T0253148

DOWNWIND OF THE ATOMIC STATE

Downwind of the Atomic State

Atmospheric Testing and the Rise of the Risk Society

James C. Rice

NEW YORK UNIVERSITY PRESS
New York

NEW YORK UNIVERSITY PRESS
New York
www.nyupress.org

References to Internet websites (URLs) were accurate at the time of writing. Neither the author nor New York University Press is responsible for URLs that may have expired or changed since the manuscript was prepared.

Please contact the Library of Congress for Cataloging-in-Publication data.

ISBN: 9781479815340 (hardback)
ISBN: 9781479815371 (library ebook)
ISBN: 9781479815364 (consumer ebook)

New York University Press books are printed on acid-free paper, and their binding materials are chosen for strength and durability. We strive to use environmentally responsible suppliers and materials to the greatest extent possible in publishing our books.

Manufactured in the United States of America

10 9 8 7 6 5 4 3 2 1

Also available as an ebook

For Bailey J. Rice and Sydney A. Rice

And in memory of my parents, Clifford and Joyce Rice

CONTENTS

ABBREVIATIONS

ACHRE Advisory Committee on Human Radiation Experiments
AEC Atomic Energy Commission
ANT actor-network theory
AFSWP Armed Forces Special Weapons Project
AWS Air Weather Services
BEENW *Biological and Environmental Effects of Nuclear War*
CDC Centers for Disease Control
CNI Committee for Nuclear Information
DOD Department of Defense
DOE Department of Energy
FCDA Federal Civil Defense Administration
FTCA Federal Tort Claims Act
GAC General Advisory Committee
GAO General Accounting Office
HA High Altitude
HASL Health and Safety Laboratory
H-DIVISION Health Division
HELLR *Health Effects of Low-Level Radiation*
JCAE Joint Committee on Atomic Energy
J-DIVISION Testing Division
LASL Los Alamos Scientific Laboratory
LLREH *Low-Level Radiation Effects on Health*
LRL Lawrence Radiation Laboratory
MET Military Effects Test
MLC Military Liaison Committee
MPH miles per hour
NATM National Atomic Testing Museum
NCI National Cancer Institute
NRFEM *The Nature of Radioactive Fallout and its Effects on Man*
NTS Nevada Test Site

NTSHF Nevada Test Site Historical Foundation
NYOO New York Operations Office
PHS Public Health Service
RECA Radiation Exposure Compensation Act
RFNT *Radioactive Fallout from Nuclear Testing at the Nevada Test Site, 1950–60*
RPI Rensselaer Polytechnic Institute

Introduction

The Atomic State in the Great Basin

Fallout proved unpredictable if not obstinate, and then came shot Harry. Postponed repeatedly due to the weather, the momentum to detonate was building by May 19 to avoid disrupting the remaining shots in the Upshot-Knothole series of 1953.[1] Just before dawn, thirty-two kilotons of explosive yield, twice that detonated over Hiroshima, Japan, illuminated the Great Basin Desert. The fireball drew up alluvial soil, and as the stem and mushroom cloud raced for the troposphere, ground-level dust and debris lumbered offsite. The prevailing winds were moving in the right direction. Then the prevailing winds shifted eastward. And due east, 135 miles, lay St. George, Utah. With a population around forty-five hundred, it was the largest town immediately downwind of the Nevada Test Site (NTS).[2]

Harry arrived in St. George by midmorning, a Tuesday morning, as children were running to recess and shoppers were making their rounds. Detonation was at 5:05 a.m., but Frank Butrico, a US Public Health Service (PHS) monitor, first learned of Harry's advance at 8:50 a.m. as his instruments registered radioactivity.[3] Around 9:45 a.m., he was instructed to arrange for the broadcast of a radio bulletin advising residents to shelter in place.[4] The first bulletin was at approximately 10:15 a.m., long after fallout moved over the area.

Shot Harry is depicted in figure 1.1. Following the pull of gravity, the heaviest particles, brimming with radioactivity, descended to earth the soonest, and as Harry swept through St. George and outlying areas some residents observed a metallic taste in the air, the remnants, ostensibly, of the tower and shot cab.[5] There were also scattered reports of nausea, headaches, burns on exposed skin, and hair loss, indicative of radiation sickness.[6] Such ailments were a stark contrast to assurances of safety circulating through the downwind communities in the form of Atomic

1

Figure I.1. Shot Harry on May 19, 1953. (Photo courtesy of the US Department of Energy)

Energy Commission (AEC) press releases and pamphlets. Planning for the contingencies of radioactive contamination was lethargic, but the commission's public-relations effort was in full swing.

At a meeting conducted by the Department of Energy (DOE) in 1980 in preparation for civil litigation, *Irene Allen et al. v. United States*, Butrico reviewed a memo to his superior, submitted twenty-seven years earlier, but did not recognize his own signature.[7] He surmised that somebody, at some point, revised downward the levels of radioactivity he detected in St. George and shortened the length of time before the radio bulletin advising residents to take cover. He repeated this assertion two years later, under oath, during the *Allen* trial. His recounting of May 19, 1953 provides a snapshot of organizational failures of foresight. Butrico confided, "I got the very distinct feeling that if there was a plan, it sure wasn't made known to us as monitors. If something went wrong, what would we do? Because never were there any instructions to us on that. Just Monitor. If a problem occurs, you can get in touch with Mercury, and so on. I mention this because it was quite obvious in these phone calls that everybody was playing it by ear."[8]

During the first two years of testing, fallout dispersed downwind in a bewildering manner, but prior to Harry no town experienced a direct hit. The offsite threshold in 1953 was 0.3 roentgens of whole-body external gamma radiation per week, or 3.9 in a thirteen-week period. This approximated the standard for workers in the nuclear industry, but such workers voluntarily placed themselves at risk, received training to minimize exposure, and did not include infants or children.[9] Gamma radioactivity in St. George was as high as 6 roentgens and in Cedar City, to the northeast, 5 roentgens, but AEC officials stressed that the average for the Upshot-Knothole series was less than the 3.9 guideline over thirteen weeks.[10] A conceit lay embedded in the commission's assurances regarding average radiation exposures: averages obscured the disconcerting unevenness of fallout. John G. Fuller observes, "It fell in globs, like splattered paint on a Jackson Pollock canvas. An average fallout figure meant nothing. It was the hot spots—the unpredictable spots—that counted. And no one could tell where all of them were, locally or a long distance away."[11]

Harry was hardly the sum of the commission's missteps during the Upshot-Knothole series. Nancy, a twenty-four-kiloton tower shot, swept over sheep and their human caretakers trailing east to lambing grounds at Cedar City, Utah. The first signs of trouble included burns, blisters, and scabs around the mouth and lesions extending up the muzzle, along the head, and down the back underneath the wool, which "slipped" or pulled off in chunks during shearing.[12] At lambing time, the ewes were lethargic and often died in a strikingly abrupt manner, and some lambs were stunted and undersized at birth.[13]

Kern Bulloch, along with his brother McRae, encountered fallout while trailing sheep after Nancy in March and from April to May witnessed dead and dying animals. Kern Bulloch testified before congressional investigators in 1979:

> When they started to lamb, we started to losing them, and the lambs were born with little legs, kind of pot-bellied. . . . And we just started to losing so many that—my father was alive at that time—that he just about went crazy. He had never seen anything like it before. Neither had I; neither had anybody. . . . They put Geiger counters on those sheep, and they said, "This is running the needle off the post. This counter is going off the post. These are really hot. They're really hot." . . . I think we figured we lost between 1,200 and 1,500 head. Mostly—most of them were lambs.[14]

An interorganizational effort was hastily assembled and then consolidated under the auspices of the AEC's Division of Biology and Medicine. Investigators considered three scenarios: (1) radioactivity was a direct cause; (2) radioactivity represented a contributing factor in tandem with poor range conditions and the stresses of trailing, or moving long distances, and pregnancy; (3) radioactivity played no role in the deaths of about forty-four hundred sheep. "Almost immediately, the investigation began to unearth damaging information," R. Jeffrey Smith notes.[15] After seven months of inquiry, however, AEC officials settled on the least plausible explanation: fallout played no role in the grisly and anomalous event.[16] The problem was malnutrition. With decades of experience raising sheep in the hardscrabble ranges of southern Nevada and Utah, the ranchers were skeptical of the commission's explanation.[17] Nonetheless, they confronted an institutional definition of the situation buttressed by the rational-legal authority of an organization charged with atomic development while ensuring public safety, a dual mandate that the AEC never could balance.

Nevada offered a host of advantages for a continental proving grounds: it was sunny, dry, and close to the nerve center of the atomic state at Los Alamos, New Mexico, and the federal government already controlled a large expanse of land. It also exemplified key disadvantages. Management of fallout was predicated on "dispersal and dilution" in such a manner as to avoid Los Angeles to the southwest, Las Vegas to the southeast, and Salt Lake City to the northeast.[18] In practice, the optimal direction was a tight northeasterly corridor over southern Nevada and Utah. Any deviation could expose residents in the many small towns and ranching communities within a roughly 150-mile direction. Regardless of initial trajectory, it held the potential to expose a large segment of Americans to radioactive debris as it dispersed across the country before reaching the Atlantic.

AEC officials promised precision and control, but fallout showed little regard for such wishful thinking. It moved in diverse directions, at a differing pace, at variant altitudes, came to earth unexpectedly, or remained aloft for an extended period, even as hot spots appeared here but not there. And while the commission's public reports were reassuring the empirical signals were not. Terrestrial redeposition was beset by endemic uncertainty. The burgeoning atomic state possessed offices, manufacturing and engineering facilities, laboratories, and economic and political gravitas stretching from one end of the country to the other. It was a paragon of scientific and technological rationality, but fallout was conspicuously intractable, as Harry made clear.

The Upshot-Knothole series of 1953 was a turning point. In its wake, there was discussion within the AEC of abandoning atmospheric testing in Nevada, and there were proposals to better inform the public of the radiological risks so people could take steps to protect themselves and their families. Abandonment was quickly dismissed, as money had already been invested in the NTS, and better informing the public gained little traction because of a deep-seated conviction among AEC and military officials that the public was prone to unreasoning panic.[19]

Country of Lost Borders

The Great Basin is a land of extremes. It is starkly arid, and despite sweltering summers, the winters are cold, as much of it lies above 4,000 feet in elevation. It encompasses approximately 200,000 square miles, and it is composed of a series of mountain ranges running north-south and interspersed by sprawling valleys.[20] To traverse it laterally is to ascend, descend, and repeat. "If the Great Basin was a song, it would be a combination of 'The Bear Went Over the Mountain' and 'A Hundred Bottles of Beer on the Wall.' If it were a movie, it would be *Groundhog Day* with tumbleweeds," Chip Ward notes.[21] Its few streams and rivers

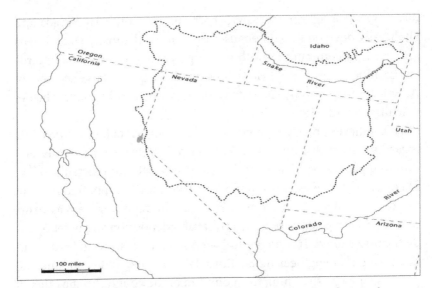

Figure 1.2. Location and extent of the Great Basin. (Figure based on floristic character-istics and courtesy of Donald K. Grayson from his book *Giant Sloths and Sabertooth Cats: Extinct Mammals and the Archaeology of the Ice Age Great Basin* [Salt Lake City: University of Utah Press, 2016])

refuse to run to the sea. Water only escapes through evaporation. Much of its flora and fauna either bites, stings, or impales. At Badwater Basin in Death Valley, one can stand on a briny lakebed 282 feet below sea level, the lowest point in North America, while the weight of 13,000 feet of Telescope Peak looms overhead. The Sierra Nevada and Cascade mountain ranges capture moisture moving eastward from the Pacific, and as the air is stripped of water it has little to offer on the leeward side. The Great Basin is an enormous "rain shadow," in meteorological parlance. And while the upward arching arms of the Joshua tree is the emblem of the Mojave Desert, the saguaro the hallmark of the Sonoran, and the ocotillo the denizen of the Chihuahuan, the Great Basin is a sagebrush "ocean": sagebrush, jackrabbits, and coyotes.[22] "God's dogs," Terry Tempest Williams remarks.[23]

Figure 1.2 depicts the location and extent of the Great Basin Desert. To immerse oneself in this sagebrush ocean is to encounter a primordial feeling of spaciousness. Then there are the winds—unruly if not mer-curial. Then there is the silence, "a great spatial silence."[24] As Stephen

Trimble observes, "This silence makes for clear thinking about how one fits into the world."[25] Then there are the views. It is not uncommon to peer down an alluvial slope, across a gleaming bone-dry playa, and up the pinnacle of a snowcapped peak fifty miles distant. From Wheeler Peak in eastern Nevada, the view extends for seventy miles.

The conviction that the desert is a barren wasteland feeds into a moral void that powerful actors find inviting. Land perceived as devoid of life can hardly then be understood as stripped of it.[26] Mary Austin, an early critic of western ambivalence, observed, "East away from the Sierras, south from Panamint and Amargosa, east and south many an uncounted mile, is the Country of Lost Borders. . . . Desert is a loose term to indicate land that supports no man; whether the land can be bitted and broken to that purpose is not proven. Void of life it never is, however dry the air and villainous the soil."[27]

The Great Basin is not easily understood from a mind-set rooted east of the hundredth meridian, the hydrological transition point meandering down the Dakotas, western Kansas, Oklahoma, and western Texas. It has long confounded the brash and ill prepared. Interlopers have often sought to remake the Great Basin in their own image, and those who are rooted in this harsh land have too often acquiesced to imposed parochial interests in return for short-term economic gain.[28] Even for the residents of the Mormon towns scattered far and wide in the Great Basin, militarism too often passes for patriotism. They constitute a "peculiar people," desperate for recognition as real Americans, and too often they have accepted the intrusion of the military-industrial complex onto a landscape providing refuge from religious persecution in the mid-nineteenth century—east of the hundredth meridian.

As nineteenth-century emigrants moved westward to California and Oregon, the Basin was an impediment to be endured. Extremes in temperature and aridity were foreboding from the buckboard of a wagon. Misjudging its expanse could be tragic.[29] Death Valley derives its name from such misjudgment, the snowcapped peaks of the Sierras lying just beyond. Like distant waves on hot asphalt, things are not always what they seem in the Great Basin. Hubris frequently prefigures disaster here.

Figure 1.3 depicts the location of the NTS. It begins where the Mojave Desert, reaching eastward from California, fades into the Great Basin, which envelops much of Nevada before stretching into parts of Oregon,

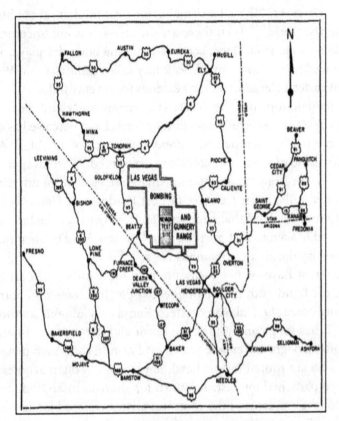

Figure 1.3. The location of the Nevada Test Site. (Atomic Energy Commission, "Atomic Tests in Nevada," March 31, 1957)

Idaho, and Utah. To the midcentury architects of the atomic state, this comparative blank spot on the map was a barren wasteland. Here they carved out an outdoor laboratory, and between 1951 and 1962 over one hundred open-air atomic devices were detonated. The southeast entrance of the NTS is just sixty-five miles northwest of Las Vegas, and in total it encompasses 1,375 squares miles.[30] This is, it is often remarked, larger than the state of Rhode Island. "The Nevada Test Site (NTS) is big on a scale possible in few parts of the world, and in a way that only the West of the United States is big," Rebecca Solnit observes.[31] Along the western edge of the NTS, at Yucca Mountain, the DOE has spent $14 billion building a deep geologic repository for the entombment of high-level radioactive waste, and just beyond the northeast edge of the NTS lies the mythical airbase known as Area 51.

Much of the Great Basin is dedicated to military training, weapons development, and sequestration of hazardous waste. One cannot travel northward from Las Vegas far before having to swing west or east due to the "present absence" of militarism, which is both ubiquitous and yet largely ignored by the public.[32] "Until the Second World War, the Southwest was a failure on all the economic maps drawn up by the nation," Charles Bowden notes.[33] But as militarism expanded in complexity and scale, the need arose for sacrificial spaces, bringing federal funding to some but radiological contamination and illness to others. Indeed, the Great Basin bears the visible scars of environmental violence dispersed far and wide. "The problem with defense is how far can you go without destroying from within what you are trying to defend from without," Dwight D. Eisenhower lamented.[34]

Few people understood the American West as did the writer Wallace Stegner. His childhood was marked by movement across the terrain, as his father was convinced that making it big was always around the next corner.[35] The everyday minutia of farm life, nonetheless, is replete with responsibilities come rain or shine. Stegner recounts as a youth tying a series of lariats from house to barn before an impending blizzard, the rope serving as a guide during the chaos.[36] It is an analogy that Stegner extends to scrutinizing the past. Like a rope strung from backdoor to barn, the willingness to confront uncomfortable lessons is a bulwark against the disorienting influence of mythology. We need to rig up a line between past and present lest we lose our way into the future.

Major Claims Advanced

As Dina Titus notes, the AEC was not a "villainous" organization unconcerned with public health.[37] And yet the monitoring of radioactivity and efforts to educate and inform downwind residents of the potential risks were pursued in a haphazard manner. Detonations in Nevada accelerated weapons development during the tensions of the Soviet arms race, but the AEC was negligent in confronting the hazards of radioactive contamination—as they were understood at the time. This was an iterative, evolving process, however, as opposed to crass, amoral indifference.

In turn, the argument herein rebuts two commonplace explanations: (1) AEC officials understood the risks but simply did not care about the public health consequences; (2) AEC officials did the best they could to mitigate and minimize the risks given the state of knowledge at the time. What occurred is more multifaceted than these assertions admit, and it is within this complexity that lessons are found applicable to the social control of large organizations managing risky sociotechnical systems in contemporary society. AEC operational conduct during the era of continental open-air atomic testing is a case study in organizational deviance. As Diane Vaughan argues, "organizational deviance" refers to conduct that "deviates from both formal design goals and normative standards or expectations."[38] This is the conundrum to be explained.

Table 1.1 outlines the argument developed in the pages that follow. Organizational deviance was predicated on assumptions that proved erroneous over time and contributed to failures of foresight, leading to mistakes that served as the impetus to misconduct and cover-up. These heuristics had systemic implications such that AEC officials were continually chasing the unanticipated material and biophysical characteristics of fallout rather than proactively initiating measures necessary to protect the health of downwind residents. In turn, the AEC neglected to take actions that could have made testing safer, and adherence to unfounded organizational heuristics helps explain why. These key assumptions include that (1) fallout is subject to predictable atmospheric dispersion and dilution; (2) the public is prone to unreasoning panic; and (3) external gamma radiation exposure is the primary hazard of testing. Organizational actors often rely on mental shortcuts amid complexity and uncertainty, but such heuristics can obscure information that is at variance with established beliefs or policy.[39] Without outside accountability and internal reflection, heuristics underpinning operational conduct can become blinders.

Moreover, open-air detonations were predicated on the diffusion premise: nature is so vast and robust that it can absorb the harmful provocations of humanity. That is, amid nature's limitlessness, there is room for diffusion, dilution, and dissipation. In practice, atomic testing in Nevada and the Pacific Proving Grounds made a mockery of

TABLE I.1. Major Claims Advanced

Organizational-Level Dynamics

Erroneous operational assumptions:
• Fallout exhibits predictable atmospheric dispersion and dilution patterns.
• The public is prone to unreasoning panic over fallout and has little role to play in the management of offsite radioactive contamination.
• External gamma radiation is the primary hazard of atmospheric testing.
• *Diffusion premise*: Nature is so vast and robust that it can absorb the harmful provocations of humanity without deleterious side effects for people or nature.

AEC officials had unprecedented power to restrict or classify information, and this inhibited broader and more informed dialogue and debate.

The ability to restrict information contributed to an overreliance on public relations and the effort to sustain a preferred definition of the situation, to the exclusion of prudent management of fallout.

Institutional-Level Dynamics

Pressure from the armed services led to an increasing scale and complexity of activities— including reliance on tower shots—and rendering management of fallout more difficult.

The Nevada Test Site was a key node in a larger sociotechnical system with strong expansionary tendencies, generating momentum and path dependency dynamics.

Lack of accountability, particularly in the early years of testing, provided the space for AEC and military officials to engage in operational conduct that they may otherwise have avoided.

Atmospheric testing was conducted by a technocracy often pursuing its own parochial objectives, as dictated by technical expertise and bureaucratic priorities.

the diffusion premise. Increasingly powerful devices spread long- and short-lived radionuclides far and wide, and instead of dilution there was bioaccumulation and movement up the food chain.

Public apprehension did not precede testing but arose in response to evidence of AEC duplicity. Fear of the public, however, was entrenched in AEC culture from the beginning and shaped decisions made and those neglected in ways that undercut the management of fallout. Rather than possessing an endemic irrationality regarding radioactivity, what the downwinders feared and the public stills fears is that people in positions of power will not inform them of salient hazards in a forthright manner. They worry that their concerns will be abandoned for the interests and objectives of a technocracy that is beyond public reproach. As William Freudenburg notes, what the public fears is recreancy, or a "failure of institutional actors to carry out their responsibilities with the degree of vigor necessary to merit the societal trust they enjoy." It is a breach of fiduciary conduct, a "retrogression or failure to follow through on a duty or trust."[40] Open-air atomic testing illustrates that such concerns are not without precedent.

AEC officials primarily focused on external gamma radiation and the occurrence of acute health effects, while internal emitters (radioactive material that has been inhaled or ingested) and long-term biophysical effects never garnered substantive concern. Radioiodine was one of the most hazardous contaminants dispersed downwind, typically ingested through milk, but was consistently overlooked by AEC officials—even as the scholarly literature over the course of the 1950s documented its carcinogenic effects. Radioiodine seeks out the thyroid gland, and the latency period between exposure and the expression of cancer may be measured in decades. During the era of open-air testing, AEC officials never established a systematic survey of radioiodine in fresh milk downwind of Nevada. There was a conviction among AEC and military officials that the absence of acute injury due to external gamma exposure exemplified the absence of *any* public health effects.

All the major nuclear powers conducted open-air detonations at the "extremes of empire," or remote edges of their political and military influence.[41] In electing to commence with continental detonations, however, the AEC did so in an internal peripheral space. It was a landscape largely ignored by the rest of the country but nestled deep within the homeland. Those who lived in the Great Basin were primarily displaced Native Americans and scattered Mormon outposts. They were marginalized segments of the population but still US citizens, and their concerns could not be brushed aside as easily as the indifference accorded inhabitants of the Marshall Islands. In turn, AEC and military officials worked to sell the public on the safety and necessity of testing in Nevada and continually reinforced this narrative when events undermined their overly optimistic definition of the situation.

Due to entrenched, specious organizational heuristics in response to the fluidity of fallout, the AEC's public-relations program consistently outpaced pragmatic management of radioactive contamination. The engineering of public opinion was vigorously pursued through press releases as well as speeches, the distribution of pamphlets, and films aired in the downwind communities. Public relations were deployed to curb dissent, even as radioactive fallout could not be contained. In turn, there was increasing divergence between the discourse and the biophysical and material effects that were at odds with the definition of the situation that AEC officials labored to sustain. We give meaning to nature through the social construction of reality, but nature is more than language and continually challenges our socially negotiated conceptions of it, particularly when they are out of step with repeatable, observable dynamics. AEC and military officials wrestled with this tension over the course of the 1950s. They possessed unprecedented powers of classification, however, and many reports and memoranda were sequestered from public view due not to national security but public-relations concerns.

In addition to the adoption of erroneous organizational-level heuristics, the AEC was caught in a maelstrom of institutional dynamics. In particular, the increasing pace and complexity of activities over time were spurred by the armed services' effort to wrest control over the NTS. When testing began in January 1951, Nevada was a backyard laboratory for Los Alamos Scientific Laboratory (LASL), but LASL soon began to lose control because of a push by the military to get in on the atomic action.[42] Moreover, testing began in a tepid manner and over time constituted a more significant hazard with the adoption of tower shots and higher explosive yields. Tower shots increased the risk of fallout, due to the updraft of alluvial soil, but were essential to the diagnostics of weapons testing as well as troop maneuvers and effects testing—the examination of detonations on military clothing and hardware—conducted at the behest of the armed services. Responding to the excessive demands of the armed services undercut efforts to concurrently manage offsite fallout.

The AEC's proving grounds in Nevada lay at the back end of a larger sociotechnical system with government bureaucrats, military officials, academics and affiliated universities, and civilian contractors who did the commission's bidding. Despite public assurances to the contrary, it was generally anticipated within the AEC that there could be risks to public health, but such awareness grew over time in tandem with economic investment in the NTS and its importance within the larger system of weapons design and production. The degree to which AEC officials sought to address the risks of radioactive fallout was circumscribed from the beginning, but as concern became more prominent, the larger sociotechnical system, in which the NTS was enmeshed, exhibited momentum that was difficult to arrest. And as the risks became more recognizable, it also became apparent that comprehensive management of fallout would render continental detonations impractical. To enforce differential exposure limits by gender and age would have been laborious and necessitated exposure standards so restrictive as to limit the number of shots conducted in a series and the frequency of successive series. This would have had a ripple effect throughout the larger atomic-industrial complex. In turn, over time, AEC officials embraced a Faustian trade-off: expedited weapons testing but some measure of public sacrifice.

Further, the AEC's dual mandate to pursue weapons development in tandem with public safety was compromised, in the early years of testing, by a lack of outside accountability. In 1946, the US Congress established the Joint Committee on Atomic Energy (JCAE). The JCAE had exclusive jurisdiction over all bills related to military and civilian aspects of atomic technology and was tasked with congressional oversight. For nearly a decade after its initial conception, the committee was more a booster club than a watchdog. This changed in the wake of the Castle Bravo detonation in the South Pacific in 1954. Lacking robust outside accountability but possessing the power to sequester information ensured that AEC officials could dictate the terms of the debate over radioactive fallout, particularly in the early years of testing.

Lack of outside accountability along with the capacity to restrict and classify information reinforced groupthink as an indelible aspect of AEC organizational culture. This contributed to the secretive character of AEC operational conduct. Not only were AEC and military officials terrified of public opinion, but they were convinced that the public had no productive role to play other than quiescence. One of the most remarkable aspects of atmospheric atomic testing entails the effort, as the 1950s wore on, to temper and discipline an entrenched technocracy that was resistant to outside influence.

The conduct of continental atmospheric atomic testing was fundamentally undemocratic, as public oversight and participation were avoided at every turn. The counterargument stipulates that this was necessary because of a looming Soviet threat. In times of national crises, in turn, the rights of the people and obligations of institutional structures in society may be suspended temporarily for the good of the collectivity. Whether this Faustian trade-off was necessary to the degree it was enacted is a difficult question to answer. The 1950s were a remarkably tense period geopolitically, and a continental test site was key to expedited weapons design and development. What is much clearer is that absent public accountability, organizational conduct may serve a technocratic order with its own objectives and values. "Technocracy" refers to the suppression of public debate in favor of technical expertise and bureaucratic priorities.[43] Scientific and technological expertise, rather than the will of the people, is at the core of technocratic control. The power to dictate how weapons testing was carried out was centralized within the AEC but subject to the ambitions of the armed services. This technocratic order was remarkably insulated from society while imposing on the public an unprecedented degree of risk.

The public health impacts of atmospheric testing in Nevada are difficult to clearly specify, but this does not mean that such impacts are nonexistent. The downwind residents were exposed to the greatest risk, but hot spots occurred throughout the intermountain west, midwestern, and northeastern states. Further, low-level radioactivity produces effects that are discernable only over time, that vary by age at exposure as well as gender, and that mirror diseases that already exist within the general population. The challenge is to distinguish illness induced by fallout from other factors. These complexities worked to the advantage of the atomic-industrial complex, and a sobering implication of open-air testing is that the impact on public health may never be known with certainty.

Atmospheric testing in Nevada was an ontological wake-up call. "Ontology" denotes inquiry into how it is that things exist—the study of being or substance. As Andrew Pickering observes, it involves "questions of what the world is like, what sort of entities populate it, how they engage with one another."[44] Open-air atomic detonations illustrated that the world is not as we generally assume it to be. This is ironic, of course, as physics is typically viewed as paving the way for the command and control of nature. And yet we live in a world brimming with nonhuman activity and change that we manipulate but do not master. Despite our having unleashed the power of the atom, dominion remains illusionary, while heedless provocation is ubiquitous. Such things matter, as the contemporary world is replete with large, complex organizations managing risky technological systems. The legacy of open-air detonations in

Nevada raises important questions as to the balance between militarism and national security, on the one hand, and the right of the public to be informed of risks to which they may be exposed, on the other. Moreover, it highlights the importance of understanding the factors, including the erroneous ontological assumptions, shaping organizational failures of foresight and the drift into recreancy.

Preview of the Chapters

This book's chapters are analogous to lariats strung to and from the past, and they are divided into three overarching sections: (1) The commencement of testing in January 1951 through the end of 1952 was characterized by recurring indications that fallout was unpredictable, in tandem with institutional momentum to utilize the NTS for an increasing number of objectives while conducting higher-yield tower shots. (2) The Upshot-Knothole series of 1953 was pivotal, as the incubating dangers of fallout came to fruition. This incited consternation within the AEC that testing could be curtailed in Nevada. Institutional momentum and the economic benefits accruing to southern Nevada, however, pushed AEC and military officials and local boosters to double down on the vital role played by detonations in Nevada. This was accompanied by an increasing emphasis on public relations, the withholding of information from the public, and the rerationalization within the AEC that testing was necessary to ensure national security. (3) The period 1954–1962 is illustrative of more robust efforts to monitor offsite contamination but also more expansive testing in Nevada—all while the AEC worked to sustain its definition of the situation, stipulating that testing did not present a public health hazard.

Chapter 1: Uranium and the Agency of Nature in the Risk Society

Uranium is the most consequential rock in the world.[45] It exemplifies everything that modernist pretense neglects: independent movement, change, force, and transformation. Uranium is indicative of the epistemological and ontological conundrums of modernity, as articulated by Ulrich Beck's risk society thesis and Andrew Pickering's conception of the performativity of nature.[46] Perhaps the most surprising aspect of the AEC's conduct in Nevada was a reluctance to recognize that fallout was dynamic and fluid and that radiological contamination could reach further geographically and deeper into the food chain than anticipated. The subatomic vibrancy of uranium did not translate into a conception of the performative and restless contours of the nonhuman world more broadly. This speaks to the myopia of Western rationality and issues that Beck and Pickering confront and that AEC officials grappled with, shot after shot, while chasing fallout across the country.

Chapter 2: Trinity and Lessons Not Learned

Trinity foreshadowed the challenges inherent in open-air testing in Nevada. Moreover, Trinity heralded not simply the dawn of atomic military technology but a shift in modernity, as the contradictions of science and technology now exist on an unprecedented scale; exhibit temporal distance between cause and effect, inhibiting accountability; and lack salient social and scientific cues from which a firm definition of the situation can be established. In important respects, the risk society as sketched out by Beck began on a desert playa in New Mexico. And the unanticipated efficacy and intractability of nature exhibited here was repeated, again and again, downwind of Nevada.

Chapter 3: The Emerging Atomic Spectacular, 1950–1951

In the aftermath of Hiroshima, there was debate regarding military versus civilian control of atomic technology, and the Atomic Energy Act of 1946 stipulated that the armed services defer to civilian control. This meant little, in practical terms, during the early years of atomic testing. Indeed, the armed services sought to utilize the AEC's outdoor laboratory for a variety of activities for which it was not intended when it was founded in 1950. The first two series, one early in 1951 and again in the fall, revealed the following: (1) The lessons of Trinity did not resonate with AEC officials. This includes awareness of the variable character of the prevailing winds and the occurrence of hot spots of gamma radiation. (2) Absolute safety was assured, although the dynamics of fallout were far from certain. From the outset, AEC officials backed themselves into a discursive corner. They promised the command and control of nature, and this made it difficult to admit problems arising over time for fear it would erode the commission's legitimacy. (3) By the fall of 1951, the momentum to expand the scale and complexity of activities at the NTS already undercut offsite radiological monitoring. Troops and equipment were assembling onsite, and changes to the testing schedule came with little notice to those who were burdened with radiological safety operations. (4) The public did not panic with the commencement of detonations but did find them entertaining. This clashed with the commission's initial "low-key" public-relations approach. It was not possible to detonate atomic devices in Nevada without the public's knowledge, and as fallout proved intractable AEC officials increasingly relied on public relations to sustain a preferred definition of the situation, which was at odds with the fluidity of fallout.

Chapter 4: The Road to Upshot-Knothole, 1952

The problems encountered in 1951 grew more pressing one year later, and the empirical cues indicating the threat of radioactive contamination elicited concern within the AEC but scarce changes in operational protocol. The prevailing winds twisted and turned from the meteorological forecasts, and hot spots arose with abandon, while the rhetoric within the AEC clashed with the sunny picture presented to the public. Indeed, the AEC assembled a committee tasked with assessing such problems. The committee recognized the increasing complexity and scale of operations and that the many "fringe" projects proposed by the armed services strained existing resources, including those devoted to management of offsite fallout. The committee concluded that testing in Nevada was essential to weapons development, however, and reiterated that no town downwind had been subjected to a significant fallout event.[47] Just days after the committee report was released within the AEC, one town experienced just that.

Chapter 5: Dirty Harry and the Material-Discursive Bind, 1953

Prior series suggested that fallout was obstinate, but the Upshot-Knothole series of 1953 made this fact abundantly clear. Shot Simon rained out over New York State, Nancy swept over ranchers and their sheep herds just offsite, and Harry landed atop the largest town in the downwind region. These events demonstrated that AEC operational conduct was lethargic in response to the fluidity and recalcitrance of nature. The series was so jarring that AEC officials suspended testing in Nevada until a second internal committee could, again, examine the issue. In the interim, however, the commission was embroiled in an interorganizational effort to determine what killed over four thousand sheep. If it was radioactivity—and everything pointed to just that—then it held the potential to derail continental testing.

Chapter 6: Dead Sheep and the Fluidity of Fallout

For AEC and military officials, dead sheep presented a public-relations problem, but it should have sparked a recognition of the vibrancy of internal emitters of radiation. Nibbling and grousing as the sheep trailed eastward, when the herds reached Cedar City, Utah, they were hot enough to peg a Geiger counter needle and exhibited unmistakable symptoms of radiation injury. In a remarkable instance of "believing is seeing," AEC and military officials came to an inexplicable conclusion: the sheep probably suffered from malnutrition. This illustrated both willful and strategic ignorance or the avoidance of some information and overreliance on other sources, as well as the active production of doubt and uncertainty, to arrive at a conclusion preferred from the beginning.[48] This was possible due to the AEC's power to restrict or classify information along with a lack of outside accountability. The investigation, conducted over seven months, found a way to conclude that which the empirical evidence least suggested.

Chapter 7: Respite, Reconfiguration, and Operation Teapot, 1954–1955

Testing resumed with the Teapot series of 1955, and operational conduct was reconfigured because of the recommendations of a second internal committee, which documented failures of foresight that the AEC never publicly admitted. The committee insisted that investment in the NTS precluded abandonment of the tests but recommended taller shot towers and more robust offsite radiological monitoring.[49] The AEC also embarked on a stepped-up public-relations campaign, and the centerpiece was a film shot in St. George, Utah, and unveiled with the commencement of Operation Teapot. The film depicted an idealized scenario of the day Harry swept through town in May 1953. It was fiction. Despite such efforts, by 1955 the public was growing weary of fallout. Moreover, the Castle Bravo detonation of 1954 dispersed radioactive contamination over thousands of square miles of the Pacific and introduced "fallout" to the American public. Cracks were showing in the AEC's hegemonic narrative and eliciting pushback from scientists and everyday citizens in response to the excesses of the atomic state.

Chapter 8: Bulloch et al. v. United States: Deception and Dirty Science, 1956

Just as the AEC resumed testing in Nevada, after nearly two years, another bomb dropped: ranchers from the Cedar City area were not convinced that malnutrition killed their sheep and in 1955 filed a lawsuit in federal district court in Salt Lake City. *Bulloch et al. v. United States* went to trial in 1956 and constituted a dire threat to continental testing. The government countered with duplicity and dirty science. The sheep investigation of 1953 embodied the organizational production of willful and strategic ignorance, but the *Bulloch* trial necessitated more overtly unscrupulous machinations, such that mistake evolved into misconduct. Judge A. Sherman Christensen ruled for the government, but more than twenty years later and still on the bench, he was presented with documents that were not offered at trial in 1956 and declared that the government committed "fraud upon the court."[50] He vacated the original judgment and ordered a new trial, but this was rejected on appeal. It was a remarkable instance of citizens suing the government, lawyers representing the government prevailing in the courtroom, the judge issuing a scathing indictment of their conduct twenty-six years later, and an appeals court issuing what many observers consider a decision influenced by political, as opposed to legal, standards.

Chapter 9: Operation Plumbbob: Accelerated Testing in the Shadow of a Moratorium, 1957

Despite prevailing in *Bulloch et al. v. United States*, AEC and military officials understood that a moratorium on atmospheric testing was on the horizon. This was primarily due to thermonuclear detonations conducted by the US in the South Pacific and the Soviets in eastern Kazakhstan. In the shadow of a moratorium, testing accelerated in Nevada. Operation Plumbbob in 1957 encompassed twenty-four shots, the most of any series, conducted over four months, the longest of any series, releasing over 340 kilotons of explosive yield into the atmosphere, the highest of any series.[51] Six years after testing began in Nevada, and many lessons learned the hard way, 1957 was a year for the record books. Moreover, while Plumbbob was unfolding in Nevada, the JCAE was conducting hearings in Washington, DC. Of particular concern were the uniformity thesis, which insisted that fallout exhibited uniform terrestrial redeposition, and the question of stratospheric holdup of radioactive debris. AEC officials argued for both uniformity and stratospheric sequestration, as long as ten years, so that by the time debris returned to earth long-lived radionuclides constituted less of a hazard. Their testimony was contradicted by concurrent studies, and in subsequent years research documented neither uniformity nor considerable stratospheric holdup. The testimony of AEC officials, however, helped to delay a temporary moratorium for another year.

Chapter 10: Baby Teeth, Project Sunshine, and a Moratorium on Testing, 1958–1962

By 1958, a groundswell of public and scientific critique confronted AEC officials, and it came not downwind of Nevada but from national and international concern over thermonuclear detonations in the South Pacific. What happened in Nevada was overshadowed by broader concerns, even as each detonation drifted across the continent. This groundswell was driven by nongovernmental organizations and scientific associations whose members were dissatisfied with the paternalistic conduct of the atomic state. In particular, the Greater St. Louis Citizens' Committee for Nuclear Information embarked on a data-gathering project garnering considerable attention: the gathering of baby teeth for analysis of strontium-90. Baby teeth were a means for charting ingestion of strontium-90 through milk and a reminder of what was at stake in allowing the AEC to dominate the national debate. Further, details of the commission's Project Sunshine leaked to the press. Project Sunshine began in 1953 and was shrouded in secrecy, as it involved the surreptitious collection of bodies and body parts for radiological analysis. Particularly prized were stillborn children. The data indicated that the uptake of strontium-90 was more prodigious than AEC officials were comfortable admitting publicly. In turn, by 1958 the AEC no longer controlled the narrative, and in November, the US, the United Kingdom, and the Soviet Union embarked on a temporary moratorium lasting nearly three years.

Chapter 11: Sedan, Silent Spring, and the Reenchantment of Nature

Despite a three-year moratorium, AEC and military officials were busy promoting the peaceful uses of atomic devices under the auspices of Project Plowshare. The goal was to demonstrate the viability of atomic devices for the construction of canals, ocean ports, mountain passes, open-pit mining, rerouting of rivers, creation of lakes, and even underground to enhance the recoverability of oil and natural gas deposits. The most memorable dimension of Project Plowshare was the Sedan shot of 1962. It left behind an enormous crater at the NTS. Sedan was not an open-air test, but the designation means little as it was one of the dirtiest detonations ever conducted in Nevada. Two months later, Rachel Carson's book *Silent Spring* arrived on bookshelves nationwide. Sedan and publication of *Silent Spring*—months apart in 1962—were not simply a historical curiosity but indicative of a burgeoning tension: the former a testament to the capacity of science and technology to remake nature and the latter an indictment of the myopia underlying its reckless modification. Indeed, Sedan and *Silent Spring* offer divergent ontological lessons—differing conceptions of the relationship between the human and the nonhuman—that we still seek to comprehend.

Conclusion: The Legacy of Open-Air Testing in the Contemporary Era

AEC and military officials were embroiled in a performative give-and-take with nonhuman nature that they found difficult to conceptualize. Detonations in Nevada comprised a tacking back and forth between human agency and the materiality or force and efficacy of a variety of entities, processes, and things both organic and inorganic—things with the propensity to provoke transformation and change. It remains a blind spot in Western culture: the nonhuman world acts as well as reacts to human provocation. Nature does not act with intention or malice, but nature acts. This is one of the key lessons of atmospheric testing that AEC and military officials never came to terms with, but we scarcely have the latitude to hold onto antiquated mythology much longer. Nature is restless and efficacious. Reticence to appreciate this vibrancy underlies the mistakes committed by AEC and military officials as well as other instances of organizational mistake, misconduct, and disaster in the contemporary era.

PART I

Fallout and Failures of Foresight

1

Uranium and the Agency of Nature in the Risk Society

A rock would seem to embody matter as passive, but uranium is quite the opposite. Its propensity to provoke transformation and change has reshaped human societies in ways both deliberate and unintended, and it remains a challenge controlling its startling efficacy. Human social organization, moreover, is predicated on the force and vibrancy of the material and biophysical nonhuman world in a myriad of ways. Rather than a clear delineation between the social and the natural, there is entanglement and varying degrees of socionatural hybridity.[1] As Noel Castree argues, "hybridity" refers to things "not quite natural, not quite social" but a little of both.[2] And in lieu of mastery, there is emergence, surprise, and recalcitrance in the encounter between humanity and non-human entities, processes, and things.[3]

If one were to continuously divide elemental matter, what would be the smallest unit arrived at that still comprises its identifying character-istics? In the language of quantum mechanics, the answer is the atom. At the heart of an atom is the nucleus, which is composed of a varying array of protons and neutrons. A neutron possesses a neutral charge, a proton a positive electrical charge, and circling the nucleus is one or more negatively charged electrons swarming as if "bees around a hive."[4] Uranium is distinctive, as its nucleus is packed with protons and akin to a drop of water teetering on an edge.[5] Albert Einstein sketched out the equivalence of mass and energy—$E = mc^2$—wherein mass (m) multi-plied by the speed of light (c) squared is convertible to energy (E), and since the speed of light is 186,000 miles per second, even a small quan-tity of mass has vast energetic effects. The task confronting Manhattan Project scientists was to evoke such equivalence. Tom Zoellner notes, "A uranium atom is simply built too large. It is the heaviest element that occurs in nature, with ninety-two protons jammed into its nucleus. This approaches a boundary of physical tolerance. The heart of uranium, its nucleus, is an aching knot held together with electrical coils that are

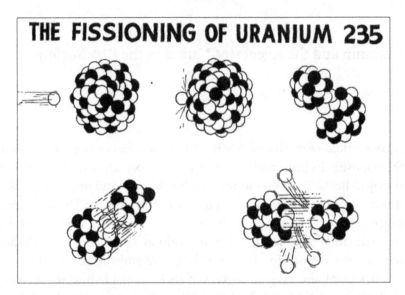

Figure 1.1. Fissioning of uranium U-235. (Armed Forces Special Weapons Project, *Radiological Defense Indoctrination Course for Armed Forces Schools* [Edgewood Arsenal, MD: Chemical Corps School, 1950], NTA Accession No. NV0752882)

as fragile as sewing thread—more fragile than in any other atom that occurs in nature. Just the pinprick of an invading neutron can rip the whole package apart with hideous force."[6]

That a uranium atom is a drop teetering on an edge is the key to the atomic age. Of the $2 billion invested in the Manhattan Project, the bulk was spent on facilities at Hanford, Washington, and Oak Ridge, Tennessee, for capturing U-235.[7] Uranium U-238 has 92 protons and 146 neutrons and is abundant in nature, while U-235 has 92 protons but 3 fewer neutrons and is scarce.[8] The difference is crucial, as U-235 is better at capturing an incoming neutron, fracturing the drop wobbling on the edge, such that the nucleus splits into two smaller, unevenly composed nuclei or fission products, there is a loss of mass, and an average of 2.4 new neutrons are ejected—disrupting surrounding nuclei in a cascading manner.[9] This process is depicted in figure 1.1. Zoellner further explains, "The subatomic innards of U-235 spray outward like the shards of a grenade; these fragments burst the skins of neighboring uranium nuclei, and the effect blossoms exponentially, shattering a trillion trillion atoms within the space of one orgiastic second. A single atom of uranium is strong enough to twitch a grain of sand. A sphere the size of a grapefruit can eliminate a city."[10]

Uranium is exceptional, but many things dug from the ground, arising from the soil, and roaming or razed from the landscape play a role in reshaping human societies due to particular material or biophysical characteristics. Much of organic and inorganic nature provokes effects that are subtle or jarring. Nonetheless, what is missing in the modern worldview is the "force of things," or recognition that the nonhuman world is not simply pliable, inert stuff subject to human manipulation but conjures activity by itself.[11] Humans are assumed to be the "motor of history," but the nonhuman materiality through which social organization is built and sustained is ignored.[12] As Karen Armstrong observes, "For Descartes, the universe was a lifeless machine, the physical world inert and dead."[13] Uranium is not inert and dead. Moreover, the nonhuman world acts independently and reacts to human provocation in a manner that René Descartes and other luminaries of the Scientific Revolution did not envision.

Materiality is the capacity of the organic and inorganic to engage in action-reaction and therefore affect, or indeed resist, transformation and change. That uranium or plutonium is alive in some allegorical manner was often expressed by Manhattan Project scientists. In quantities too small to invoke a chain reaction, it is docile. The physicist Leona Marshall Libby remarked, "Plutonium is a very dense metal. When you hold a lump of it in your hand, it feels warm, like a live rabbit. This is because it is emitting energetic alpha particles at a very high rate."[14] Building a bomb necessitated research into the mass necessary to invoke a sustained chain reaction—the sudden rush of materiality. And as U-235 or plutonium approaches criticality, it is more akin to a mythical creature. "Like tickling the tail of a sleeping dragon," the physicist Richard Feynman suggested.[15]

The Restless Atom

At the turn of the nineteenth century, many physicists doubted the existence of atoms, and few believed in transmutation, whereby one element converts to another. Things shifted, Alfred Romer notes, with the discovery of an invisible source of energy: "Early in the new year of 1896, all over the world, people opened their newspapers to read a story from Vienna. The report said that a German professor named Routgen had discovered a way of photographing hidden things, even to the bones within a living, human hand."[16]

Wilhelm Roentgen stumbled upon X-rays. The "X" is a placeholder for the mysterious source of energy.[17] Roentgen's discovery prompted the French physicist Henri Becquerel to conduct experiments in which he discovered that uranium possessed a similarly invisible energy source. Soon thereafter, Marie and Pierre Curie demonstrated the propensity of uranium to shed electrified bits of matter and coined the term "radioactivity." In 1905, Einstein postulated the equivalence of mass and energy, and nearly a decade later, with war looming in Europe, H. G. Wells published the novella *The World Set Free*, envisioning a fearsome new weapon exploiting such equivalence: an "atomic" bomb. Whoever controlled this weapon, Wells cautioned, controlled the world.[18]

In subsequent years, the physicist Ernest Rutherford identified proton and electron particles as well as differing types of radiation—alpha, beta, and gamma. The neutron was detected in 1932—completing the schematic of the atom—and in 1933, while crossing a London street, the physicist Leo Szilard envisioned what a neutron could provoke: a chain reaction.[19] Due to the neutron's neutral electrical charge, it is ideal for cutting through the outer shell of an atom and disrupting the nucleus. Speaking of Szilard, Zoellner explains, "He was excited and horrified by this insight. The fictional had become the suddenly possible, even likely."[20] Nine months prior to Szilard's vision, Adolf Hitler became chancellor of Germany, and advances in physics collided with fascism in Europe and the specter of another world war. This confluence provided an unprecedented urgency, and three decades after Wells coined the term, Manhattan Project scientists at Los Alamos, New Mexico, possessed the weapon he envisioned.

This account is the conventional tale. It is incomplete, however, as it does not do justice to how science works in the real world. The conventional tale places human cognition and agency at the center of the explanatory account, and yet the history of atomic physics is better depicted as the push and pull of human provocation and nature. It is punctuated by trial and error, serendipity, and the iterative, evolving recognition of how subatomic particles act under a variety of conditions. It is less an illustration of human cognition, by itself, uncovering the mysteries of a hidden world than repeated human provocation to assess how matter responds and then altering thought accordingly. Science is more performative than it is typically portrayed.[21] Romer argues, "Everything that we know in physics we have learned from experiments, and not from a small handful of them either. Behind even a tiny scrap of knowledge there may lie dozens and dozens. Of them all, the ones we have learned the most from in the end are those that went wrong."[22]

Physics evolved less through cognitive innovation than through *performative engagement* with the material world: deploying a given experimental design, stepping back to see how matter reacts, using another tactic, measuring the response, and then adjusting expectations, yet again. Recognition that U-235 is fissionable was less a discovery than a tacking back and forth over an extended period between researchers employing a variety of experimental setups to study the action and reaction of uranium.[23] Such provocation and response revealed that uranium is continually shedding subatomic particles and transforming into something else entirely. It becomes radium and then radon and then polonium and then eleven other elements, until finally reaching a stable, nonradioactive state as lead.[24] Along the way, it loses mass and emits energy. Radiation is the transport of energy, whereas radioactivity refers to the subatomic particles and gamma waves associated with unstable elements undergoing decay.[25] Radioactivity is less a thing than incessant change.[26]

Quantum theory and the delineation of atomic structure are one of the most remarkable achievements of modernity, and yet they undercut prevailing assumptions of fixed and stable nonhuman entities. They depict a world bustling with movement and efficacy. At the commencement of the Scientific Revolution, Francis Bacon envisioned that humanity could attain an "empire over things," but he could scarcely have anticipated the vibrancy of many of these things.[27] What Bacon, Descartes, and others left behind was awareness of a dynamic organic and inorganic world. Prior to the Scientific Revolution, nature was viewed in contradictory ways, but it was not viewed as passive.[28] A central challenge is to recapture this sense of the nonhuman realm as anything but docile.

Radioactivity does things. Subatomic particles collide, shatter, and provoke change. Gamma waves can strip an atom of one or more of its electrons such that it becomes "ionized." An ionized atom exhibits an electrical charge as the balance between electrons and protons is disrupted, and instability renders them disposed to form novel combinations with other atoms.[29] And liberated electrons cause cellular damage. "In living tissue, ionization sets off a chain of physical, chemical, and biological changes that can result in serious illness, genetic defects, or death," Catherine Caufield notes.[30] And while somatic injury is limited to the exposed individual, genetic damage is passed on to successive generations. Radioactivity possesses movement, force, and pernicious temporal persistence.

Bruno Latour argues that laboratory research entails constructing a context that allows the object of scrutiny to defy the prevailing story line used to talk about it to date.[31] It is a particularly focused setting in which concept and deed, language and behavior, are juxtaposed and "the greatest degree of intimacy between words and things can be achieved."[32] But as performative engagement with uranium moved from self-contained laboratories to outdoor testing sites, reciprocal give-and-take expanded in scale. "Fallout" was coined in 1945 to describe the radioactive debris descending to earth after an atomic detonation, and uranium generates a myriad of new radioactive particles as it fissions. Radioactive materials exhibit varying half-lives and differing chemical composition.[33] Strontium-90 has a half-life of twenty-nine years and seeks out the bones, causing leukemia. Cesium-137 has a half-life of thirty years and locates in soft tissues, causing various tumors. Radioiodine exhibits a half-life of a scant eight days, but once ingested, typically through milk, it concentrates in the thyroid gland. All provoke effects that often become explicit only over time.

Andrew Pickering and the Agency of Nature

For Andrew Pickering, science as it is practiced in the laboratory presents to us in microcosm our relationship to the world. Drawing from Latour and actor-network theory (ANT), he sketches a world of emergent entanglement where we live "in the thick of things," as agency lies not simply with humanity.[34] In turn, ontology matters, Pickering suggests, as "how we understand the world and how we act in it are systematically bound together."[35] Raymond Murphy explains, "Human agents dance with the moves of nature's actants to form hybrid constructions, with both influencing the other and both having some autonomy. The metaphor of a dance focuses on the joint movements of purposive human actions and non-oriented actions of nature. . . . Human agents and nature's actants are performers in movements influenced by the other's creative actions."[36]

Tuning refers to the effort to discern and then capture observable, repeatable biophysical-material dynamics. It is akin to turning a radio dial to and fro in search of the clearest signal possible. Whether in the laboratory or society writ large, human agency is dependent on harnessing biophysical-material properties, processes, and reactions. This is characterized by *resistance and accommodation*, the former referring to biophysical-material feedback at odds with human intent and the latter referring to reflective human responses to biophysical-material intransigence. This dialectic of resistance and accommodation is rarely guided by instrumentally rational control but is instead characterized by unpredictable transformation and change in which cause and effect are often difficult to fully anticipate beforehand. Tuning contributes to instrumentally rational ends but embroils humanity in a reciprocal dance of agency wherein thought and language rarely proceed but rather follow from the feedback generated through human provocation.[37] Whether in the laboratory or society writ large, we now live in a world where cognition is less anterior to performative engagement with the nonhuman realm than struggling to keep pace.

The ultimate objective of tuning is to stabilize and routinize the harnessing of nonhuman force and change and suppress the dance of agency to within acceptable boundaries. The goal of science and engineering, in turn, is to construct *interactive stability*, reflecting cultural assumptions of duality and mastery.[38] This is exemplified in the mechanistic capture of nature embodied in technological artifacts, big and small, that are reliably employed to achieve particular objectives. Science does not uncover the objectivity of the world so much as it seeks to remake the world by harnessing the performativity of nonhuman entities, processes, and things arrayed in reliable machines and extensive sociotechnical systems that are the backbone of societal complexity.

"Our dealings with the environment are a catalogue of dances of agency writ very large," Pickering suggests, and Cartesian duality is not how the world is but how it is made to appear.[39] Cartesian duality insists on a fundamental separation between the human and the nonhuman, but it is a superficial gloss that we work to construct and sustain. Underneath is hybridity and reciprocal entanglement. A well-functioning nuclear power plant is an *island of stability* generating electricity flowing into a distribution grid, which itself is the focus of procedures and protocols designed to ensure interactive durability. And when the electricity grid goes down, so too does the interactive stability of many other systems, like ripples across a pond.

In turn, once interactive stability has been constructed it must then be sustained, and this is a challenge, as anomalies and mistakes are commonplace over time in large organizations managing complex sociotechnical systems. The social becomes bound up with nonhuman materiality, demanding maintenance and repair, lest things come undone.[40] Institutional structures arise to provide oversight, expertise is promoted and codified, procedures are formalized, and sanctions are leveled to discourage deviance from established protocol. This dynamic demands a mutual disciplining of nature and social organization. "Just as the material contours and performativity of new machines have to be found out in the real time of practice, so too do the human skills, gestures, and practices that envelope them," Pickering notes.[41] Rather than dominion, there is *anthropogenic commitment* and reciprocal reconfiguration. Social organizational patterns predicated on nonhuman agency create path dependencies and begrudging acceptance of novel risks. Given that societal complexity is predicated on the capture of nonhuman performativity, humanity itself becomes captured in fashioning islands of stability demanding oversight and maintenance.[42]

Pickering's conception of islands of stability or sociotechnical arrangements exhibiting durability suggests where risk emerges: the inability to maintain the boundaries of human-nonhuman assemblages. It is a failure to contain the material force and efficacy of nonhuman entities, processes, and things according to preconceived plans. Society is composed of people and things, ideas and matter, and agency is derived from the way these heterogeneous elements are configured. Assemblages of the human and the nonhuman are often fluid, uncertain, and prone to breaking down in surprising ways.[43] Interactive complexity can generate emergent, nonlinear dynamics that are tightly coupled or occur quickly within a sociotechnical system—undercutting interactive stability.[44] As Latour suggests, things often deviate from the "straight path of reason and control" to create a "labyrinth" of unexpected connections and pathways.[45] The history of the US atomic weapons complex reveals numerous near misses and scarcely averted disasters.[46] This highlights the challenges of sustaining interactive stability in a system riddled with complexity and nonhuman materiality. On the surface, atomic technology stands as a testament to dualism and mastery, but its hidden history reveals quite the opposite.

Toward a Performative Conception of the Risk Society

"What would happen if radioactivity itched?" Ulrich Beck ponders.[47] If it did, the public would not be dependent on organizational actors, both corporate and governmental, to intercede on their behalf. Therein lies the conundrum: absent expertise expressed in a fiduciary manner it is impossible for the layperson to gauge the effects of radioactivity. Sociotechnical prowess introduces hazards that the individual, family, or community cannot readily negotiate, while experts often disagree about the degree of risk. Instrumental rationality now exceeds the boundaries of knowledge and containment of the unintended side effects of innovation, while leaving the public in a state of *anthropological shock* as our "personal access to reality" is undermined.[48] "Instrumental (formal) rationality" refers to the conscious, methodical calculation of the most efficacious and efficient means and procedures in the accomplishment of a particular end.[49] In turn, despite brilliant advances in science and technology, in some ways we are more vulnerable as modernity progresses rather than increasingly liberated and free. Commenting on the Chernobyl nuclear accident in 1986, Beck argues, "In the face of this danger, our senses fail us. All of us—an entire culture—were blinded even as we saw. We experienced a world, unchanged for our senses, behind which a hidden contamination and danger occurred that was closed to our view—indeed, to our entire awareness."[50]

Advanced industrialized societies are lurching toward "self-annihilating progress," Beck insists, as innovation is accompanied by a host of unanticipated contradictions.[51] The impacts on health from exposure to low-level radiation, antibiotic resistance, genetically modified food, pesticides, herbicides, synthetic chemicals, global viral outbreaks, and ozone depletion and climate change are emblematic of such dialectical forces. It is a shift in the objective character of risk and the prominence of public and political debate regarding such changes. In turn, Beck argues that we are witnessing the transformation from an industrial to a reflexive or second modernity predicated on *manufactured uncertainties* that prior generations did not confront.[52] "[Manufactured uncertainties] are distinguished by the fact that they are dependent on human decisions, created by society itself, embedded in society, and thus *not* externalizable, collectively imposed, and thus individually unavoidable."[53]

Industrial modernity is characterized by bounded risks primarily centered around the workplace and limited in time between the exposure and/or event and the expression of injury, but reflexive modernity is beset by manufactured uncertainty extending across time and space. Wealth shielded the privileged from the brunt of natural and workplace risks in industrial modernity, but such privileges are eroding as risk increasingly cuts across class distinctions.[54] Who could truly find refuge from radioactive contamination dispersing from Nevada, shifting with the winds, and raining down hundreds, if not thousands, of miles away?

Substances dispersed far and wide through wind, water, and food or present in everyday consumer goods present nearly incalculable exposure and effect on any given individual, elicit consequences that mirror diseases already existing in society, and traverse political and corporeal boundaries with impunity. Within industrial modernity, actuarial analysis and insurance premiums are key mechanisms for managing risk. In late modernity, there are manufactured uncertainties that insurance companies simply do not want to confront. These include accidents involving civilian nuclear power as well as genetic and chemical innovations for which the scope, complexity, and ubiquitous characteristics make indemnification an exercise in absurdity.

Beck's *Risk Society: Towards a New Modernity* was published coincident to the Chernobyl accident, and the timing was prescient, as radioactive debris seamlessly transgressed national borders and evoked effects over an extended period and in an indiscriminate manner that was impossible to mitigate or assess with precision. This dual edge of science and technology is accompanied by *organized irresponsibility* as government bureaucracies and private corporations are rarely held to account for the imposition of hazards that are difficult to empirically verify or dismiss. Contamination is thus normalized, as risks are neither substantiated according to existing scientific protocol and legal norms nor proven to be fallacious and nonexistent.[55] In turn, faith in expertise and institutional actors in society erodes. "Modern society has become a risk society in the sense that it is increasingly occupied with debating, preventing, and managing risks that it itself has produced," Beck observes.[56] "We face the epistemological challenges of a 'changing order of change.'"[57]

Self-annihilating progress, due to manufactured uncertainties, and organized irresponsibility are but a handful of conceptual building blocks in Beck's depiction of the shift from industrial to reflexive modernity, and while he acknowledges the epistemological conundrums, he does not examine prevailing ontological assumptions regarding the interrelationship of technoscience and nature. Indeed, despite Beck's evocative perspective, he does not question the validity of prevailing Cartesian presuppositions as to the duality and separation of humanity and nature. Cartesian dualism insists that the essence of humanity is immaterial and separate from the passive and fixed character of nature. Culture, thought, and language are presumed to drive historical change. The "changing order of change" that Beck describes is not simply epistemologically daunting but ontologically turbulent.

Both Beck and Pickering focus on science but in different ways. Beck is concerned that science outpaces our understanding of self-induced hazards. "By opening more and more new spheres of action, science creates new types of risks as well."[58] Novel risks also signify a shifting ontological landscape. Beck understood that the forward march of instrumental rationality folds back on itself as it is accompanied by hazards that defy traditional risk assessment, logic, and technique, but his provocative critique fails to recognize what Pickering describes as the dance of agency between humanity and nonhuman entities, processes, and things. That is, Beck's risk society thesis does not fully envision the performative interplay of human sociotechnical prowess and nature. The world that Pickering elucidates is characterized by push and pull, as the nonhuman world is active and vibrant and has effects that are both conspicuous and subtle. Absent consideration of this performative dance, it is difficult to fully conceptualize the potential implications of the new era of modernity that Beck sketches out. Indeed, Beck does not articulate an overarching concern with nature-culture dualism, nor does he explicitly locate the contradictions of the risk society as emanating from this false dichotomy.

The Trinity detonation of July 16, 1945, in south-central New Mexico depicted the startling efficacy of plutonium, but more broadly, it prefigured a new era of socionatural hybridity as novel forms of social organization and destruction became possible. We create technological infrastructure but are, in turn, shaped by both the harnessing of nonhuman materiality and those instances in which such forces escape containment. Chernobyl, Fukushima Daiichi, and Three Mile Island are examples of sociotechnical breakdown that, upon closer inspection, reveal the tenuousness of constructed dualism and serve as a repudiation of the belief that only humans act.[59]

Environmental risk is derived from failure to maintain the stability of human-nonhuman configurations such that the force and efficacy of entities, processes, and things exceed that for which they were originally engineered. Manufactured uncertainty is a depiction of the coupled becoming of human and nonhuman agency in unpredictable ways, and environmental hazards are expressed through the vibrancy of nature arrayed in complex assemblages that are productive but not necessarily durable and circumscribed in space and time. *Environmental risk is matter and material agency set in motion in an unbounded manner.* And it is difficult to fully envision what may go wrong, and why, when the agency of nature is bracketed from view.

Both Beck and Pickering offer valuable insights into the "changing nature of change" since midcentury, and a greater melding of their ideas opens up new theoretical and empirical possibilities. Both sketch out not a postmodernist account but a critical modernist perspective in which undisciplined instrumental rationality is at the root of the problems they describe. Together, their scholarship highlights both the epistemological and ontological challenges of late modernity and, ideally, a more prudent path forward.

The Road Ahead

Modernity promises the stepwise evolution of human societies based on rational thought and behavior, and science and technology are crucial symbolic and institutional components of this evolution. Indeed, they would alleviate human suffering through the command and control of nature. Science and technology have bequeathed to humanity miraculous inventions but rarely without accompanying trade-offs. This is the untoward development that Ulrich Beck and Andrew Pickering sketch out. It is the paradoxical and incalculable amid breathtaking sociotechnical achievements. The ontology of modernity presupposes that only humanity acts and that the world is, in principle, mechanistic and thus decipherable.[60] In turn, we bracket from view our performative intertwining with the nonhuman world. Accelerating unintended side effects, however, point to the need to scrutinize how science is practiced and how the world is composed.

Open-air detonations did not entail the diffusion and dilution of substances within a natural world separate from humanity. Things continually came back around in ways illustrating that nature is not an empty container but responsive to human provocation. In important respects, nature continually challenged prevailing discursive constructions through which it was understood once the flash and shockwave subsided. What is later construed as history typically focuses only on human activity while failing to account for the resistance and accommodation of the nonhuman world.[61] Lessons that could be learned, in turn, rarely resonate. And it is this dance of resistance and accommodation that AEC and military officials were not fully cognizant of despite being repeatedly buffeted by the unexpected.

Together, Beck and Pickering underscore what humanity least wants to hear: humility and prudence in our engagement with nature are virtues to be embraced. Apace with instrumental rationalization in the modern age is the devaluation of nature, and we must seek greater accommodation between the social and the nonhuman world. Ecological rationality, Murphy argues, embodies the reformulation of instrumentally rational social action to account for the complexity, uncertainly, and feedback of nature, in its various guises.[62] This encompasses greater awareness of ecological processes in and of themselves and as inter-twined with human activity. It encompasses moving beyond the "natural ontological attitude of modernity" wherein humanity presumably acts through the fixed properties of the biophysical-material world. The biophysical and material, then, is assumed to be the malleable substrate subject to human intentionality.[63] In contrast, we scarcely master but prodigiously manipulate, and the legacy of open-air atomic testing illustrates the liabilities of confusing manipulation with mastery, knowledge for wisdom, and of being reticent to recognize the hybridity and entanglement of social organization and the nonhuman world.

Methodological Considerations

This study is a qualitative and single-case analysis. A single-case study is particularly suitable to inquiry regarding the *how* and *why* of a unique event with the potential to foreground important dynamics beyond the case under examination.[64] The data are derived from archival research and consist of internal AEC documents, the commission's publications intended for the public, and court opinions and congressional inquiries.[65] The analysis is structured chronologically and focuses on key decisions and events shaping the commission's actions and nonactions relative to the management of fallout. Of particular concern is how the risks of fallout were defined and redefined, or inhibited from redefinition, because of dynamics outside the organization, organizational structure, and the assumptions of key organizational actors. This comprises three levels of analysis: (1) the broader context in which the organization is situated; (2) the characteristics of the organization itself; (3) the cognitive, sense-making practices of key officials or units within the organization. Although research often privileges a given level of analysis, organizational mistake, misconduct, and disaster typically emerge from a confluence of dynamics at all three levels.[66]

The protocol guiding the archival research was informed by Diane Vaughan's model of identifying patterns, assessing their validity relative to primary and secondary data sources, and striving to articulate the essential factors shaping the dynamics under examination. It is a process of remaining aware of contradictory data and assembling and reassembling the factual and theoretical account as new information is encountered. Further, one must not force existing conceptual schemes onto the case or make definitive conclusions before examining the data in depth, and it is crucial to elucidate the dynamics of interest as situated within the organizational and historical context in which they occurred.[67] As John Scott notes, "Texts must be studied as socially situated products."[68] This entails coming to "understand how people in another time and place made sense of things," Vaughan cautions.[69]

The Nuclear Testing Archive contains more than 375,000 documents. Confronting voluminous archival documents necessitates a strategy for searching the materials contained therein.[70] I began, first, by searching for those documents illustrating how officials in the AEC came to construct an operational protocol regarding the hazards of fallout before testing commenced. Second, the research then focused on the manner whereby officials successively worked to make sense of the intensity, extent, and characteristics of fallout as testing proceeded. Third, I searched for archival materials illustrating the propensity for officials to substantively alter operational protocols or indeed reluctance to do so. Fourth, as the research progressed, the focus shifted to the contrast between the organizational construction of risk articulated within internal AEC reports and memoranda and that expressed in the commission's documents intended for the public.

Theoretical Considerations

In qualitative research, there is often a balance between the search for emergent theoretical constructs, that is, the production of grounded theory, and reliance on existing theoretical ideas lending coherence to the interpretation under development. This study relies on both emergent theoretical constructs derived through an inductive focus and existing theoretical scholarship consistent with key dynamics uncovered within the archival data.

Theory is typically defined as a series of interconnected statements that define, describe, and explain some aspect of the social world. The goal is to impart order and meaning onto "empirical uniformities" or observable, repeatable dynamics.[71] Theories exist at different levels, ranging from a focus on concrete social and organizational behavior to abstract, macrostructural statements. It is akin to flying in an airplane.[72]

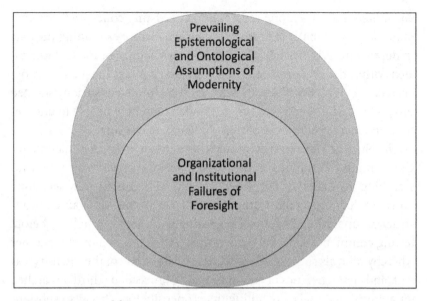

Figure 1.2. Theoretical focus.

At a given height, some things are visible, while other things neces-
sitate a change in elevation before they come into focus. In turn, this
study draws from the literature on the dark side of organizations, which
is middle-range theory and similar to flying at fifteen thousand feet.
"Middle range" refers to theory pitched at the level of specific empirical questions.[73] I also draw from the theoretical scholarship of Ulrich
Beck and Andrew Pickering, and this is analogous to flying at thirty-five
thousand feet.

Figure 1.2 articulates this distinction. The literature on the dark side
of organizations is concerned with failures of foresight and blind spots
in decision-making. At a more abstract level is the scholarship of Beck
and Pickering. AEC and military officials engaged in the mismanagement of fallout due to a variety of blind spots but particularly those
rooted in an erroneous conception of the interrelationship of the human
and the nonhuman, the social and the natural. AEC officials understood
that open-air detonations generated a connection between the organizational, technological, and ecological, and yet in substantive terms, this
intertwining was remarkably invisible. It was understood that radioactive debris ascended into the atmosphere, but the idea that radioactivity

intertwined with grass, cows, milk, and babies was less clear. The failure of AEC officials to anticipate and plan for radioactive contamination stemmed from organizational failures of foresight rooted in a reticence to acknowledge that the nonhuman world is teeming with complexity, movement, and pushback.

There are significant liabilities to formal-instrumental rationality premised on the assumption that nature is inert and malleable. Large, complex organizations are the carriers of such liabilities, and the hazards born of organizational failures of foresight are often expressed through ecological disruption, contamination, and change. Nature is a "medium of interaction" that powerful actors in society seek to manipulate, and they often impose on others the injurious effects of such manipulation, particularly when things go wrong.[74] Open-air testing involved the unprecedented production of environmental risk over a broad expanse of geographic space. Officials were enmeshed in an uncertain entanglement, with winds moving in different directions at variant altitudes, varied topography, scattered rainfall hundreds of miles from Nevada, patchwork patterns of radioactivity upon fallout, and the migration of radionuclides through the food chain. To overlook such vibrant and bewildering socionatural dynamics is to overlook the central conundrums that AEC and military officials encountered.

2

Trinity and Lessons Not Learned

The Manhattan Project was founded in June 1942 to build an atomic bomb before the Third Reich could do so. Army Major General Leslie R. Groves served as the project's military head, while the theoretical physicist J. Robert Oppenheimer led the scientific effort. The US Army ran the project, under the guidance of Groves, while the scientists at Los Alamos reported to Oppenheimer. The effort included construction of industrial facilities at Oak Ridge, Tennessee, and Hanford in Washington State to produce enriched uranium and plutonium. At the project's peak, it employed 130,000 people and overall cost $2 billion (approximately $30 billion today).[1]

The Pajarito plateau, forty miles northwest of Santa Fe, New Mexico, was an area that Oppenheimer knew well, having spent summers in New Mexico growing up. Construction at Los Alamos, code-named Site Y, began in the spring of 1943, and the population grew from several hundred to nearly six thousand by the end of the war. It was a city that did not exist officially, and a mix of civilian and military personnel worked under strict secrecy while focusing on a singular objective that many were not fully aware of until the bombing of Hiroshima.[2] Soldiers, physicists, construction workers, engineers, chemists, mathematicians, generals, spouses, and babies all made for a remarkable community. Rebecca Solnit explains, "Los Alamos is beautiful. It is a mesa 7,000 feet above sea level, even now accessible only through narrow roads that wind along steep slopes and precipices, a fortress of rock thrust up into the deep blue ever-changing New Mexico sky."[3]

Scientists first focused on a gun assembly device in which one subcritical mass of uranium is propelled into another to initiate a sustained chain reaction.[4] By 1944, attention also turned to an implosion device in which criticality is achieved through squeezing plutonium by way of conventional high explosives. And after years of labor, things took a turn. "During the spring of 1945, the pace at Site Y quickened, and the

Figure 2.1. Trinity site, pre-detonation, 1945. (Photo courtesy of Los Alamos National Laboratory)

tension mounted. Work for many became frantic, and many of the staff disappeared into the desert for weeks at a time," Jon Hunner notes.[5]

Jornada del Muerto (Journey of Death) playa in south-central New Mexico constituted an arid, risky shortcut for sixteenth-century Spanish settlers traveling north to Santa Fe or south to El Paso. The ancestral lands of the Mescalero Apache, it lies at the upper reaches of the Chihuahuan Desert extending south into Mexico. Los Alamos is defined by both a spot on this playa and an event named Trinity. Oppenheimer coined the code name, and it is thought to be inspired by select lines of a John Donne poem referencing Christian theology, which begins, "Batter my heart, three-person'd God."[6] Here humanity embarked on a productive and destructive entanglement with nature in unprecedented ways.[7]

The photograph in figure 2.1 was taken prior to development of the site. The vantage point is looking east toward the Oscura Mountains. The area is characterized by a hard physicality. As Gerard DeGroot observes, it "looked blasted even before it was bombed."[8] Richard Rhodes notes, "The Jornada was host to gray hard mesquite, to yucca sharp as the swords of samurai, to scorpions and centipedes men shook in the morning from their boots, to rattlesnakes and fire ants and tarantulas."[9] The Oscura, or "dark," range rises four thousand feet and frames an imposing eastern boundary; to the west is the Rio Grande River, running from one end of the state to the other; to the south are the rugged San Andreas Mountains; and northward lie scattered ranches and towns amid intermittent mesas and canyons.

Planning for a test of the more complex plutonium device began in March 1944. There were questions whether it would work. In the gendered language at Los Alamos, failure was designated a "girl" and success a "boy."[10] It was an immense undertaking accompanied by scientific experiments, diagnostic measurements, photographic documentation, and distinguished visitors from around the country. Construction included barracks, a technical stockroom, a mess hall, an infirmary, and heavily fortified concrete bunkers ten thousand yards (5.7 miles) to the north, south, and west of ground zero. N-10000 housed recording instruments and searchlights, W-10000 was equipped with searchlights and high-speed cameras, and S-10000 served as the control bunker from which the countdown sequence was initiated.[11] Base camp served as Project Trinity headquarters and lay five miles beyond S-10000. In the final weeks of preparation, at least 250 personnel from Los Alamos were working at Trinity, and more than 100 soldiers were guarding the site.[12] Film badge records indicate that 355 people accessed the test site on July 16.[13]

Prior to the test, Army personnel searched the countryside and identified sixty-three ranches within thirty miles of ground zero, the closest just twelve miles north.[14] Surrounding ranches and towns were not notified beforehand, to maintain secrecy; but more than 140 soldiers were stationed north of ground zero in the event evacuations were necessary, and this contingent included one five-person radiological safety monitoring team stationed at Trinity to assist with offsite monitoring. The evacuation detachment was prepared to move up to 450 people. Twenty Army intelligence personnel were positioned in towns up to one hundred miles away and given barographs, seismographs, and radiation meters to record earth shock, the blast wave, and fallout intensity—data needed if the Army confronted lawsuits due to property damage—and to observe public reaction to the detonation.[15]

In mid-June, Joseph Hirschfelder and John L. Magee, physicists focusing on postshot radiological effects, drafted a memo suggesting that offsite fallout could be more intense over a wider area than anticipated.[16] It was met with skepticism, particularly by General Groves, but prompted creation of the evacuation contingent. The peak threshold of gamma radiation exposure triggering evacuation was 15 roentgens per hour, but any evacuations would garner public attention and be enacted only if deemed absolutely necessary. "Secrecy cloaked the Manhattan Project from the outset. In the hierarchy of project goals it stood second only to making bombs that worked," Barton Hacker observes.[17]

Four two-person teams were prestationed in the outlying towns of Socorro, Nogal, Roswell, and Fort Sumner and tasked with monitoring fallout. They were in contact with Base Camp by radio or telephone.[18] Among hundreds of participants, only four two-person teams and one five-person contingent were directly involved in measuring ground-level fallout offsite. These thirteen people were assisted by Army intelligence agents stationed in outlying towns, but their primary role was to collect data to safeguard the Army's interests if civil lawsuits should arise. "Safety never commanded topmost concern at Los Alamos. Getting the job done came first. In testing the bomb, however, safety may have ranked even lower than normal," Hacker notes.[19] Japanese surrender was inevitable but not necessarily imminent. Moreover, the Manhattan Project had advanced scientific and technological innovation, as evidenced by thousands of new patents, but after spending $2 billion, there was pressure to deliver an end product.[20]

Trinity as a Coupled Performance

Trinity was an intricate tacking back and forth between human ingenuity and the materiality of nature—a coupled, socionatural performance. Physicists, chemists, engineers, metallurgists, meteorologists, soldiers, and military officers were entangled with nature small enough to hold in one's hand but constituting forces vast and impersonal. And although the scientists and engineers could control virtually all aspects of the test, the weather they could foresee in only a tentative manner. Another uncertainty was the explosive force of the device.[21] "Fallout would be governed by two great unknowns: the yield of the bomb and meteorological conditions," DeGroot observes.[22] Both exceeded expectations.

The prevailing winds at Jornada del Muerto are from the west, but an overarching concern was a shift in direction carrying fallout over the town of Soccoro to the northwest or Albuquerque and Santa Fe to the north.[23] There were little historical data focusing on the Jornada del Muerto area to work with, and the nearby Oscura Mountains caused variation in local wind direction and speed.[24] Jack M. Hubbard served as the meteorological forecaster for the Trinity test, and he sketched a list of ideal weather conditions. Temperature, humidity, wind speed and direction, high pressure, low pressure, precipitation, upper-level troughs, lower-level troughs, temperature inversions—all were important considerations. The ideal scenario was clear skies, low humidity, high visibility, no inversions, and light surface winds moving east-northeast from ground zero.[25]

The test was scheduled for July 4, but the "gadget" was not quite ready. The test was rescheduled for July 16, but it coincided with a forecast of inclement weather.[26] July is monsoon season in New Mexico. The hottest time of year is also when much of the desert Southwest receives what little rainfall there is each year, and it often comes as a torrent or not at all.

Hubbard had no discretion in the scheduling of the test, and by mid-July President Harry Truman would be attending the Potsdam Conference alongside Soviet Premier Joseph Stalin and Winston Churchill. A "boy" at Trinity would improve Truman's bargaining position and potentially shape the postwar geopolitical landscape in a manner congruent with US interests.[27] As Groves later confided, there was immense pressure to give Truman what he wanted.[28] The test would not be synchronized around the optimal weather conditions but geopolitical considerations.

Hubbard could foresee problems on the horizon and appealed for a date after July 16 when the forecast was more favorable, but he confronted a "fait accompli."[29] The test would not be delayed. And on July 15, few of the items on Hubbard's wish list were likely to be met in the next twenty-four hours, and as the day wore on, things only got worse. Ferenc Szasz observes, "Everything he did not want was present: rain, high humidity, inversion levels, and unstable winds. Rain was undesirable because it might 'scrub' the cloud and bring down a high concentration of radioactive particles in a small area. Safety lay in spreading the material as widely as possible. The only item on which everyone agreed was that the bomb should not be exploded in the rain."[30]

The gadget was so novel that some people harbored nagging concerns that the intense heat might interact with nitrogen and set the atmosphere on fire. Mathematical calculations indicated that this was not likely, but the very idea was unsettling. There was also concern regarding "looping."[31] If the device had only a moderate yield and the mushroom cloud encountered decreasing atmospheric pressure, it might spread out and drop in temperature, and if it then encountered a strong inversion or layer of warmer air, it could bounce back to earth and inundate the test area with radioactivity. And then there was the possibility that it could produce a thunderstorm.[32] Loud explosions interacting with atmospheric instability and high humidity, it was theorized, might induce precipitation. That an atomic detonation could change the weather was an idea that would later be discounted, but at the time, it was anything but clear. If the detonation did not set the atmosphere on fire, create its own thunderstorm, get scrubbed by a distant storm, head for where the people were, or bounce back to earth soon after liftoff, then everything would go according to plan.

At 4:00 p.m. on July 15, the key players assembled at McDonald Ranch, two miles from ground zero, to debate whether to proceed with a 4:00 a.m. detonation time. This group included Groves, Oppenheimer, Hubbard, General Thomas Farrell, Trinity project director Kenneth T. Bainbridge, and others.[33] The McDonald Ranch was built in 1913 and had been a cattle and sheep outfit before the family was forced to evacuate in 1942 to make room for a gunnery range.[34] The subcritical pieces of the plutonium core were transported here in the backseat of an Army sedan, "like a distinguished visitor," and assembled in the master bedroom prior to insertion into the gadget.[35] The house is built of adobe mud bricks. The interior floors are wood, as is the front porch. A windmill sits outside, and the roof is covered with tin. It would look at home in a John Wayne movie and is a quaint backdrop for the conduct of innovative science and technology.

Groves pulled Oppenheimer aside. The long, thin, chain-smoking physicist and rotund military man retreated to an adjacent room. When they returned, it was announced that a final decision would await a

2:00 a.m. conference to look over the latest weather data. At 11:00 p.m., the wind was blowing toward N-10000, and a heavy air mass hung over Trinity; three hours later, a thunderstorm pounded the site with rain and wind gusts up to thirty miles per hour.[36] Dignitaries were assembled twenty miles northwest of ground zero, Truman was half a world away at Potsdam awaiting word, resources were assembled to evacuate and triage causalities incurred onsite or offsite, and the device was perched atop the shot tower, nestled in a rectangular metal shed; but the weather was abysmal. The one thing that Jornada del Muerto was known to have little of was all around.

At 2:00 a.m., many of the most brilliant theoretical minds of their generation and results-oriented, war-weary military officers stood arguing the situation at Base Camp as lighting struck overhead, the wind howled, and rain threatened. It was as if God were in attendance. Szasz explains, "No dramatist could have staged the setting for the Trinity atomic detonation with more power than it had in fact. The tension could be felt everywhere. Cloud-to-could lightning lit up the tower at intervals, and the rain fell in torrents. The ground was strewn with puddles, and the numerous scientists worried about possible electrical shorts."[37]

Oppenheimer was nervous, pacing. Some Los Alamos scientists advised him to cancel the test.[38] Groves was angry about the weather, even though Hubbard clearly forecast unfavorable conditions.[39] The 4:00 a.m. schedule was abandoned, but Hubbard suggested that conditions would improve around dawn. The test would commence at 5:30 a.m.—just before morning light. It would be delayed, again, only if there were rain at ground zero but would commence even if there were rain showers in the area. It was going to happen, that is, even if sacrifices had to be made.[40]

Military officials and scientists were on the cusp of a giant leap forward but were plagued by that which has long confounded humanity. Wind and weather exhibit materiality and agentic force. Indeed, it is hard to imagine a more ubiquitous example of the propensity of nature to act as well as assist, resist, frustrate, and wreck human plans. Marq de Villiers notes,

> The search for an understanding of wind and the weather it brings has been a constant of human history, for wind is a changeling that can bring

blessings but also hard times. Wind can be soft and beguiling, seductive; the caress of a gentle breeze stroking the skin is one of the great pleasures of the human adaptation to our natural world. But sometimes wind can be deadly, intensifying violently into a kind of personal malevolence. Like a short-tempered and belligerent god, the wind has a power that can appear arbitrary, excessive, overwhelming, devastating, uprooting trees, wrecking houses, sinking ships, battering people, scarring psyches.[41]

Just after 4:00 a.m., the clouds were breaking, and the wind was stable and moving toward the northeast. Stars were visible in the night sky.[42] At around 4:45 a.m., Hubbard informed Bainbridge the weather was not ideal but good enough.[43]

The timing sequence was initiated at 5:10 a.m., and the countdown began. At five minutes to detonation, a green rocket launched skyward amid the short wail of a siren; the two-minute signal rocket fizzled, but there was a longer wail of the siren. At one minute, another signal rocket ascended into the darkness. At thirty seconds, four red lights flashed on the console at S-10000 as the needle on a volt meter jumped left to right. At ten seconds, a gong sounded in the control bunker.[44] As time approached zero, Tchaikovsky's *The Nutcracker Suite* unexpectedly filled the air as a distant radio station crossed wavelengths with the frequency at Trinity.[45] At the most inopportune time, the light, airy, jocular cadence intruded on the most audacious effort in history to pry into the secrets of primal nature: big science, exotic technology, and the brusque and determined logic of military prerogatives amid oboes, flutes, and violins.

At 5:29:45 a.m., there was heat, blinding light, rolling thunder, a jarring shockwave, and a towering mushroom cloud. In the masculinist jargon of the scientists and engineers, it was a boy.

The gadget employed conventional high explosives shaped into a series of lenses and surrounding two hemispheres of plutonium. When the plutonium was squeezed from roughly the size of an orange to that of a lime, it jump-started a self-sustaining chain reaction and the era of atomic military technology. General Thomas Farrell, deputy to General Groves, remarked, "The effects could well be called unprecedented, magnificent, beautiful, stupendous, and terrifying. No man-made phenomenon of such tremendous power had ever occurred before. The lighting

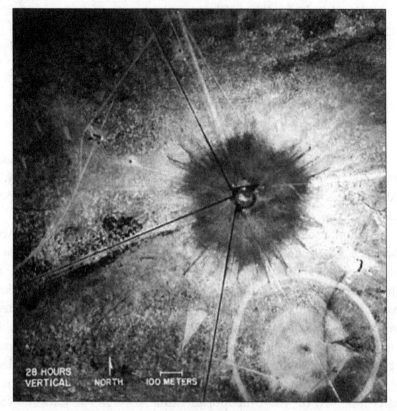

28 HOURS
VERTICAL   NORTH   100 METERS

Figure 2.2. Trinity site, post-detonation. (Photo courtesy of Los Alamos National Laboratory)

effects beggared description."[46] The physicist Isidor Rabi proclaimed, "It blasted; it pounced; it bored its way right though you. It was a vision which was seen with more than the eye. . . . A new thing had been born; a new control; a new understanding of man, which man had acquired over nature."[47]

Figure 2.2 is a photo of the Trinity site twenty-eight hours after the detonation. The fireball scorched the desert as it lifted alluvial soil within the stem and mushroom cloud. The one-hundred-foot tower was all but gone. A greenish, glassy substance, Trinitite, coated the crater surface, and mangled jackrabbits lay scattered more than eight hundred yards from ground zero.[48]

Twenty-eight months of frenzied and secretive work, high on a remote mesa, 210 miles away, jarringly came to fruition on the Jornada del Muerto. The same model was detonated over Nagasaki, Japan, less than a month later. That this place in south-central New Mexico is indelibly linked to a Thursday morning in Nagasaki is hard to imagine standing amid the creosote and ocotillo, the sun and cheat grass. At Trinity, everything feels impossibly distant. The force and scale of atomic military technology, however, readily transgress conventional conceptions of time and geographic space. Humanity harnessed a primal force of nature but was captured, in turn, by the newborn capacity to provoke the equivalence of mass and energy.

The Fluidity of Fallout

The rearrangement of matter lies at the heart of the atomic age but in tandem is the introduction of "fallout" into the vernacular. Fallout demonstrates the liabilities of atomic military technology not simply in war but in peacetime.[49] Trinity depicted technique and precision, but the aftermath was fluid and indeterminant. The Manhattan Project moved beyond the orderly industrial facility, the cleanliness of the laboratory, and the pacing of the seminar room to entanglement with the world. And one of the most remarkable achievements in physics and engineering was followed by chasing radioactive debris across the desert.

At first, debris hung over ground zero.[50] As the upper level of debris raced to forty thousand feet, lower levels turned in a corkscrew fashion.[51] Such unbalanced pirouettes were evidence of divergent winds. Soon, the tempo picked up. From Base Camp, Cyril S. Smith, a metallurgist, observed, "There was a dust cloud over the ground, extending for a considerable distance. . . . The obvious fact that all of the reaction products were not proceeding upward in a neat ball but were lagging behind and being blown by low altitude winds over the ground in the direction of inhabited areas produced very definite reflection that this is not a pleasant weapon we have produced."[52]

Fallout lumbered toward the northwest, where the town of Socorro lay thirty-five miles distant, but around 6:15 a.m., the wind veered toward the north-northeast.[53] Offsite monitors detected a "skip" of fifteen

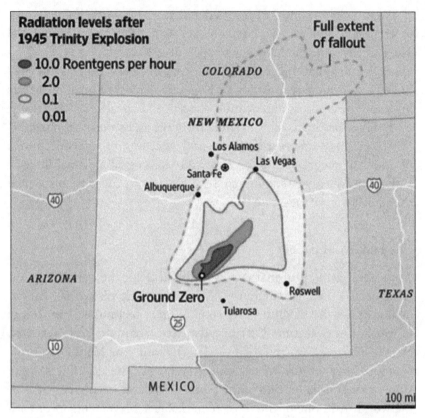

Figure 2.3. Trinity detonation fallout trajectory. (Courtesy of Los Alamos Historical Document Retrieval and Assessment Project)

miles from ground zero before they encountered ground-level gamma radiation.[54] They also encountered a breakdown of radio contact, inhibiting coordination, and at about twenty miles from ground zero, they exceeded the range of radio contact with Base Camp.[55] "Vagaries of wind and terrain produced anomalies in the fallout pattern, a problem compounded by lack of radio communication between monitors. They could trade information only when their paths crossed or at prearranged meetings," Hacker notes.[56]

Figure 2.3 charts the trajectory of the Trinity fallout. At 8:00 a.m., a reading of 15 roentgens per hour was recorded outside Bingham, New Mexico, and then higher readings were detected further east in Hoot

Owl Canyon, twenty miles from ground zero.[57] Gamma radiation registered 15–20 roentgens per hour. As detailed in a top-secret memo by Colonel Stafford Warren, chief of the Manhattan Project Medical Section, radiation was intense enough to "cause serious physiological effects."[58] Such readings were surprising but not alarming, as there were no residential dwellings thought to be in the area.[59]

The evacuation detachment was dismissed at 1:00 p.m., as it was assumed that no further problems were likely to arise and despite gamma readings from the Bingham area as high as 7 roentgens per hour around 3:30 p.m.[60] "Army officials seemed to be in a hurry to sound the 'all clear' after the Trinity blast," Thomas Widner and Susan Flack explain.[61] Later in the afternoon, however, a change in wind direction pushed radioactive debris westward and back toward the town of Carrizozo, New Mexico. Evacuation was hastily debated. It was decided to conduct more monitoring, and by the time such efforts were concluded, debris had dispersed from the area.[62] By evening, it was thought that fallout was no longer a pressing issue, but a surprise was awaiting the next day.[63]

On Tuesday, July 17, monitors returned to Hoot Owl Canyon and stumbled on a ranch that was not listed on a map that Army intelligence personnel had created. The Ratliff ranch was occupied by an elderly couple and their grandson. Mr. Ratliff described fine ash littering the ground as if it were snow.[64] A second ranch, not listed on any maps, was also discovered in the area.

Rural residents downwind were not advised to avoid local produce and milk. As radioactivity settled on pastures and gardens and entered the water supply, there was the potential for internal ingestion in addition to simply gamma radiation exposure. Many ranches had a cistern collecting run-off from metal roofs, and there was rain throughout the area on the evening of July 16. The sole focus of the offsite monitoring effort was external radiation exposure and the possibility of acute effects. There was no concerted effort to assess the internal dose from inhalation or ingestion of alpha or beta radiation, although such considerations were well known at the time.[65] Los Alamos personnel were routinely subject to nasal-swab analysis to assess the inhalation of radioactive material, but such procedures were not employed among residents downwind.[66]

Indeed, after Trinity, there was no systematic effort to assess the long-term health effects of residual radioactivity. There were periodic studies of flora, fauna, and soil but only a clandestine monitoring of select ranches known to have encountered high gamma radiation. Personnel revisited several families over time, particularly in the Hoot Owl Canyon area, but never disclosed why they were there, asked only vague questions regarding health status, and performed only cursory visual examinations.[67] Because of this, in tandem with the problems associated with radio communication among the monitors as well as the scarcity of offsite monitors, it is uncertain the full extent of the exposure of downwind residents to transient energy in its various guises.[68]

The visible segment of fallout dissipated around Vaughan, New Mexico, approximately one hundred miles northeast of ground zero.[69] Less visible elements traveled further. A paper mill in Indiana that manufactured strawboard used in the packaging of Eastman Kodak industrial X-ray film took in contaminated river water. When the film was developed, it was marred by mysterious pinpoints and blotches, and a little detective work concluded that the cause must have been Trinity, an event made public after the bombings of Japan in August. As open-air testing began in Nevada, in 1951, the film industry often detected what the US public was only scarcely informed of: fallout traveled far and wide.

The device exceeded expectations with regard to explosive yield, but it was inefficient. Of the 6 kilograms of plutonium, around 4.8 kilograms did not fission and dispersed offsite.[70] It has been detected on Chupadera Mesa, twenty-eight miles from ground zero.[71] Plutonium exhibits a restless materiality, and with a half-life of twenty-four thousand years, it epitomizes environmental persistence. It is the "world's most dangerous" element, and it is one of the most enigmatic.[72] Glenn Seaborg, who identified plutonium in 1941, attested, "Plutonium is so unusual as to approach the unbelievable. Under some conditions it can be nearly as hard and brittle as glass; under others, as soft as plastic or lead. It will burn and crumble quickly to powder when heated in air, or slowly disintegrate when kept at room temperature. . . . It is unique among all of the chemical elements. And it is fiendishly toxic, even in small amounts."[73]

In a top-secret report to Groves dated July 21, Stafford Warren revealed that Trinity fallout formed a "very dangerous hazard" thirty miles wide and extending nearly ninety miles north-northeast of ground zero. "The distribution over the countryside was spotty and subject to local winds and contour." Radioactivity was detectable two hundred miles away four days after the test, and there was "still a tremendous quantity of radioactive dust floating in the air." Warren concluded with a pointed observation: "It is this officer's opinion that this site is too small for a repetition of a similar test of this magnitude except under very special conditions. It is recommended that the site be expanded or a larger one, preferably with a radius of at least 150 miles without population, be obtained if this test is to be repeated."[74]

Power is rooted in appropriated nonhuman materiality. Power is also the capacity to shift the deleterious side effects onto others without fear of accountability. Trinity was meticulously mapped in the months leading up to mid-July, but the aftermath was sketched in only broad terms. If the detonation was akin to the tango, precise and tightly choreographed, management of fallout was distinctly improvisational. The evacuation detachment did not travel with the debris cloud but stood idle unless called on, by which time the heaviest fallout would have long passed overhead. The range of the car radios employed by offsite monitors was not delineated until they gave out. There was no systematic effort to measure the internal dose received through inhalation or ingestion. The ranches scattered across the countryside were not carefully accounted for prior to the detonation. And a long-term survey of the potential health effects due to external or internal radioactive exposure never occurred.

In 2020, the National Cancer Institute (NCI) released a series of studies assessing the potential health effects of Trinity in response to growing concerns that it left behind a legacy of disease.[75] The NCI surmised, "Some people probably got cancer."[76] The NCI stressed that it is impossible to assess with certainty how many people were affected or to conclusively link any individual cancer to Trinity, given the lack of data.[77] Illness and death from cancer were not recorded for decades after Trinity in the counties most affected by fallout. Nonetheless, Elizabeth Cahoon and colleagues estimated that "as many as 1,000 or as few as 290 cancers have already occurred or are projected to occur in the future that would not have occurred in the absence of residual radiation exposure from Trinity fallout." The most likely illness is thyroid cancer from ingestion of radioiodine in fresh cows' milk.[78] Moreover, Harold Beck and colleagues estimate that the majority of unfissioned plutonium came to earth in a band stretching eighteen to sixty-two miles northeast of ground zero and that approximately 80 percent was deposited within the state of New Mexico, but they argue that this dispersal, primarily over uninhabited terrain, does not constitute a hazard to public health.[79]

The Lessons of Trinity

Trinity illustrated that wind direction and speed are capricious and that the mixing of the fireball with alluvial soil ensures changes in terrain create hot spots that are impossible to predict beforehand. This necessitates a large and flexible offsite monitoring effort to account for the vagaries of nature. In Nevada, however, offsite monitoring began in a tepid manner and grew more robust only after fallout proved a substantive threat. Moreover, an exclusive focus on external gamma radiation neglects how uneven redeposition interacts with local food production. Many ranches downwind of Trinity collected rainwater, raised gardens, and drank cow or goat milk from their own stock or local sources.[80] Numerous residents downwind of Nevada did the same. Living close to the land accentuated accumulation of internal emitters of radiation. At Trinity, it was assumed that internal emitters were not a public health concern, except when external gamma radiation levels were high. This was an unwarranted oversimplification. But everything was new at Trinity. This assumption, however, persisted over eleven years of testing in Nevada. Trinity was accompanied by a policy of strict secrecy. The detonation was seen and heard for hundreds of miles, and the Army distributed a press release insisting that it was an accidental detonation of an ammunition cache.[81] Short-term disinformation may be justified when a reasonable threat to national security can be demonstrated. Strict secrecy about testing in Nevada, however, made it nearly impossible for downwind residents to protect themselves by engaging in simple decontamination procedures. There would be no more detonations in south-central New Mexico, but Trinity was small in yield compared to what was to come. And within a radius of 150 miles downwind of the NTS lived thousands of people scattered in rural towns, far-flung ranches, mining outposts, and Native American communities.

At Trinity, a new ontological stance between humanity and nature was achieved, and it is as demanding epistemologically as the science from which it is derived. Further, it is an ontology of emergence in which new elements and old are taken up by the individual body, and it is characterized by invisibility amid vibrancy and change. The phrase "permissible dose" is indicative of this new relationship between the state and its citizens. No longer is practicing for war conducted in isolation from the individual, family, and community, and "permissible dose" signifies the degree of imposition defined as acceptable—as dictated by coalitions of experts and formally recognized institutional bodies. Trinity was a rupture with regard not simply to destructive military capabilities but also to the entanglement of science, technology, and the body. One did not have to be on the battlefield any longer to witness the devastation of military technological prowess. Residents of rural downwind towns throughout the Southwest came to literally embody the priorities and decision-making criteria of civilian and military officials.

Uranium was once the overburden or waste sifted through in the search for more reputable things, such as silver.[82] Now it plays a role in a multitude of potentially catastrophic as well as practical activities. And as uranium has become entangled with the human and the social, it has contributed to a dialectical rebound effect. Many of the most vexing problems of modernity are rooted in the side effects of advancement in science and embodied in technological innovation. They are the unintended consequences of productive, instrumental activity. They are new risks, for which there are few historical reference points to guide present decision-making.[83] They constitute a "new species of trouble."[84]

Environmental risk is matter and material agency set in motion in an unbounded manner, and fallout was a central actor during the era of open-air testing. It did not act with intention but nonetheless forged connections over space and time between the animate and the inanimate, the nonhuman and the human. Brimming with energy in a variety of chemical concoctions and the materiality to provoke biophysical change, it encompassed the potential not simply for crescive organizational disaster but for disasters, plural. Fallout is the dialectic of the atomic age. It is the material, ecological, and social contradiction arising at the historical moment when the power of humanity to transcend the natural world appeared certain. Despite repeated promises, a "clean" atomic bomb has never been achieved.[85] And while physicists became more adept at predicting the radioactivity that a given design would produce, the way transient energy spread across the landscape remained resistant to calculation and control.

Society does not stand apart from the nonhuman world but is an expression of its force and vibrancy, and instrumental rationality premised on a fundamentally erroneous conception of the interrelationship between the human and the nonhuman, society and nature, elicits unintended and increasingly destructive outcomes. The vision animating the Scientific Revolution of the sixteenth century, however, largely continues apace with increasing evidence that nature is not passive and mechanical.[86] To envision the social in the thick of things is to adopt a perspective that is cognizant of the push and pull, action and reaction, between humanity and nonhuman nature. Few things so dramatically embody the inadequacies of conventional ontological conceptions of the world as uranium. We are embroiled in a productive but perilous entanglement with a rock. This dance must be carefully choreographed to remain within specified boundaries. It is a rock that demands discipline and restraint.

Today, Trinity is demarcated by a lone stone obelisk standing twelve feet high. The site is accessible to visitors two days a year. It is as if this allows the public to meander and commemorate but not spend too much time dwelling on the abrupt alteration in the capacity to inflict destruction, utilizing nature as the medium through which it is enacted.

At a recent visitor's day, representatives from White Sands Missile Range sat on folding chairs, resting their elbows on a folding card table, underneath a sign proclaiming, "Free Answers."[87] In truth, there are more lingering questions than definitive answers at Trinity. Iver Peterson argues, "In that flash the major technical questions about atomic fission were instantly resolved, only to be replaced by moral and ethical questions that have proved infinitely harder."[88] Ellen Meloy observes, "Here was a desolate void on which to stage a mesmerizing spectacle of engineering, the grand drama of abstract science made fire. If there was love here, it was not for this desert but for perfection, success, the splendor of man's handiwork."[89]

Jornada del Muerto has a history of people testing fate. Fallout could have moved in the direction of Socorro, Tularosa, Alamogordo, or, worse still, Albuquerque. If it had come down on Los Alamos, it surely would have appeared biblical in its reckoning. There was some measure of serendipity or providence that the most intense ground-level gamma radiation dispersed across scarcely populated countryside. The arid expanse of south-central New Mexico would be reserved not for further atomic detonations but for rocket testing and development. Open-air detonations moved on to paradise and then back to the continent, nestled within the expanse of the Great Basin. Here the architects of the postwar atomic state would again test the fate of people living downwind.

The conviction that the desert is barren underpins a "moral void" in which powerful actors assume that irredeemable, deleterious consequences are of little import. Meloy notes, "[The military-industrial complex] inherited an ethos that says arid lands are wastelands, not merely marginal but submarginal, places where nothing rusts quickly and the land seems a parched void. Here few people can be harmed by hardware that is intended for maximum harm. Everything is out of sight, yet nothing is hidden. Both the spectacle and the scenery can become intoxicating."[90]

Few segments of state activity so consistently elude critical public debate as does military technological development. "National security" is akin to a magic potion disabling democratic oversight. It is not the oldest trick in the book, but it is one of the most reliable. There may be valid reasons to withhold information from the public, but the latitude afforded the military-industrial complex in the years after Trinity was wide indeed. "Paramount objectives throughout the era were weapons production and the utmost secrecy. A land dominated by silence and sky, dust and time, held these secrets well," Meloy suggests.[91]

3

The Emerging Atomic Spectacular, 1950–1951

The Atomic Energy Act of 1946 (McMahon Act) stipulated civilian control of postwar atomic development while providing for military input. On January 1, 1947, all materials, equipment, infrastructure, and information were reassigned to the Atomic Energy Commission. The AEC possessed research laboratories, weapons production plants, and administrative offices dispersed from one end of the country to the other and a monopoly on nearly all research and development utilizing fissionable materials, while contracting out operations to private contractors. And it was granted exceptional powers of classification and restriction of information. "The size and scope of the AEC's operations, together with the sweeping powers granted by Congress, clearly indicated that it was intended to be a powerful agency," Robert Duffy observes.[1]

Figure 3.1 outlines the organizational structure of the AEC. Five civilian commissioners appointed by the president sat atop the organizational pyramid, one designated as chair. Due to the complexity of atomic technology, Congress was reticent to appoint just one person to lead the organization and opted for individuals characterized by varying occupational positions and experience. A general manager, also appointed by the president, assisted the commissioners and served as the chief executive concerned with day-to-day operations. The commissioners focused on wide-ranging policy decisions, while the general manager was concerned with the implementation of policy.[2]

The commission consisted of program divisions of research, production, biology and medicine, and military application, with a director for each appointed by the commissioners, and each program director reported to the general manager. The Division of Military Application was concerned with experimentation and development for military purposes, and the director was required to be a member of the armed forces. The Division of Biology and Medicine was tasked with assessing the effects of radioactivity on health and making recommendations

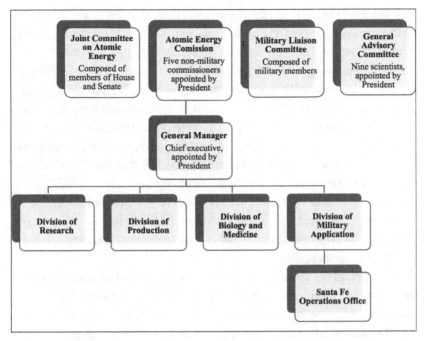

Figure 3.1. Organizational structure of the Atomic Energy Commission, as it relates to atmospheric atomic testing.

based on unbiased expert opinion. As the 1950s wore on, top officials in the Division of Biology and Medicine, however, often appeared more worried about threats to public relations than public health.

Three major advisory bodies flanked the commissioners. The Joint Committee on Atomic Energy was composed of nine individuals from the US Senate and nine from the House of Representatives. The JCAE had jurisdiction over congressional legislation pertaining to atomic weapons and energy development. The General Advisory Committee (GAC) included nine distinguished scientific advisers appointed by the president to counsel the commissioners on scientific and technical matters. The Military Liaison Committee (MLC) was composed of retired or active members of the armed services, and its objective was to advise the commissioners on matters related to the military applications of atomic technology. The Atomic Energy Act gave the MLC authority to make written recommendations to the commissioners and appeal directly to the secretary of war, or after reorganization, the secretary of defense, if the commissioners were acting in a manner the MLC deemed detrimental to weapons development.

In commenting on the Atomic Energy Act of 1946, Byron Miller explained, "An unusually clear example of the role which democratic processes can play in the framing of legislation is presented by the history of the law controlling atomic energy within the United States." On the other hand, he conceded, "A law at best establishes general propositions which influence but do not determine the effect it will have."[3] Power shapes the real-world employment of even the most carefully crafted legislation, including subsequent legislative amendments. Despite the intent of Congress to yield control over atomic development to civilian authority, the "pendulum" had swung in favor of military control in practice soon thereafter.[4]

That President Truman supported civilian control of atomic technology was indicated in his appointment of David E. Lilienthal as the first chairman of the AEC, and it was a choice met with suspicion by those in Congress who advocated exclusive military control of atomic technology.[5] From the beginning, then, the AEC was beset by the momentum of the atomic-industrial complex established during the Manhattan Project, in tension with the demands of broader democratic oversight. Lilienthal was appointed in September 1946 to a five-year term, but due to political pressure, he would be gone after three years.[6] His chairmanship of the AEC conflicted with a rising political tide in Congress as the US embarked on a "permanent war economy."[7] Detection in September 1949 of the first Soviet atomic detonation, moreover, enhanced the political position of those who were committed to a resolutely militarized policy approach to atomic development.

Looking back from the vantage point of 1963, Lilienthal suggested that the commission's role as civilian atomic custodian had long "evaporated." He asserted, "The AEC functions chiefly as a designer, developer, maker, and tester of atomic weaponry. . . . The AEC as weaponeer has in fact become very much part of the military establishment, serving the needs and goals of that military establishment as defined by the military."[8] This is reminiscent of President Eisenhower's warning regarding the military-industrial complex just prior to leaving office, two years earlier. Eisenhower cautioned against the scale of the US defense establishment and its "unwarranted" effects on democracy.[9] It was a startling admission by a man who had been president since January 1953 and grappled with some of the most daunting challenges of the Cold War.

The immediate postwar debate on civilian versus military control of atomic technology centered on the question of how to rationally adjust affairs in the new atomic age. In 1950, Eugene Rabinowitch, the founding editor, along with Hyman Goldsmith, of the *Bulletin of the Atomic Scientists*, confessed that the *Bulletin* was intended to "preserve our civilization by scaring men into rationality."[10] Writing in 1949, Lilienthal cautioned against science and technology being unhinged from democratic principles: "Unless the application of research and technology are consciously related to a central purpose of human welfare, unless technology is defined and directed by those who believe in people and in democratic and ethical ends and means, it could be that the more research money we spend the further we miss the mark. It is like driving in an automobile that is going in the wrong direction; the faster you drive the farther away from your destination you will be."[11]

Regarding the fervor in the years following World War II, Paul Boyer inquires, "What was the result? Except for the ambiguous McMahon Act, practically nothing. The international-control movement collapsed; the nuclear arms race began."[12] The US state embarked on *atomic technological rationality*, or the tendency, as it becomes integrated into society and central to the parochial aims of increasingly influential interest groups, to reshape what is more broadly considered an acceptable means for securing preferred ends and to the detriment of alternative courses of action. This influence is both discursive and material, as atomic technology is increasingly interwoven within society. The means become the focal concern, while the ends are overshadowed. Instrumental rationality applied in and through technology shapes what appears reasonable, and as technology is misconstrued as existing beyond political debate, society is pushed to bend to the needs of technology and its advocates; protest appears irrational, provincial.[13] The problem is not technology per se but social relations of control and domination. William Leiss suggests, "The fault lies not with technological rationality itself, but with the repressive social institutions which exploit the achievements of that rationality to preserve unjust relationships among men."[14]

During the 1950s, the atomic-industrial complex exhibited remarkable momentum, and AEC officials capitulated to the demands of the armed services to expand and accelerate atomic weapons experimentation and development. Such capitulation was not simply motivated by evidence of Soviet military expansion but in the service of domestic interest groups that benefited from undisciplined defense spending, as Eisenhower so evocatively warned in 1961.

Pushing for a Continental Test Site

The atomic-industrial complex desired space: political, social, and geographic. The McMahon Act provided the institutional structure whereby the AEC possessed considerable autonomy from the president and members of Congress outside the JCAE. Bureaucratic powers of restriction of information imparted a wide measure of insulation from public oversight. What remained was a need for space measured in acres, and Project Nutmeg was established in 1948 to examine possible continental testing sites. The central criteria were suitable meteorological conditions and the absence of nearby population centers.[15] The idea was greeted with skepticism by the AEC commissioners and set aside for reconsideration in the event of an overriding emergency.[16] The startling recognition in September 1949 that the Soviets had detonated an atomic device one month prior and war in Korea in 1950 provided the impetus to revisit the idea.[17] Five candidates were identified: Alamogordo-White Sands Missile Range in New Mexico, the Las Vegas Bombing and Gunnery Range in Nevada, Dugway Proving Grounds in Utah, Camp Lejeune in North Carolina, and the area between the towns of Eureka and Fallon in central Nevada.[18]

Southern Nevada offered better prospects for reducing offsite radiological contamination than Dugway or White Sands, due to more sparsely populated areas downwind, but Camp Lejeune along the North Carolina coast was the most advantageous, with prevailing winds moving east over the Atlantic.[19] The federal government did not possess all the land needed, however, and there were residents who would have to be relocated. It was quicker to carve out a section of the existing Las Vegas Bombing and Gunnery Range, an expanse of land covering five thousand square miles. In choosing a continental proving ground, radiological safety was a central consideration, but it was not the only consideration. Site location relative to air and road travel from Los Alamos and the difficulty of establishing a viable site were also key criteria. In this regard, Nevada looked better than Camp Lejeune.

On December 18, 1950, President Truman lent his signature, and a section of the Las Vegas Bombing and Gunnery Range was ceded to the AEC. "The essential need now is for a site at which a few relatively low order detonations may be done safely and with minimal non-productive cost in time, effort and money at the earliest possible dates, preferably within the next two to three months," states the memo that Truman signed.[20] Five weeks later, the first atomic device was detonated. It was the "backyard" laboratory that LASL earnestly desired.[21] It also gave the Army and Navy the opportunity to influence atomic weapons research and development. The logistics of detonations in the South Pacific, five thousand miles from the West Coast, were daunting and discouraged development of lower-yield, tactical battlefield weapons and testing of effects or of the impact of detonations on soldiers, equipment, animals, and built infrastructure such as bridges and buildings. Testing exclusively in the Marshall Islands practically ensured an emphasis on large-yield, aircraft-deliverable devices such that the Air Force remained the dominant military service working in conjunction with the AEC. "By the late 1940s both the Army and Navy brass were beginning to feel left out of the nuclear action," Miller notes.[22]

Fallout and the Social Construction of Risk

In August 1950, at a conference at Los Alamos, attendees discussed the feasibility of accurate meteorological forecasting in Nevada. The consensus was that there was minimal risk to the public except for "rainout," wherein radioactivity is deposited on the ground in concentrated form due to precipitation. That dry fallout was not a significant threat was a tacit assumption, as it was assumed that changes in wind direction and velocity were predictable with "reasonable accuracy." However, "people will receive perhaps a little more radiation than medical authorities say is absolutely safe," the conference report notes. Enrico Fermi, the Noble Prize–winning physicist, injected a note of caution into the proceedings, observing "the extreme uncertainty of the elements" they had to go on: "we did our best with these."[23] Within the report, there is no discussion regarding the effects of radiation on children or pregnant women or the problems endemic to internal emitters of radiation.

The conference, assembled five months before testing began, is noteworthy as central assumptions underlying AEC operational conduct were not vigorously debated but mutually confirmed by attendees. And yet, as Fermi observed, there was considerable uncertainty regarding redeposition of fallout. This is not in itself novel, as organizations frequently confront complexity and uncertainty. What is crucial is the degree to which the institutional context and organizational culture encourage explicit scrutiny of taken-for-granted assumptions underpinning operations. While Fermi sensed the degree of uncertainty, AEC officials often did not exhibit such reflexivity in the years to come.

A central dynamic of modernity is belief that the future is not dictated by fate or the whimsy of the gods but by goal-directed activity in the present. We need not wait passively for a future that there is no hope of shaping, but through rational decision-making failure can be reduced and the probability of success enhanced—thus driving people to new opportunities and achievements. Peter Bernstein explains, "The word 'risk' derives from the early Italian word *risicare*, which means 'to dare.' In this sense, risk is a choice rather than a fate. The actions we dare to take, which depend on how free we are to make choices, are what the story of risk is all about."[24]

Eugene Rosa offers a less romantic definition: "Risk is a situation or event where something of human value (including humans themselves) is at stake and the outcome is uncertain."[25] Risks are ontologically real, but individual and collective perceptions nevertheless matter in regard to what is viewed as a salient concern. This is the "dual reality" of risk.[26] Probabilities, odds, the likelihood that something will or will not occur in the future—all depict the degree of uncertainty one confronts. Moreover, when confronting uncertainty, individuals and organizations often refer to "patterns of the past" in making sense of how to proceed in the present.[27] The past is not a foolproof guide, but it is someplace to start.

Trinity was a place to start working through the problems discussed at Los Alamos, and yet it scarcely seemed to register. The lessons learned in south-central New Mexico lacked institutional momentum. One crucial lesson: detonation atop a tower pulls up alluvial soil, and as the cloud disperses, the heaviest particles descend first and in a concentrated and capricious manner. Prevailing wind patterns can be fickle, and in tandem with topography, hot spots can occur with abandon. This lesson is less important when conducting airdrops, but even after the adoption of large tower shots, AEC officials continued to ignore the significance of hot spots. Another crucial lesson: where there are hot spots of gamma radiation, there may be high levels of internal emitters of radioactivity ingested through water or food.

In 1979, a group of downwinders filed a lawsuit in US federal district court. *Irene Allen et al. v. United States* encompassed three years of discovery processes and legal motions, three months of testimony, and seventeen months later, a "groundbreaking legal statement."[28] Judge Bruce S. Jenkins concluded that the AEC misled downwind residents as to the potential health effects of exposure to fallout and failed to adequately monitor radioactivity, release information in a comprehensive and timely manner, and educate residents as to the simple precautions necessary to minimize exposure.[29] For the first time, a federal judge ruled that the AEC failed to exercise due care in the conduct of open-air testing in Nevada *and* determined that low-level radioactivity was a proximate cause of injury in ten of the twenty-four "bellwether" cases under examination.

Allen et al. paved the way for additional lawsuits, and this ensured that government lawyers vigorously contested the memorandum opinion rendered by Jenkins.[30] Ultimately, the downwinders learned that "the king can do no wrong." That is, in 1987, *Allen et al.* was overturned on judicial appeal as it was determined that government actions were protected from tort claims under the discretionary function exemption of the Federal Tort Claims Act (FTCA) of 1946.[31] The US adopted the English legal doctrine of sovereign immunity, holding that the monarch (state) could not be sued without permission and arguing that the absence of this doctrine would unduly inhibit enactment of governmental activities.[32] The FTCA was intended to ease such restrictions, but the discretionary function exemption has been interpreted so broadly as to constitute "an impermeable barrier."[33] Enactment of the Warner Amendment by the US Congress in 1984, moreover, transferred all legal liabilities of the numerous contractors involved in weapons testing to the federal government.[34]

In contrast to the generally optimistic assumption established at Los Alamos in August 1950, in Jenkins's detailed judicial opinion, he flatly rejected the assertion that officials at or working under the auspices of the AEC paid sufficient attention to the scientific knowledge *at the time* in making key decisions affecting the health and safety of downwind residents. Jenkins called into question a single maximum threshold of radiation exposure without regard to age or gender, arguing that there was no evidentiary basis justifying such a policy, and he questioned adoption of the *threshold hypothesis*, wherein assessment of radiological risk is premised on noncumulative impacts such that no permanent health effects ostensibly arise if an established dose is not exceeded within a given time frame. Moreover, Jenkins argued that the threshold concept was not widely embraced among relevant scientific associations and that there existed extensive debate regarding genetic alterations due to *linear dose-effect* exposure to radiation.[35] Indeed, such debate continues to reverberate.

In turn, continental testing was premised on a sanguine appraisal of the predictability and uniform effects of fallout, which contributed to mistakes and thus demonstrated the need to increase safety measures beyond those originally envisioned. Moreover, operational protocol was formulated neither in callous disregard nor rigorous concern but to ensure the viability of continental testing. To establish variant threshold levels by age or gender—lower for children and women—would have been time and resource intensive in practice. To establish a policy of cumulative, nonthreshold radiation exposure would have ensured the built-in obsolescence of LASL's outdoor laboratory. Pushing the boundaries may be daring and, indeed, may contribute to the accomplishment of desired objectives, but it appears a lofty and romantic endeavor only when the unintended effects, should they arise, are imposed on others.

Operation Ranger: January–February 1951

In 1950, the population of the United States exceeded 154 million, of which 160,083 resided in Nevada—just a tenth of 1 percent of the total. Clearly, Nevada possessed abundant open space. In 1950, Clark County, Nevada, including Las Vegas and surrounding communities, was home to 48,289 people, or approximately 30 percent of state residents. By 1960,

Figure 3.2. Guard station at the Nevada Test Site. (Photo courtesy of the Nevada National Security Site)

the population of Nevada increased to 285,278, of which 127,016 people lived in Clark County, now home to over 44 percent of the state's residents.[36] In turn, between 1950 and 1960, Las Vegas and Clark County were growing, were increasingly the political and economic epicenter of the state, and were hardly a dusty desert outpost anymore.

Figure 3.2 shows the entrance to the NTS in the mid-1960s. Money and people poured into the state during the years of atmospheric atomic testing. "For Nevada, the pork barrel took the shape of a mushroom cloud," Gerald DeGroot insists.[37] Detonations proceeded when the prevailing winds were moving east-northeast to preclude fallout over Las Vegas to the southeast. Testing brought in federal dollars and tourists eager to gamble and catch a glimpse of an early-morning atomic flash or ascending mushroom cloud, but while radioactivity drifted toward Las Vegas on occasion, the city was generally spared the risks of testing while cashing in on the economic benefits. This is noteworthy because as the perils of fallout became apparent to rural inhabitants far from Clark County, the two major newspapers in Las Vegas vigorously defended testing, at times castigating and mocking rural Nevada and Utah residents raising concerns over fallout.[38] Indeed, soon after testing began, the Clark County official seal was redesigned to depict a billowing mushroom cloud.

On January 11, 1951, the AEC released a terse statement: atomic testing would soon begin and would occur intermittently on an unannounced basis, for an indeterminate period, and "under controlled conditions." "Health and safety authorities have determined that no danger from or as a result of AEC test activities may be expected outside the limits of the Las Vegas Bombing and Gunnery Range."[39] AEC officials were committed to releasing little information and anticipating that the public would scarcely notice the detonations. Nonetheless, they were fearful of "unthinking public reaction."[40] To the relief of AEC officials, the announcement did not evoke public concern.[41]

Illumination of the predawn sky on January 27 signaled the inauguration of the commission's new proving grounds. As outlined in table 3.1, Able was small, timid even.[42] At one kiloton, it hardly rivaled the yield unleashed on Japan, but low-intensity fallout unexpectedly came down with snow in Rochester, New York. Fallout from the eight-kiloton Baker shot the day after Able passed over Bunkerville, Nevada, where schoolchildren were called out of class to watch the pinkish-orange cloud pass overhead, and, like Able, it intertwined with precipitation in the northeastern US, was detected at university facilities, and interfered with film manufacturing at Eastman Kodak. Like Able, shot Easy was small, at just one kiloton. Baker-2 was eight times bigger, however, and the pressure wave broke a large plate-glass window at a Las Vegas storefront seventy miles away, while fallout passed near the city and drifted toward Phoenix, Arizona. The final detonation of the series, code-named Fox, was the largest shot and produced a jarring shock wave, while the flash was visible as far away as Oakland, California, and Boise, Idaho.[43]

TABLE 3.1. Operation Ranger Detonations

Code name	Type	Yield (kilotons)
Able	airdrop	1
Baker	airdrop	8
Easy	airdrop	1
Baker-2	airdrop	8
Fox	airdrop	22
Total yield		40

Source: DOE, *United States Nuclear Tests: July 1945 through September 1992* (Las Vegas: Nevada Operations Office, 2000).

Operation Ranger included five airdrops over a ten-day period deto-nated above the desert. Four of these were less than ten kilotons, while Fox was twenty-two kilotons. Trinity, by comparison, was twenty-one kilotons, Hiroshima fifteen kilotons, and Nagasaki twenty-one kilo-tons.[44] A major objective of Operation Ranger involved the test firing of fission triggers for the still theoretical hydrogen bomb, and there was little military involvement. Operation Ranger focused on LASL weapons development priorities.[45] The US Air Force Air Weather Services (AWS) was tasked with monitoring and assessing meteorological conditions be-fore each detonation—a crucial role to ensure management of fallout through "dispersal and dilution."[46] Offsite fallout was monitored by the Rad-Safe Group from LASL, which conducted scattered ground surveys within a two-hundred-mile radius downwind and oversaw air sampling up to six hundred miles from the NTS.

Crowds assembled on Mt. Charleston and along roadways to catch a glimpse of the mushroom clouds. Gladwin Hill, the *New York Times* Los Angeles bureau chief, wrote many of the stories in the paper detailing atmospheric testing in Nevada. Reporting from the outskirts of the NTS during Operation Ranger, Hill mused, "Then suddenly the surrounding mountains and sky were brilliantly illuminated for a fraction of a second by a strange light that seemed almost mystical. It was strange because, unlike light from the sun, it did not appear to have any central source but to come from everywhere. It flashed almost too quickly to see. Yet the image that persisted for a fraction of a second afterward in the eye was unforgettable."[47]

The *Times* has long been one of the most prestigious newspapers in the country, with political connections and material resources that few other papers can approach. *Times* reporters and editors were ideally situ-ated to hold AEC officials accountable for the management of fallout, and internal memoranda and reports illustrate that AEC officials were sensitive to news coverage. However, in the early years of testing, *Times* reporting was oblivious to the AEC's increasing struggle to contain fall-out, even as local and regional newspapers began questioning the integ-rity of the commission's conduct over the next two years.[48]

The public did not greet testing with consternation but embraced the novelty of bombs on the outskirts of Las Vegas, and national and local media were eager to get in on the story. It is more difficult to sustain a position of rigid information management, William Kinsella notes, when material containment is incomplete.[49] The blast wave, flash, and detection of radioactivity at industrial and university facilities around the country undercut the AEC's "low key" discursive approach.[50] During Operation Ranger, Hill proclaimed, "The silence of the cold gray dawn on the mountain-dotted eastern Nevada desert, punctuated only by the faint chirping of crickets in the sagebrush, was shattered this morning by the blast of the third of the Atomic Energy Commission's new nuclear explosion tests. Experiencing it gave one, despite the scientific explanation about atomic fission, a strange feeling of fleeting contact with eternal power."[51]

Ranger proceeded with few problems related to offsite fallout, but AEC officials were troubled by the detection of radioactivity at university laboratories and its impact on film manufacturing.[52] The concern stemmed less from the levels of radioactivity, which were low, but from potential lawsuits and negative press. AEC officials decided it was imperative to reach out to Eastman Kodak and the National Association of Photographic Manufactures, the industry trade association. This encompassed distribution beforehand of the starting date, number of detonations, and approximate length of a test series so that precautions could be taken to forestall economic losses from ruined film, information that residents of the downwind communities were not privy to and could not employ in taking their own precautions.[53]

In preparing the public-relations plan for the next series, Buster-Jangle in the fall of 1951, the AEC concluded, "The final reaction of the Nevada and general site area public is best thumb-nailed by saying the people felt they had been participants in a very major national enterprise, and were rather proud of the whole thing." Nevertheless, the AEC report noted the need to preempt "latent local fear or radioactivity," despite evidence that downwind residents were prone to such things.[54]

Prior to the inaugural detonation at the NTS, AEC officials in Washington, DC, transferred administrative responsibility for continental testing to the Santa Fe Operations Office and LASL's Testing Division (J-Division), tasked with management of atomic weapons testing while contracting out services to private firms as well as to military service laboratories.[55] Offsite radiological safety (rad-safe) operations for the first two series in Nevada were the responsibility of LASL's Health Division (H-Division), situated within the Division of Biology and Medicine, headquartered in Washington, DC.

The organizational structure of the AEC was distinctly decentralized. The five politically appointed commissioners in Washington, DC, set the policy agenda, supported by the general manager, also located in DC, who relayed information to the commissioners from five field managers dispersed around the country. Each field manager possessed considerable autonomy in carrying out the broad policy directives they were given—including the field manager at the Santa Fe Operations Office.[56] The Los Alamos–Santa Fe node was at the heart of testing activities at the NTS, in consultation with the AEC's Division of Military Application. "Operationally, the chain would begin and end at the local site. The scientists at Los Alamos, and the field manager in charge of the operation, would initiate the series of tests and would complete the test series," Howard Ball notes.[57] The AEC commissioners, in turn, were distant geographically from Nevada and by virtue of organizational structure. All test series conducted at the NTS first required presidential approval. But even though President Truman voiced concerns beforehand about the safety of continental testing, from its commencement in January 1951 to his leaving office two years later, he was notably removed from direct involvement—preferring to delegate responsibility to AEC chairman Gordon Dean.

Operation Buster-Jangle: October–November 1951

Operation Ranger presented minimal radiological risks, but in May 1951 AEC and military officials proposed a subsurface detonation, which was guaranteed to generate fallout. Indeed, the objective was to bury a one-kiloton device fifty feet below the surface to intentionally create a cloud of dust and debris for analysis. It would be the world's first subsurface detonation of an atomic device. So significant were the uncertainties that a mandatory public evacuation within a forty-four-mile radius of the NTS was considered. At a conference at LASL, supporters argued that the detonation would advance understanding of fallout characteristics on the battlefield, but others feared it could turn public opinion against the commission's new outdoor laboratory, if not contaminate much of the test site.[58] It was a "calculated risk," but one those who were in attendance unanimously deemed acceptable.[59] It was, without irony, designated as a "safety shot." In the weeks and months after the LASL conference in May, however, the idea came under increasing scrutiny and was abandoned.[60] Rather than a deep subsurface detonation, the "Jangle" portion of the upcoming Buster-Jangle series included a one-kiloton surface and one-kiloton shallow subsurface shot—both generating the first instances of notable offsite fallout. The commission, of note, was already under pressure from military officials to take risks at their nascent proving grounds.

Moreover, in preparing for Buster-Jangle, AEC officials lamented the attention that Ranger garnered. AEC chairman Dean remarked, "it is impossible to set off a nuclear explosion within the continental United States without a lot of people knowing about it."[61] In turn, the public information plan for Buster-Jangle sought to desensationalize testing, emphasizing that it was "not particularly novel or exciting."[62] Such efforts were futile.

Attention is one thing, but "unreasoning panic" is another, and AEC officials were relieved that there was little public apprehension. Nonetheless, the expectation that downwind residents would react in an unreasoning manner to fallout was a foundational assumption shaping AEC conduct, and it elicited crucial implications with regard to operational conduct. Fear of the public was invoked throughout the entirety of atmospheric testing in Nevada—even in the absence of observable public concern.

As outlined in table 3.2, Buster-Jangle consisted of airdrops and the first tower, surface, and cratering shots, as well as the first troop maneuvers. The Army was impatient to conduct exercises, and the other branches of the armed services wanted to send observers. By the summer of 1951, the Military Liaison Committee submitted such plans for approval, and AEC officials scrambled to draft radiological safety guidelines for troop exercises.[63] AEC officials consented to the exercises but, to the dismay of Army officials, insisted that troops remain at least seven miles from ground zero at detonation time. An estimated nine thousand Department of Defense (DOD) personnel engaged in various activities during Buster-Jangle. Some observed detonations, approximately sixty-five hundred personnel engaged in battlefield maneuvers, while others conducted damage effects testing on military equipment,

TABLE 3.2. Operation Buster-Jangle Detonations

Code name	Type	Yield (kilotons)
Able	tower	less than 0.1
Baker	airdrop	3.5
Charlie	airdrop	14.0
Dog	airdrop	21.0
Easy	airdrop	31.0
Sugar	surface	1.2
Uncle	subsurface	1.2
Total yield		71.9

Source: DOE, *United States Nuclear Tests: July 1945 through September 1992* (Las Vegas: Nevada Operations Office, 2000).

participated in scientific data gathering, or acted in a supporting role for the aforementioned.[64]

Shot Able was the first of the series. Perched atop a one-hundred-foot tower, it "fizzled."[65] Shot Baker sent radioactive debris toward the California coast, where it hovered for some time, and given the millions in the area, it provoked consternation in the AEC. The upper levels of the Baker cloud, however, migrated east and were detected in Topeka, Kansas, and parts of Oklahoma and Texas before drifting over Massachusetts. Next was Charlie, a fourteen-kiloton airdrop. In describing the radioactive cloud, Richard Miller observes, "Meteorologists at the site tried in vain to guess the direction the cloud would take. Conflicting wind patterns twisted the stem into a spiral shape almost 6 miles high. The mushroom hung over the test site in that configuration for almost two hours before beginning a grand—and rainy—tour of the United States."[66]

Dog was next, and it was the first shot to include troop maneuvers. It was a comparatively large, twenty-one-kiloton airdrop detonated at fourteen hundred feet above the desert. After observing the detonation, troops moved toward ground zero accompanied by radiation monitoring teams. The objective was to train soldiers in atomic battlefield maneuvers, to test military tactics, and to assess the effects on military equipment and fortifications.[67] A further goal entailed assessing the psychological effects of atomic warfare on participating troops.[68]

Shot Easy was the fifth in the series, and at 31 kilotons, it was the largest detonation to date. The bulk of the fallout migrated east and encountered a cold front, diverting its path in a long, sweeping arc from southern Arizona to Texas and then across Florida to the Atlantic.[69] Buster-Jangle concluded with Sugar, a 1.2-kiloton surface shot, and the subsurface Uncle detonation, also estimated at 1.2 kilotons. Despite their comparatively lower yields, both Sugar and Uncle produced substantial radioactivity at ground zero, a challenge since both included troop maneuvers. Moreover, Sugar, the surface detonation, generated a noticeable increase in offsite fallout, and Uncle, the device buried seventeen feet down, generated even greater offsite radioactivity.[70]

AEC officials were satisfied that radiation exposure to DOD personnel had not exceeded the guidelines established beforehand; but many service members were not wearing film badges, and a comprehensive exposure record was not catalogued.[71] Further, AEC officials were convinced that while there was greater offsite fallout than during Ranger, the intensity of radioactivity was not a cause for concern.[72] Several Buster-Jangle detonations, nevertheless, displayed bewildering atmospheric dynamics. Divergent and capricious winds aloft, temperature inversions, regional topography, and precipitation far from the NTS were confounding factors. The AEC had yet to launch a concerted offsite radiological monitoring program in the downwind communities—a reflection of the assumption that dry fallout was not a substantive threat. Not until 1953 did the AEC, in conjunction with the US Public Health Service (PHS), establish seventeen fixed monitoring stations in the NTS region, in addition to mobile monitoring teams, to assess intensity and duration of fallout.

After the conclusion of Buster-Jangle, in December 1951, Harry F. Schulte of the LASL Heath Division submitted a classified report to division leader Thomas L. Shipman illustrating the growing pains facing those who were charged with ensuring offsite radiological safety. Schulte listed the difficulties that mobile rad-safe monitoring teams confronted in setting up sampling stations downwind of the NTS due to the vast distances and scarcity of roads to move about efficiently. Meteorology in tandem with varied terrain created diverse distributional patterns of airborne radioactivity. Schulte suggested, "It is difficult to see, at present, how the fall-out program can be conducted on a much smaller scale and still yield useful results. One of the difficulties of interpreting data obtained at Buster or Ranger is in the uncertainty whether the location sampled was in or near the point of maximum concentration. For this reason, a large number of sampling stations must be employed."[73]

Just twelve days after receiving Schulte's report, T. L. Shipman submitted a classified report of his own to Alvin C. Graves, the head of LASL's J-Division and the test director for continental detonations. Shipman stressed that the unit was understaffed for key tasks, equipment and supplies were lacking, transportation was inadequate, and the close timing of detonations, one to the next, strained rad-safe operations. Further, the lines of authority among the 187 members participating in the unit's operations were convoluted and confusing—a mix of civilian and military personnel. Of these 187, the mobile monitoring group patrolling for radioactivity within a two-hundred-mile radius of the NTS comprised only ten two-person teams. In turn, Shipman expressed dismay with the abrupt expansion of Buster-Jangle to include military effects testing and troop exercises, which took place during the "Jangle" phase. What he was expressing is institutional momentum, a bird's-eye view of the different interest groups jockeying to utilize the NTS. Shipman contended,

> In the early part of 1951, Operation Buster looked like a reasonably simple affair. It grew alarmingly with the inclusion of a Military Effects Program, and then the entire appearance of the operation was changed by the addition of Jangle. No firm facts were available on which the Rad Safe organization could estimate its requirements; changes and additions appeared and reappeared up to, and in some cases, after the various detonations. Some way must be found of insisting on an early cut-off date, so that one can say "This is the operation and specifies the requirements; additions or amendments must wait for a future operation."[74]

Both Schulte's and Shipman's reports depict the increasing tempo and scale of activities at the NTS and the struggles of rad-safe personnel to keep pace. The reports were classified, however, and were a contrast to the AEC's public pronouncements that safety was always the primary focus. Indeed, Shipman made a crucial observation foreshadowing the problems arising in later test series: "In Operation Jangle, thanks to the kindness of the winds, no significant activity was deposited in any

TABLE 3.3. Disqualification Heuristics Guiding AEC Operational Conduct

Heuristic	Systemic implications
1. Predictable dispersion and dilution patterns exhibited by fallout	• Focus on meteorological forecasting to the exclusion of social-organizational preparation • Little need to inform and educate downwind populations in regard to fallout exposure
2. Focus on noncumulative, undifferentiated effects of radioactivity	• Focus on acute to the exclusion of crescive effects • Little need to monitor long-term exposure of downwind populations • Focus on intensity of external gamma radiation only
3. Expectation of an unreasoning, panic-prone downwind public	• Reticence to employ robust offsite radiological safety education and procedures • Undervaluation of the experiential effects reported by downwind residents • Preference for engineering solutions to offsite fallout to the exclusion of social-organizational solutions

populated localities. It was certainly shown, however, that significant exposures at considerable distances could be acquired by individuals who actually were in the fall-out while it was in progress."[75]

Operational protocol throughout the era of open-air testing in Nevada embodied three specious heuristics, outlined in table 3.3: (1) Fallout is subject to predictable atmospheric dispersion and dilution patterns. (2) Radioactivity has noncumulative and undifferentiated effects on the human body. (3) The public is prone to unreasoning thought and behavior regarding radioactive fallout. These heuristics had crucial implications with regard to operational conduct, as AEC officials failed to take precautions that were clearly within their power to enact. Moreover, commitment to the heuristics outlined in table 3.3 created a context in which key decision-makers within the AEC were continually chasing fallout across the Great Basin and beyond.

The expectation that fallout exhibits relatively predictable dispersion and dilution patterns contributed to a focus on meteorological forecasting to the exclusion of robust deployment of rad-safe personnel arrayed across the landscape and prepared to monitor fallout wherever it may have gone. There was little need to educate members of the public, who were viewed as ignorant and prone to hyperbole anyway, when the trajectory and redeposition of debris were relatively predictable. And if there were problems related to fallout, informing the public and providing pertinent education to minimize exposure was viewed as risky, as it may have undercut prior public-relations efforts insisting that there was no hazard. A focus on external radioactivity, moreover, and the conviction that internal emitters were not a problem absent high levels of gamma radiation tempered the perceived need to monitor radionuclides in locally grown gardens or milk consumption. Apprehension in the downwind communities arose over time but not due to some inexplicable fear of all things atomic but as residents came to suspect that the AEC was downplaying the risks of fallout.[76] Despite expectations of irredeemable irrationality, members of the public displayed a notable weighing of the pros and cons of the situation they confronted.

The disqualification heuristics outlined in table 3.3 would soon become of greater import. That the assumptions underpinning AEC operational conduct were erroneous was only hinted at in 1951. The next series, in 1952, more clearly demonstrated that continental testing was a fragile and tenuous endeavor from a public health perspective. Ranger and even Buster-Jangle were a tepid beginning. The detonation yields would get larger, effects testing and troop maneuvers more grandiose, and the hazards more pressing.

The Emerging Atomic Spectacular

New York Times coverage quickly settled into an *atomic spectacular* frame of reference, organized around the novelty of testing. Vivid descriptions of the sensory experiences of witnessing a detonation were prominent. Gladwin Hill intermittently broke into first-person narrative in his reporting, to give readers a sense of the excitement and trepidation.[77] Rather than detailing the AEC's protocol for monitoring fallout and responding should problems arise, *Times* coverage was concerned with the aesthetics of one detonation relative to the one prior and expressing disappointment at less visually spectacular shots. Testing as a tourist attraction, moreover, was a consistent refrain, reinforcing the entertainment value of detonations on the outskirts of a city predicated on tourist activities that were distinct from those offered elsewhere. Generally shielded from the imposition of fallout, economic boosters in

Las Vegas embraced testing as another spectacle in addition to casinos and elaborate stage shows, and the *Times* advertised the atomic spectacular before a national audience.[78]

Within the *Times*, there was also a distinct *atomic fetishism*, whereby the technology was imbued with mystical qualities. Concrete operational policies of the AEC as well as broader questions about state-society relations regarding a technology with known risks to public health were neglected in lieu of commentary as to the color and shape of the most recent mushroom clouds. Glenn Feighery observes that newspaper coverage focused on the "awe" factor, but the idea of the atomic spectacular goes further in suggesting that the *Times* was so enamored by the beautiful, exciting, and at times monstrous that AEC organizational dynamics were not considered in depth.[79] Hideous, alluring atomic entities were arising northwest of Las Vegas, and *Times* reporters were mesmerized.

By the fall of 1951, however, Gladwin Hill was becoming more critical of the AEC. He was irritated that the press was forced to report on detonations while parked on the side of roadways or atop a mountain.[80] It was unacceptable, he declared, in a democracy valuing freedom of expression that the commission would remain so secretive. What Hill wanted was a closer view.

Amid novel events, journalists often reflect the official organizational or institutional interpretation in news accounts.[81] And it is understandable that *Times* reporters would adopt an atomic spectacular interpretive framing, given the geopolitical tensions of the early 1950s. Nevertheless, the inattention to offsite fallout is striking. Not only was the issue not considered in 1951, but assurances of safety by AEC officials were so frequently reiterated that questions pertaining to fallout seemed unnecessary. *Times* reporting was framed to encourage a particular interpretation, and the risks endemic to atmospheric testing were not a suggested consideration.

4

The Road to Upshot-Knothole, 1952

During the Buster-Jangle series of 1951, the armed services begrudg-
ingly agreed to the AEC's insistence that troops remain seven miles from
ground zero at shot time, but in planning for the Tumbler-Snapper series
of 1952, they launched a counteroffensive, arguing that this hampered
their ability to engage in battlefield training. They wanted troops as close
as seven thousand yards (3.9 miles). Indeed, the Marines refused to
participate in Tumbler-Snapper if the seven-mile limitation were again
imposed.[1] AEC officials were reticent to acquiesce to such demands given
the larger-than-anticipated yield of many shots. Moreover, the margin of
error inherent to airdrops could put troops as close as seven thousand
yards in danger. Shields Warren, a medical doctor and director of the
Division of Biology and Medicine, authored a forthright letter to Brigadier General K. E. Fields, director of the Division of Military Application,
in opposition to the proposal. Warren was worried about the safety of the
troops but also the negative press that injuries or death might evoke, and
he argued that any accident would be "magnified" by the press and would
threaten continued use of the NTS, noting that the test site was "accepted
by the public as safe."[2] AEC chairman Gordon Dean and the other commissioners expressed concern but agreed to allow troops to be stationed
closer to ground zero—if the military agreed to assume full responsibility
for their safety.[3] Such a retreat was an illustration of the military's growing influence over activities at the NTS.

An estimated 7,350 troops participated in Tumbler-Snapper as observers or engaged in battlefield maneuvers, while another 3,250 assisted
with scientific experiments or supporting services such as food, housing,
transportation, and communication.[4] Richard Miller notes, "By early
1952, tactical weapons were a very big item with everybody but air force
officers. They saw the whole thing as nonsense and an attempt to weaken
the strategic arsenal. . . . But to the other services, it was as though they
had been given a fantastic new toy to use as they saw fit. And with the

AEC safely out of the way, the generals rushed to the chalkboards. There would be war games to end all war games. Troops. Tanks. Artillery. *Paratroopers*. It would be a busy year."[5]

Tumbler-Snapper also included the first tower shots. By 1952, it was understood that tower shots generate localized fallout, but as weapons testing became more sophisticated over time, it necessitated stationary diagnostics instruments and therefore tower detonations.[6] Effects testing and troop maneuvers only increased reliance on tower shots, as airdrops were too inaccurate—particularly as troops were positioned ever closer to ground zero.

An Atomic Plague

As reliance on tower detonations arose, the AEC confronted difficult decisions with regard to how much to tell the public about fallout intensity and location. Increasing "close-in fallout" should have prompted AEC officials to consider educating the public downwind as to the simple precautions necessary to reduce exposure should a heavy fallout event occur. Officials were reticent to take such steps, and this reticence predates testing in Nevada. The US Army and the War Department went to extraordinary lengths to control the images of death and illness at Hiroshima and Nagasaki.[7] What Allied reporters saw and what they reported was scripted and controlled, newsreel footage was classified or disappeared altogether, and news stories were censored. "From the very start, the visual record of the atomic bombings would be limited to structural effects, while the human dimension would be evaded or ignored," Robert Lifton and Greg Mitchell explain.[8]

The Australian war correspondent Wilfred Burchett refused to acquiesce to US Army censorship and boarded a train from Tokyo to Hiroshima, alone. Arriving a month after the bombing, Burchett was the first Allied journalist to view the devastated city, but he was equally astonished by a lingering malady afflicting many who survived the initial blast. He referred to it as an "atomic plague" in London's *Daily Express* newspaper on September 5, 1945. "In these hospitals I found people who, when the bomb fell, suffered absolutely no injuries, but now are dying from the uncanny after-effects," Burchett wrote.[9] The story got him expelled from Japan.

"Fallout" describes radioactive debris descending back to earth, but over time it has also come to denote the transference of the hazardous by-products of the activities of some onto others. A conventional chemical explosion produces a blast wave, heat, and a flash of light, but an atomic detonation elicits all of the above in extravagant abundance and something more. While the blast wave, heat, and flash are easily recognizable through normal human senses, radioactivity is invisible, odorless, and tasteless. A conventional detonation is acute and definable in geographic space and time, but the damage that results from the violent reshaping of matter is open-ended and difficult to define geographically or temporally. The distinction between the conventional and the atomic is more than simply increased scale but lies in the character of the effects persisting in the aftermath. The atomic plague is unlike any other plague that humanity has confronted.

Radioactivity displays variant forms with different impacts on the human body. *Alpha* particles are the least penetrating. They do not pass through the outer layer of skin. *Beta* particles pass through the outer layer but cannot reach deep into the body. *Gamma* radiation, akin to an X-ray, passes through the human body. To be in the vicinity of an external source of gamma radiation can be dangerous, but the impacts of alpha and beta particles are significant when inhaled or ingested.

From Operation Ranger in January 1951 through the end of the Upshot-Knothole series of 1953, offsite radiation monitoring within two hundred miles of the NTS was conducted by Los Alamos Scientific Laboratory, the US Public Health Service, or various military units.[10] Monitors with two-way radios were stationed at strategic locations before each shot and directed to locales corresponding to where the debris cloud seemed to be heading according to aircraft monitoring from above. Before the Tumbler-Snapper series of 1952 began, however, responsibility for radiation safety operations was transferred from Los Alamos to the Armed Forces Special Weapons Project (AFSWP), as LASL officials felt that the task was burdensome and strained existing resources.[11] AFSWP comprised members of each of the armed services and was concerned with stockpile maintenance, storage, handling, and testing of atomic devices.[12] The threshold adopted from Operation Ranger in 1951 through the end of Upshot-Knothole in 1953 was 0.3 roentgens of whole-body external exposure per week, or 3.9 in a thirteen-week period. A roentgen is a measure of ionizing radiation relevant to the induction of physical effects.[13] A DOE report notes, "In the absence of formal standards, the AEC applied occupational standards to offsite populations."[14] That is, the AEC adopted the standard for workers in the nuclear industry at the time.[15]

Moreover, radioactivity did not come back to earth in a smooth, even manner. That terrestrial redeposition was akin to the smattering of paint on a Jackson Pollack canvas suggests that comprehensive monitoring of radioactivity necessitated substantial effort.[16] Indeed, it required more effort than the AEC intended at the outset. Absent an accounting of the hot spots, moreover, it was impossible to fully assess the degree to which the data collected depicted the radioactive burden imposed on the downwind communities.

After Trinity, Stafford Warren authored a memo to General Leslie Groves documenting ground-level radioactivity far from ground zero, and he discouraged further detonations in New Mexico: "It is recommended that the site be expanded or a larger one, preferably with a radius of at least 150 miles without population, be obtained if this test is to be repeated."[17]

As delineated in table 4.1, downwind of the NTS—to the east-northeast—lived tens of thousands of people in a menagerie of towns, ranches, farming communities, and Native American settlements. None loomed large on a map, but they added up to a substantial number of people. In 1979, the Radiation Exposure Compensation Act (RECA) was introduced in the US Congress by Senator Ted Kennedy of Massachusetts, but it failed to garner sufficient support. It was again considered in 1990, cosponsored by Kennedy and Senator Orrin Hatch of Utah, and signed into law in 1991. This legislation provides financial compensation for injuries sustained due to radioactive fallout from weapons testing in Nevada. Table 4.1 outlines the counties in the original RECA bill, which had a greater focus on central Utah, while the 1990 version excludes seven Utah counties but includes six counties in Arizona. As table 4.1 shows, approximately 19,000 people lived in the five Nevada counties that were most at risk, and around 130,000 people lived in seventeen Utah counties directly downwind of Nevada. It adds up to nearly 150,000 rural residents.

Table 4.2 outlines the counties in the RECA legislation of 1990. In 1950, there were over 19,000 people in five Nevada counties, more than

TABLE 4.1. Population of Original RECA-Designated Counties (1979) in 1950 and 1960

County/town	1950 population	1960 population
Nevada County		
Eureka	896	767
Lander	1,850	1,566
Lincoln	3,837	2,431
Nye	3,101	4,374
White Pine	9,424	9,808
Total	19,108	18,946
Utah County		
Beaver	4,856	4,331
Carbon	24,901	21,135
Duchesne	8,134	7,179
Emery	6,304	5,546
Garfield	4,151	3,577
Grand	1,903	6,345
Iron	9,642	10,795
Juab	5,981	4,597
Kane	2,299	2,667
Millard	9,387	7,866
Piute	1,911	1,436
San Juan	5,315	9,040
Sanpete	18,891	11,053
Sevier	12,072	10,565
Uintah	10,300	11,582
Washington	9,836	10,271
Wayne	2,205	1,728
Total	138,088	129,713
Total all counties	157,196	147,659

Source: US Census Bureau, *U.S. Census of Population: 1960*, vol. 1, *Characteristics of the Population* (Washington, DC: US Government Printing Office, 1963).

TABLE 4.2. Population of Enacted RECA-Designated Counties (1990) in 1950 and 1960

County/town	1950 population	1960 population
Nevada County		
Eureka	896	767
Lander	1,850	1,566
Lincoln	3,837	2,431
Nye	3,101	4,374
White Pine	9,424	9,808
Total	19,108	18,946
Utah County		
Beaver	4,856	4,331
Garfield	4,151	3,577
Iron	9,642	10,795
Kane	2,299	2,667
Millard	9,387	7,866
Piute	1,911	1,436
San Juan	5,315	9,040
Sevier	12,072	10,565
Washington	9,836	10,271
Wayne	2,205	1,728
Total	61,674	62,276
Arizona County		
Apache	27,767	30,438
Coconino	23,910	41,857
Gila	24,158	25,745
Mohave	8,510	7,736
Navajo	29,446	37,994
Yavapai	24,991	28,912
Total	138,782	172,682
Total all counties	219,564	253,904

Source: US Census Bureau, *U.S. Census of Population: 1960*, vol. 1, *Characteristics of the Population* (Washington, DC: US Government Printing Office, 1963).

61,000 people residing in ten counties in Utah, and over 138,000 people living in the six Arizona counties listed in table 4.2.[18] All were rural areas, but they added up to more than 219,000 people in 1950. The counties in Arizona experienced robust growth over the course of the 1950s. By 1960, there were more than 250,000 people living just downwind of the AEC's outdoor proving grounds. If the goal was to establish a test site with nobody around, per Stafford Warren's advice, then south-central Nevada hardly fulfilled this mandate.[19]

Further, the NTS is located within the traditional lands of the Western Shoshone and Southern Paiute tribes. As demarcated in figure 4.1,

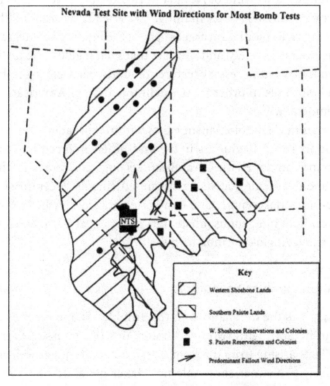

Figure 4.1. Native American communities downwind of the Nevada Test Site. (Eric Frohmberg, Robert Goble, Virginia Sanchez, and Dianne Quigley, "The Assessment of Radiation Exposures in Native American Communities from Nuclear Weapons Testing in Nevada," *Risk Analysis* 20, no. 1 [2000]: 101–111; courtesy of Eric Frohmberg)

many Native American settlements were directly downwind of the AEC's proving grounds. More broadly, the mining and milling of uranium, the engineering and testing of atomic weapons, and the detritus left behind by the atomic-industrial complex have had a substantial impact on Native American lands.[20] It is an expression of the "internal nuclear colonialism" of the American West or, as Charles Pinderhughes notes, "a geographically-based pattern of subordination of a differentiated population, located within the dominant power or country."[21] Forces within a country impose their own particular objectives on the land, at the expense of marginalized segments of the domestic population. Moreover, research suggests that the impacts on Native American communities downwind of Nevada were probably exacerbated by their reliance on local subsistence hunting.[22] Valerie Kuletz argues, "The presence of internal colonialism argues against the myth of an integrated and truly democratic society, and it argues that such regional inequalities are not temporary but necessary features of industrial society—features we choose not to see in order to maintain the myth of American equality and democracy."[23]

Internal nuclear colonialism exists within a broader context of settler colonialism.[24] Beginning in the mid-nineteenth century, Mormon immigrants enveloped the Great Basin, seizing land, remaking the landscape, and displacing Native American settlements. A century later, the atomic-industrial complex and its assorted boosters pushed its own priorities on both indigenous inhabitants of the Great Basin and a menagerie of rural, Anglo-Mormon towns.

Operation Tumbler-Snapper: April–June 1952

Table 4.3 lists the detonations in the Tumbler-Snapper series of 1952. It began in April with two one-kiloton airdrops to assess variance in blast effects arising from different detonation altitudes as well as weapons effects.[25] Neither shot Able nor Baker generated notable offsite radioactivity. The next shot, Charlie, was a large, thirty-one-kiloton airdrop detonated above the desert. A roster of distinguished guests from the military, Congress, and defense industry corporations were in attendance, and for the first time, the press was onsite. Situated at a rock outcropping designated "News Nob," the press had a view looking

TABLE 4.3. Operation Tumbler-Snapper Detonations

Code name	Type	Yield (kilotons)
Able	airdrop	1
Baker	airdrop	1
Charlie	airdrop	31
Dog	airdrop	19
Easy	tower	12
Fox	tower	11
George	tower	15
How	tower	14
Total yield		104

Source: DOE, *United States Nuclear Tests: July 1945 through September 1992* (Las Vegas: Nevada Operations Office, 2000).

north toward Yucca Flat, the site of many atmospheric and then under-ground detonations. The AEC also allowed Charlie to be televised live to a national audience. Recognizing that Americans in general and the press were unlikely to ignore testing in Nevada any time soon, the AEC began to transition from a low-key public-relations approach to treating atomic detonations as a spectacle of technological-instrumental ratio-nality, and by the mid-1950s, the public relations of continental testing were the inverse of the way they began.

A frequent presence at News Nob was the *New York Times* science reporter William L. Laurence. In addition to Gladwin Hill and Hanson Baldwin, Laurence wrote many articles on continental open-air testing. He was one of the most well-known science reporters of his time. For much of 1945, he was employed by the *Times* and the War Department. It was a conflict of interest that gave him unparalleled access to the mili-tary officials and scientists working on the Manhattan Project. He was the only journalist to witness the Trinity detonation and the bombing of Nagasaki. His reporting earned him a second Pulitzer Prize. Laurence and the *Times* itself were key to the US government's postwar suppres-sion of information about the devastation at Hiroshima and Nagasaki.[26] From the beginning, the US government had problems being honest about radioactivity, and the *Times* had problems with co-optation by the US government.

Laurence displayed a penchant for religiously themed interpretations.[27] He applied this to testing in Nevada—consistent with an atomic spectacular frame of reference and at the expense of less lofty, more concrete complexities. Laurence seemed consumed by the experience. Reporting onsite during Upshot-Knothole in 1953, he mused, "Suddenly through our goggles we could see a bright light shooting out from the north. It was the light of a hundred suns rolled into one, a light not of this world. We counted three and took off our goggles. There before us, fearful to behold, was the ball of fire, a giant, iridescent sphere, a new star in the act of being created, changing shape and color at breathtaking speed."[28]

The legitimacy of the state, and democratic states in particular, rests on the exercise of rational-legal authority rather than simply the charisma of political or religious leaders or long-standing traditions.[29] "Rational-legal authority" refers to social and cultural belief in the moral validity of power exercised by the state. Elected political officials and appointees come and go, but rational-legal authority endures within structural positions in the established political order. Despite initial attempts to downplay detonations, the AEC soon embraced the atomic spectacular as a testament to rational-legal authority and expertise, propping up the commission's legitimacy while obscuring the failure to anticipate, plan for, and manage offsite radioactive fallout.

AEC officials embarked on the atomic spectacular to appease, entertain, and distance the public from atomic military development. Photographs and newsreel footage of ascending mushroom clouds by journalists operating according to strict AEC protocols gave a reassuring and decontextualized account. Such distance obscures the minutiae while emphasizing the presentation. Scott Kirsch argues that the AEC judiciously pursued "spectator democracy."[30] By controlling news coverage, AEC and military officials worked to contain debate while appearing to be open and engaged with the public. Over time, AEC officials became adept at releasing evocative images and newsreel footage but not comprehensive, timely information on fallout location, intensity, and duration. And despite finally being allowed to view detonations onsite, *Times* reporters were largely oblivious to increasing problems with offsite radioactivity or the institutional tensions between the armed services and the AEC. The line between the downwind public as spectators with a front-row seat and unwitting participants was thin indeed. And as with all illusions, there exists a fissure between superficial appearances and underlying, enduring dynamics.

Tumbler-Snapper began with four airdrops before concluding with four three-hundred-foot tower detonations with an average yield of thirteen kilotons and conducted over a four-week period. It was a significant turning point. As opposed to airdrops, Philip Fradkin observes, "May and June were different. They proved to be precursors and distinct warnings of what was to come the following year in terms of exposure of humans and animals to radioactive fallout."[31]

The first tower shot was Easy, a twelve-kiloton detonation producing radioactive debris drifting north over Ely, Nevada, before heading northeast over Salt Lake City, Utah. It then dispersed over Cleveland, Ohio, before producing scattered hot spots along the eastern seaboard.[32] About 350 miles from the NTS, Salt Lake was a city of 182,000 people in 1950.[33] Employees of a Geiger-counter manufacturer in Salt Lake detected Easy passing by and contacted the governor's office. A Geiger counter detects what humans cannot. And given the uranium deposits in and around the four corners region where Utah, Arizona, Colorado, and New Mexico abut, many people downwind owned one. It was a source of consternation for the AEC. As Easy swept over Salt Lake City, local resident Lyle Jepson won a prize, worth $10, for contacting the *Deseret News* to report levels of radioactivity higher than the natural background rate.[34] On May 8, the paper published a short, reassuring article in which a spokesperson for the AEC denied any danger.[35] The next day, an editorial in the *Deseret News* adopted a more apprehensive tone, reminding the commission of the "tremendous responsibility" it bore ensuring public safety.[36] Jepson's discovery and the subsequent *Deseret News* editorial on May 9 marks the first instance when the press in the downwind region adopted something other than a good-natured attitude toward the AEC.[37]

The incident also provoked a terse letter from Utah's Governor J. Bracken Lee to Gordon Dean. The governor was annoyed that neither he nor any other state official was contacted regarding radioactivity over Salt Lake City, and he insisted that Dean demonstrate greater cooperation with state officials in the future. Lee remarked, "Acknowledging the fact that each individual can safely receive an accumulated dose of only a certain amount of radiation, then, with malice and forethought the AEC Operations authorities subtracted from the accounts of allowable radiation exposure of about 500,000 people. Who knows how many of these citizens will later need every milliroentgen they can stand, whether it be for treatment of cancer, from atomic bombs, or even from working in necessary exposure, such as local mines."[38]

Easy was followed by Fox which produced radioactive rain over the Great Lakes.[39] Both Easy and Fox were the first of seventeen detonations between 1951 and 1958 to produce more than 1,000 "person-R" (rad) of estimated external gamma radiation exposure within two hundred miles of the NTS.[40] All seventeen were tower shots. This list indicates those detonations generating population exposure in a general manner, but it does not articulate the intensity of exposure at an individual level. A heavy dusting of gamma radiation over a sparsely populated area generates a given person-R estimate, even as a lower level deposited over a larger population may result in the same calculation. Nonetheless, Easy and Fox were emblematic of the AEC's turn toward risker operational conduct by the summer of 1952. The third tower shot, code-named George, a fifteen-kiloton detonation, produced a mushroom cloud rising over thirty-seven thousand feet along with fallout detected in Elko, Nevada, but the most intense fallout appeared in Illinois, Indiana, Michigan, and Wisconsin. The last tower detonation of the series was shot How. Fallout again moved in the direction of Elko and such disparate places as Great Falls, Montana, and Boise, Idaho.[41] A list of the twelve detonations producing the great cesium-137 fallout around the country includes Easy, Fox, George, and shot How. Together, the last four detonations of the Tumbler-Snapper series account for approximately 23 percent of cesium-137 deposition in the continental US due to testing in Nevada.[42] The Tumbler-Snapper series of 1952 was characterized by greater reliance on tower shots and, in turn, greater detection of radioactivity downwind.

Indeed, cattle and horses grazing twenty-five miles from the test site boundary were exposed to fallout from the Tumbler-Snapper series. Animals owned by the rancher Floyd Lamb displayed odd white patches of hair and ulcerations along their backs, indicative of beta radiation.[43] Thomas L. Shipman, director of the LASL Health Division and a medical doctor in charge of the Rad-Safe Group, wrote to Alvin Graves to apprise

Figure 4.2. Horses illustrating evidence of beta burns from fallout. (Photo courtesy of the Utah State Archives)

him of the problem. Shipman urged that Robert Thompsett, a veterinarian from Los Alamos, New Mexico, be part of the team examining the herd as Thompsett had personally examined injured cattle downwind of Trinity. And it was clear to Shipman that radiation was the issue: "There exists no reasonable doubt but that these cattle were injured by fall out from one of the shots in the spring series, probably one or both of the two final detonations."[44] Shipman suggested the commission confront the issue head-on: "If these cattle actually were injured by radiation, I see little choice except to tell the truth and shame the devil."[45] That never happened. AEC and military officials soon became adept at raising unreasonable doubt as fallout from the upcoming Upshot-Knothole series injured sheep numbering in the thousands.

Figure 4.2 is a photo of horses displaying evidence of beta burns. Like the telltale clicking of a Geiger counter, beta burns were a cue threatening to undercut the AEC's hegemonic narrative insisting that there were no observable offsite effects of atmospheric detonations.

Committee on the Operational Future of the Nevada Proving Grounds

In the wake of Tumbler-Snapper, AEC officials established the Committee on the Operational Future of the Nevada Proving Grounds to review operations to date. Offsite fallout was a motivating factor. The new committee was composed of AEC officials and military personnel from the commission's Division of Military Application. It was an opportunity to assess increasing reliance on tower detonations, to consider raising the height of towers to reduce localized fallout, and to discuss soil stabilization strategies at ground zero to temper the updraft of debris. Such considerations were crucial, as the upcoming Upshot-Knothole series was slated to include tower shots larger in yield than anything yet detonated. Nonetheless, such considerations were only briefly discussed. Instead, the committee is better characterized as a group of key actors from within the technocracy trying to work out their interorganizational tensions and disagreements more so than focusing on the risks of fallout.

The committee's report was released in May 1953, more than two months after Upshot-Knothole began. It reiterates the original purpose of testing in Nevada: to provide LASL with a "backyard laboratory." The idea was that as a particular design question arose diagnostic tests could be quickly arranged, and the results incorporated into weapons manufacturing in a timely manner. The first series, Operation Ranger, was identified as conforming to this overarching intent, but from there on the committee observed the increasing scope of participants and activities, even "fringe projects," and cautioned against continued expansion of operations.[46]

The upcoming Upshot-Knothole schedule was highlighted in the committee's report as evidence of the menagerie of activities and participants that were not envisioned when the NTS was created. The Federal Civil Defense Administration was staging tests to determine thermal and blast effects on housing, food, clothing, and automobiles. The Army and Marines were planning increasingly elaborate battlefield exercises, impatient to do so ever closer to ground zero, as well as effects testing. The Army was eager to test fire an atomic artillery shell from a specially built howitzer, the world's first atomic cannon shot. Moreover, the AEC was confronting requests for even more varied activities beyond the

Upshot-Knothole series: to experiment with atomic missiles, to detonate in a canyon surrounded by hills, to blow up a nuclear power reactor.[47] Military officials were even working to convince the AEC to detonate in heavy rain or snow—a proposal guaranteed to generate hazardous levels of radioactivity on- and offsite.

Norris E. Bradbury, the director of LASL and a member of the committee, was not particularly concerned with offsite fallout, but he was clearly irritated by other concerns: "I regard the tendency to use the NPG [Nevada Proving Grounds] for the purpose of weapon system tests (the forthcoming gun shot), for civil defense effects tests, for troop indoctrination and maneuvers, and for the reportorial press as quite outside the original concept of this site. Indeed this trend, if continued, can force us to abandon this site for no other reason than that the military have taken it over."[48] Institutional momentum was putting pressure on LASL's backyard laboratory, as Bradbury noted. But it was also complicating offsite rad-safe operations due to increasing reliance on tower detonations and the frenzied pace of testing.

The committee's report cautioned against tower detonations greater than thirty-five kilotons, and when the report was released in May, a forty-three-kiloton detonation code-named Simon had just rained out over New York State. The report declared, "It was the Committee's feeling that experience to date with tower and surface shots indicates there should be more care exercised rather than less care."[49] It was a remarkably unremarkable conclusion given the increasing evidence that fallout could be capricious.

By the end of 1952, downwind residents generally believed in the rational-bureaucratic authority of the AEC and trusted that it would act with diligence in the conduct of testing. There were few dissenting voices in local and regional newspapers.[50] Indeed, it was repeatedly remarked in the committee's report that the public's attitude in the downwind communities "generally is quite cooperative." Despite evidence illustrating that the public was prone to unreasoning panic, the report reads, "Radiation presents a continuing threat to public relations; while the fear is latent it can come into the open at any time; and should be anticipated by continuing public-related education, reports, etc."[51]

The Committee on the Operational Future of the Nevada Proving Grounds found it reassuring that no locality had been exposed beyond the established threshold. However, the committee conceded, "There was discussion of experience on ground and tower shots, and of the good fortune which has on occasion caused highly-radioactive clouds to wend their way in between communities."[52] When the committee released its classified internal report, in May 1953, the AEC's run of good fortune was already over.

The AEC and Failures of Foresight

Large, complex organizations are a key dimension of modernity, but things sometimes go awry.[53] This raises questions as to the dynamics contributing to organizational failure—broadly defined as deviation from expected and desired outcomes.[54] Barry Turner has argued that organizational accidents are typically not "bolts from the blue" but foreshadowed by days, weeks, even years of recurring signals suggesting that operational conduct is at variance with prevailing expectations—but these signals are ignored or underappreciated. Mistake, misconduct, and disaster, in turn, result from nonroutine activities incubating over time, as operations are at odds with what was intended and are accompanied by *failures of foresight*.[55] The burgeoning risks of the modern age are more often shaped by organizations and interorganizational relations than simply technology by itself. Turner argues, "A disaster or cultural collapse takes place because of some inaccuracy or inadequacy in the accepted norms and beliefs, but if the disruption is to be of any consequence, the discrepancy between the way the world is thought to operate and the way it really does rarely develops instantaneously. Instead, there is an accumulation of a number of events that are at odds with the picture of the world and its hazards represented by existing norms and beliefs."[56]

To understand sociotechnical disasters, then, it is crucial to reconstruct the emergent patterns that are common across organizational and interorganizational contexts characterized by looming trouble progressing into disaster. Organizational culture and structure, for example, can reduce or encourage failures of foresight. The shared, often taken-for-granted, assumptions and values developed over time as people collectively work to solve problems in pursuit of goals, to achieve internal demands for integration, and to adapt to external demands constitute organizational culture.[57] "Culture can be thought of as a set of solutions produced by a group of people as they interact about the situations they face in common," Diane Vaughan notes.[58] Culture imparts stability and identity, and it shapes individual and collective thinking, emotions, and behavior. "Structure" refers to the composition of an organization. It includes the manner in which tasks and responsibilities are delegated, the degree of centralization of power versus a more decentralized autonomy of organizational divisions, and the degree of coordination and information exchange between different divisions within an organization.[59]

The AEC was immersed in the incubation stage, as signals that were incongruous with expectations were accumulating. During the incubation stage, the unexpected may garner attention, but not necessarily to a sufficient degree to act on escalating problems that are a prelude to a more significant breakdown of operational conduct. It is at this stage that an organization with a culture of self-critique seeks to better understand those anomalies and miscues that are at odds with overarching expectations. Regarding the character of warnings, Lee Clarke suggests, "More generally, at any given moment there are messages about the relevant dangers lurking in a system. Some are specific, some general. Some are issued by people with plenty of experience and expertise, some are not. Whether those messages turn into warnings depends on the reactions to them, and on the selection processes that single out certain messages while ignoring others."[60]

Turner explained that "rigidities in belief and perception" are commonplace among organizations that are indifferent to evidence of impending problems, inhibiting appreciation of information that is inconsistent with explicit and tacit assumptions underpinning day-to-day operations.[61] Karl Weick, commenting on Turner's model, notes, "In order to act collectively, people adopt simplifying assumptions. Simplifications limit the precautions people take and the range of undesired consequences they envision. These simplifications set the stage for surprise."[62] Information, data, and cues suggesting impending problems are not simply encountered in raw form but are filtered through organizational culture and structure, which shapes the construction of meaning.[63]

Clarke makes a similar claim in referring to the pernicious effects of "disqualification heuristics."[64] Individuals and organizations rely on judgmental heuristics or mental shortcuts to aid in making inferences amid uncertainty. Time-honored heuristics serve as acceptable solutions to problems in an efficient manner. A common example is the "availability" heuristic. What seemed to work in the past often informs present-day decisions, and when future risks are considered, those that were encountered in the past get the most attention.[65]

Key decision-makers who are prone to failures of foresight are often characterized by "perceptual horizons" precluding recognition of the extent and seriousness of intermittent anomalies.[66] As Weick observes, within complex organizations, there is often a propensity toward "believing is seeing."[67] In the wake of acute disaster, organizational actors and outside regulators are invariably bewildered by the incongruence of prior beliefs and assumptions with the circumstances they now confront. In turn, risks are often ill defined and unclear, and organizations do not always possess the information needed to make crucial decisions. Nonetheless, the key question is how an organization evaluates and reacts to ill-defined problems and unanticipated events.[68] The inability to appreciate and act on signals of looming trouble often hinges on key organizational decision-makers clinging tightly to their preferred representations of reality and selective recognition of some dynamics and dismissal of others.

The incubation period with regard to atomic testing began with Buster-Jangle. By the fall of 1951, institutional momentum was accruing, and it accelerated during Operation Tumbler-Snapper, in 1952. It was not simply the number of stakeholders and variety of activities, and thus increasing reliance on tower shots, but the scale and pace of detonations over time. Soon after the inauguration of the program, Nevada was home to troop maneuvers, scientific experiments of all kinds, civil defense exercises, and elaborate weapons effects testing, in addition to simply weapons development.

Robert Gephart has reformulated Turner's model to account for interest-group dynamics and power in shaping the character of "sensemaking."[69] As Weick and colleagues note, "Sensemaking is about labeling and categorizing to stabilize the streaming of experience." Organizational actors often engage in "retroactive sensemaking" to assess the degree to which operations are at variance with what was anticipated and why, in order to formulate recommended adjustments.[70] The

Committee on the Operational Future of the Nevada Proving Grounds is emblematic of such efforts. Gephart, however, argues, "Sensemaking is inherently political: it involves the production of accounts as ground for future action in the face of alternative descriptions of reality."[71] Failures of foresight may be rooted not just in rigidities of belief and perception, constraints in information flow, limited perceptual horizons, and persistent disregard of critique from outside the organization but also enduring commitment to a particular definition of the situation. "Power is the ability to have one's account of reality perceived by others in the face of alternative claims through use of sensemaking practices," Gephart asserts.[72] In turn, mistakes and accidents are not simply due to missed signals that are at odds with expectations but are contingent on various interest groups struggling to define or indeed dismiss such signals altogether. When there are no parties at the table offering a counterweight to the needs and priorities of the technocracy, the technocracy will privilege its own needs and priorities.

Environmental risk results from failure to maintain the stability of human-nonhuman configurations such that the force and efficacy of entities, processes, and things exceed that for which they were originally designed. But risk is made meaningful through the social construction of reality, in which competing definitions jostle for public and political legitimacy. This highlights the need to focus on which actors, discourse coalitions, and rhetorical strategies contribute to *definitional power* or the ability to assert a preferred interpretation at a given point in time. Risks are real but are rendered salient through social negotiation and debate, and some definitions are more consistent with accumulated empirical evidence than others. Power shapes the production of self-evident truths, but particularly when a given definition of the situation is accepted amid contradictory empirical evidence. Moreover, the interstitial space between the world and our conceptions of it is expanding as the complexity of human manipulation of nature increases.[73] The Scientific Revolution was predicated on the successive accumulation of knowledge, even the eventual end of uncertainty; but in the risk society, uncertainty abounds, and the more deeply we delve into and rearrange aspects of nature, the wider the chasm gets.

Turner argued that sociotechnical mistakes and disasters are simultaneous examples of organizational failures of foresight. Nonetheless, some things simply cannot be known or anticipated beforehand when dealing with the properties, processes, and internal dynamics of innovative but risky sociotechnical systems.[74] Consistent with Gephart's critique, there is something missing in Turner's model, and it hinges on the fact that evidence of looming trouble—the difference between how things are defined and how they behave—often only arise over time and in tandem with the benefits accruing to a particular segment of society. Turner's model is a valuable touchstone, but it does not fully account for the ontologically turbulent shift that occurred at Trinity or the novel types of risk that arise due to advances in science and technology.

The Material-Discursive Bind

Operational conduct in Nevada evinced not simply failures of foresight but a *material-discursive bind* by the summer of 1952. There is a strong tendency in late modernity toward a bind in relation to sociotechnical risk: the hazards inherent to many materials and substances cannot be fully assessed in the laboratory and only become apparent as they are utilized in practice, but as sociotechnical innovations prove instrumentally useful and profitable, interest groups increasingly have incentives to contest evidence of any injury that arises. It is only through performative engagement that failures of foresight become evident, in turn, but this is often met with the organizational production of doubt, ignorance, and denial. Material-discursive incongruity and tension are therefore commonplace, as innovation is accompanied by assumptions and inferences that cannot be assessed mathematically or in controlled settings. The material-discursive bind is an expression of the difficulties endemic to conceptualizing the push and pull of human agency and nonhuman materiality except through practical engagement. Organizational actors, however, often feel compelled to defend those practices that illustrate accrued momentum even as the liabilities of doing so become increasingly clear over time.

The material-discursive bind of sociotechnical risk is depicted in figure 4.3. It is drawn from Ulrich Beck's conception of the risk society and Andrew Pickering's dance between human action and nonhuman efficacy and force. It recognizes that science and technology are productive and, indeed, invaluable but that they create new dilemmas in which

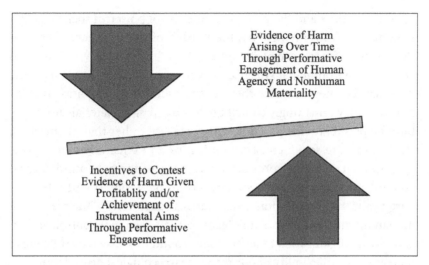

Figure 4.3. The material-discursive bind of sociotechnical risk.

differential power shapes the character of the ensuing debate. Substances and materials whose effects mirror diseases that are already common-place in society, that provoke injury only over an extended period, and that are profitable or fulfill some instrumental end best illustrate the material-discursive bind. Slow environmental violence therefore prolif-erates amid brilliant advances in science and technology as newfound dilemmas bedevil society.[75] It is a difficult balancing act, as the efficacy and force of matter become more discernable as they are entangled with instrumental human behavior, which is itself called into question by the effects it provokes but is reticent to confront. Beck articulates a situation akin to this bind, "Hazards which come into existence with the blessing of technological and state authority place authorities and policy-makers under the permanent compulsion to assert that these hazards do not exist, and to defend themselves against penetrating questions and evi-dence to the contrary."[76]

The material-discursive bind refers not simply to competing messages but also to tension between discourse and material-biophysical effects amid the competing imperatives plaguing public deliberation as power and parochial interests intercede. This dynamic erodes the public legiti-macy of the state, corporations, science, and law while invoking a more general sense of futility. And in the tension between the recognition of risk over time and the parochial interests of powerful societal actors who benefit from the activity, the burden of proof of harm lies with those who claim to have been injured. At an individual level, those who confront environmental violence typically are not fully conscious of the disordered context in which they are enmeshed, but recognition that innovation may unwittingly contribute to a tightening material-discursive bind helps situate this paradox within society rather than themselves. Retreat to pure social constructionism or an exclusively realist position, in turn, is counterproductive, as many of the most fundamental conundrums in society lie somewhere in between. The problem, in part, stems from perception. Peter Harries-Jones notes, "The premise of human domination over nature leads to a false sense of control, and, in turn, social organization of technology around this false sense of control increases the inflexibility of our response to ecological degradation."[77]

Assumptions as to humanity's mastery over the nonhuman world contribute to social-organizational patterns that are ill suited to effectively respond to problems arising when mastery proves elusive. The premise of human domination over nature leads to a false sense of control, and when things go wrong, reticence to scrutinize the pretense of control contributes to the inflexibility of institutional structures within society in addressing the contested implications of innovative science and technology. The material-discursive bind, in turn, is expressed in slow environmental violence amid competing claims to truth that are difficult to fully untangle, and political maneuvering often intercedes to ensure that things remain entangled and unclear. The optimism of modernity gives way to contradictions that pull at and fray the social fabric.

PART II

Upshot-Knothole and the Fluidity of Nature

5

Dirty Harry and the Material-Discursive Bind, 1953

Just as a magician diverts attention from one hand to another, dominant organizations engage in discursive strategies that accentuate some
things while obscuring others.[1] As the Upshot-Knothole series began in
March 1953, the AEC distributed a pamphlet in the downwind communities. "Continental Weapons Tests . . . Public Safety" articulates some
complexities of open-air testing while concealing others.[2] Of the 14,890
words in the pamphlet, "cancer" appears three times and only in regard
to the use of radiation in treating disease. "Leukemia" does not merit
a single reference. Of the words absent, perhaps none is more notable
than "milk." Cow's milk was a crucial link between the atomic state and
the public.

"No person has been exposed to a harmful amount of radiation from
fall-out. In general, radioactivity resulting from fall-out has been many
times below levels which could cause any injury to human beings, animals, or crops," the pamphlet insists. Further, it makes clear that atomic
testing is necessary to accelerate weapons development and to safeguard
the national defense and that it is efficient with regard to resources, time,
and money. Nor is radioactivity new or exotic. "Man always has lived
in a sea of nuclear radiation." The reassuring message is repeated in the
conclusion: "It is the Commission's hope that in the near future there
will be no occasion for alarm through lack of knowledge of the facts
about levels of radiation and their degree of hazard, just as there is now
no reason in actuality for alarm, since the radioactivity released by fallout has proved not to be hazardous."[3]

The pamphlet does not provide information on the precautions that
one could take should an unanticipated fallout event occur, such as dusting off or washing homegrown vegetables, clothes, hair, or children. It
does not list a telephone number for downwind residents to call to get
further information or to seek advice should a fallout event occur.[4] AEC
officials sought quiescence more than an informed public.

As testing began, the AEC was committed to a low-key public-relations approach, but two years later, the expectation that the public would not pay much attention proved naïve. The commission would have to say something. In doing so, it ran the risk of letting a preferred definition of the situation outpace pragmatic management of fallout. That is, AEC officials ran the risk of making definitive statements even as the materiality of fallout continued to elude prediction and control. There is a momentum—a *discursive momentum*—such that the longer and more robustly a given claim is put forward and oppositional claims are rejected, the more difficult it becomes to revise the established discursive position—even as empirical evidence accrues suggesting that it is wholly or partially incorrect. In turn, powerful organizations often double down on an erroneous interpretation. As Sherry Cable and colleagues observe in reference to discourse patterns at the Oak Ridge Nuclear Reservation, officials became "ensnared in their own social constructions."[5] Moreover, an organization's failure to release information earlier can inhibit its propensity to do so later—a *momentum of omission*—for fear of undercutting public perceptions of legitimacy, particularly if it is not compelled to do so by legislative or judicial action. Reticence to acknowledge an issue earlier, then, becomes a stumbling block to forthright admission later—even amid empirical evidence underscoring what organizational actors prefer not to recognize.

The AEC distributed revised pamphlets in 1955 and 1957 that were less technical in tone, include cartoons, and adopt a personal message addressed to the downwind residents.[6] The pamphlets were distributed to individuals and community groups in southern Nevada and Utah and were available at schools, post offices, gas stations, grocery stores, and local motels.[7]

The downwinder pamphlets are indicative of *fantasy documents*. Organizations outline the actions they will take should untoward events arise, and often those plans scarcely describe what occurs when things go wrong. Documents bearing the authoritative stamp of organizational expertise nonetheless serve as "rationality badges," Lee Clarke argues.[8] Planning is not simply an exercise in outlining which steps are to be taken in a given situation but is meant to communicate that the organization is attentive and competent and that there are reasons to defer to its accumulated expertise. Such documents are often an aspect of discursive momentum, moreover, as officials articulate a level of control that is not attainable in practice. However, contingency plans are not necessarily intended to deceive. They are rituals of instrumental rationality but are based on unrealistic assumptions involving problems for which there is little experience. Clarke notes, "As powerful as organizational leaders can be and as cunning as experts sometimes are, the more frightening truth is that they are often unmindful of the *limits* of their knowledge and power. We have more to fear from organizations and experts overextending their reach, propelled by forces endemic to modern society, than from conniving conspiracies."[9]

Clarke further explains, "Operational rationality is when actors transform uncertainty into risk in such a way that they can control the key aspects of the problem. And as a document, or suite of documents, garners legitimacy amongst the public it then helps structure the dialogue and debate in regard to the problems it addresses."[10] And if organizational actors do not communicate with the public in a forthright manner, discursive expressions of operational rationality can come back to haunt them when the public realizes it has been led to focus on one hand but not the other.

Discourse is the historically situated pattern of meaning(s) expressed through written or spoken language and framing some aspect of the world. Official discourse displaces knowledge born of individual experience and perception in favor of knowledge derived from large organizations and institutional arenas within society. In this sense, it is external to and imposes itself on the individual. Power and knowledge are intertwined.[11]

The atomic-industrial complex, composed of government entities and corporate contractors, has long engaged in proactive and reactive social control efforts, and the threat of job losses is a key discursive strategy employed to quell public discontent. The downwinders were not employed by the AEC or affiliated corporate contractors. As the untoward effects of the Upshot-Knothole series arose, the AEC would therefore employ novel techniques to preempt presumed "latent local fear of radioactivity."[12]

Discourse does not take place unencumbered by material and biophysical dynamics, however, and to ignore such factors is not to render their implications nonexistent, only unobserved. Sheep displaying ghastly symptoms attest to very real radiological dynamics. And although reality can never be precisely demarcated, deleterious consequences arise from adhering to distorted conceptions of reality. Raymond Murphy insists, "Some socially constructed conceptions of nature and of risk are more appropriate for their referents than others; in this sense knowledge claims are hierarchical and relativism is unconvincing. . . . Conceptions of nature and risk are socially constructed, but not in a material vacuum. They are constructed by sensory beings using as prompts observations and experiences of both nature's everyday and extreme dynamics, as well as scientific discoveries about the material world."[13]

Upshot-Knothole Series, March–June 1953

The Upshot-Knothole series included eleven detonations conducted over eleven weeks, outlined in table 5.1. It was a quick tempo. Seven shots consisted of tower detonations totaling over 130 kilotons. Upshot-Knothole was the dirtiest series with regard to radioactive contamination in the downwind communities, and it called into question the idea that

TABLE 5.1. Detonations in the Upshot-Knothole Series

Detonation	Deployment	Yield
1. Annie	tower	16 kilotons
2. Nancy	tower	24 kilotons
3. Ruth	tower	200 tons
4. Dixie	airdrop	11 kilotons
5. Ray	tower	200 tons
6. Badger	tower	23 kilotons
7. Simon	tower	43 kilotons
8. Encore	airdrop	27 kilotons
9. Harry	tower	32 kilotons
10. Grable (fired from 280mm gun)	airburst	15 kilotons
11. Climax	airdrop	61 kilotons
Total		252 kilotons

Source: DOE, *United States Nuclear Tests: July 1945 through September 1992* (Las Vegas: Nevada Operations Office, 2000).

fallout "has proved not to be hazardous." Such statements are the stuff from which discursive momentum is set in motion.

Prior to the Upshot-Knothole series, ground-level radiological monitoring relied on mobile monitoring teams documenting readings over a vast geographic area while dealing with a scarcity of roads, which inhibited efficient movement. Beginning in 1953, the AEC, in conjunction with the US Public Health Service, established seventeen fixed monitoring stations in the NTS region in addition to mobile monitoring teams. In all, ninety-five gummed-film stations—a one-foot-square sticky sheet of acetate film mounted horizontally on a three-foot stand—were placed around the country to record fallout beyond two hundred miles from the NTS. Despite such improvements, Upshot-Knothole illustrated the inadequacies of the commission's efforts to measure fallout both near and far from Nevada.

The series began with a sixteen-kiloton tower shot named Annie, accompanied by troops in trenches and effects testing on behalf of the Federal Civil Defense Administration (FCDA). Journalists as well as state and federal officials, an assemblage of more than six hundred, watched from News Nob overlooking Yucca Flat. A two-story colonial-style home was constructed thirty-five hundred feet from ground zero and another at seventy-five hundred feet. Both were furnished and included mannequins clothed in J. C. Penney attire. Fifty cars were placed around the houses. It was the FCDA's opportunity to gauge the impact of an atomic detonation on different building materials, clothing, and food supplies. "Interior shots of the house show a family of mannequins posed amidst common household items, such as lamps and tables. Without irony, photo captions explain that alert citizens who hope to survive an atomic attack should emulate the readiness of the mannequins," Andrew Kirk observes.[14]

The test was named "Operation Doorstep" to exemplify the idea that Soviet aggression was a threat reaching into the heart of the United States, but many press outlets preferred "Doom Town." Obliteration of suburban-style homes and mannequin families buried beneath rubble constituted a surreal form of national theater.[15] The FCDA sought to jar the public into taking civil defense preparations seriously without inducing a sense of nihilism. The director of the FCDA, Val Peterson, penned an appeal to the public in Collier's Weekly: "You have just lived through the most terrifying experience of your life. An enemy A-bomb has burst 2,000 feet above Main Street. Everything around you that was familiar has vanished or changed. The heart of your community is a smoke-filled desolation rimmed by fires. Your own street is a clutter of rubble and collapsed buildings. Trapped in the ruins are the dead and the wounded—people you know, people close to you. Around you, other survivors are gathering, dazed, grief-stricken, frantic, bewildered. What will you do—not later, but right then and there?"[16]

While reaching into suburbia with evocative images of atomic destruction, the test site itself remained an enigma, a suitable place for large-scale destruction, a "place without nature or history."[17] And as the FCDA was warning Americans about the dire consequences of a Soviet strike, radioactive debris was dispersing around the country. Indeed, Annie was notable not simply for the dramatic civil defense experiments. It was one of the ten dirtiest detonations with regard to population exposure to off-site fallout.[18] Situated atop a three-hundred-foot tower, Annie drew in alluvial soil that came back to earth in an irradiated state.

Shot Nancy was next. Figure 5.1 depicts the location and intensity of fallout from the twenty-four-kiloton device detonated atop a three-hundred-foot tower on the morning of March 24. Fallout often skipped before coming to earth, but Nancy came down just beyond the northeast boundary of the test site due to weaker winds aloft than predicted.[19] It came down in an area with few people but a lot of sheep, and it set in motion a series of events that the AEC would work to keep from public consciousness for decades. As debris moved further afield, it was detected in Ely, Nevada, before delivering "unusually heavy fallout" over Salt Lake City.[20] At congressional hearings in 1979, the rancher Kern Bulloch remarked, "And of course, the cloud came up and drifted over us, a little to the south of us but over us. And it was a little bit later that day that some of the Army personnel that had four-by-fours and jeeps— this road we were on was kind of a main road. It was next to the Lincoln Mine, south of the Lincoln Mine. But they came through there, and I and a companion, another sheepherder from the camp—and they said, 'Boy, you guys are really in a hot spot.'" Bulloch was advised to evacuate but would not leave the herd behind, around two thousand animals. "Well, we had to herd the sheep. We had to move as fast as they walked. That's as fast as we could move, and that's not very fast."[21]

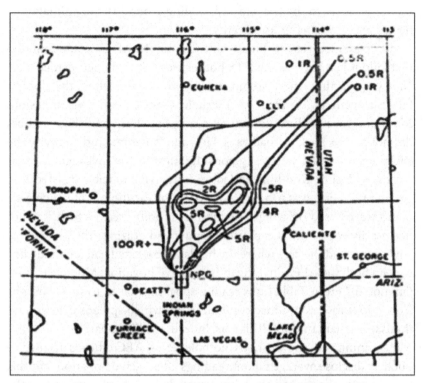

Figure 5.1. Shot Nancy fallout location and intensity. "NPG" designates the Nevada Proving Grounds. (Figure obtained from US Congress, Subcommittee on Oversight and Investigations of the House Committee on Interstate and Foreign Commerce, Health and Scientific Research Subcommittee of the Labor and Human Resources Committee, and the Senate Committee on the Judiciary, *Health Effects of Low-Level Radiation*, joint hearing, 96th Congress, 1st session, April 19, 1979, vol. 1)

The highest external gamma readings were detected around the Lincoln Mine, about forty miles northeast of ground zero, and residents were asked to take cover for two hours.[22] It was the first time that AEC officials issued a shelter-in-place directive, and due to the shielding effects of being indoors, it was concluded that residents had not been exposed beyond the established threshold of 3.9 roentgens.[23]

There were eleven herds comprising around 17,900 sheep in total grazing northeast of the NTS. Several herds were located around Pioche, several more were further west, and two were west of Caliente. One herd left the range in February and did not experience abnormal losses, but six herds closer to the test site and totaling around 11,700 animals witnessed approximately 4,400 deaths.[24]

In a classified report, Lieutenant Colonel N. M. Lulejian confided that fallout just beyond the NTS border was not well documented after Nancy, given that "there are no roads in the region where the close in fall-out occurred."[25] This was a crucial—but classified—admission. Monitors traveled dirt and asphalt roads, but sheep and sheepherders did not always stick to roadways. How much gamma radioactivity the sheep encountered was a key consideration in the investigation that began as Upshot-Knothole ended, in June. AEC officials insisted that gamma radiation was too low to account for the die-off, but such estimates were based on where the monitors monitored—which was not necessarily where the sheep were nipping and grazing. The pretense of having assessed the area where the herds were located had an undue impact on both the AEC's investigation and civil litigation. Moreover, Lulejian noted, "From Table I it is readily apparent that 10 to 50 KT bombs from 300 ft towers produce contamination which spreads over several thousand square miles."[26] This was, indeed, readily apparent.

Beginning with the Upshot-Knothole series, AEC officials decided to publicly disclose average fallout readings after each detonation, but this was abruptly abandoned after shot Nancy due to fallout approaching the 3.9 roentgen threshold at the Lincoln Mine.[27] The idea was abandoned, that is, after just the second of eleven shots in the series. Nancy ranks sixth among Nevada detonations generating the greatest fallout in the continental US.[28] It is not fallout on people for which Nancy is remembered, however, but sheep numbering in the thousands, strenuous effort invested in concluding that something other than radioactivity was the cause of injury, a civil lawsuit in which the AEC ultimately prevailed, and then twenty-six years later the same judge concluding that the government perpetuated "fraud upon the court" in 1956.

After Nancy, there were four detonations in the Upshot-Knothole series, three of which were tower shots. Ruth was a prototype of a nuclear satchel charge and small in yield. Dixie was an eleven-kiloton airdrop detonated high above the desert. Ray sat atop a three-hundred-foot tower but was a misfire. It "resembled a stack of burning tires," Richard Miller explains. Badger was a twenty-three-kiloton tower shot surrounded by thousands of Marines engaging in battlefield maneuvers "under the umbrella" of more than fifty airplanes and thirty-nine helicopters.[29]

And then came Simon. At forty-three kilotons, it was the largest tower shot to date in Nevada—although not by design, as its yield exceeded expectations by 20 percent.[30] On the morning of April 25, Simon rolled over military personnel engaging in maneuvers and effects testing as well as members of Congress and journalists assembled onsite. Philip Fradkin observes, "The VIPs watched in stunned silence as the shock wave, resembling the undulation caused by a rock thrown into a still pool, moved rapidly toward them. They were knocked backwards, their hats were blown off, their eardrums pierced by a loud riflelike crack, and all the windows of their busses were blown out."[31]

As the mushroom cloud raced to forty-four thousand feet, lower-level debris moved eastward, and radiological monitors were hastily dispatched to scattered locales. A hot spot was detected near Glendale, Nevada, and roadblocks were set up to scan vehicles.[32] While rad-safe teams were struggling to keep up with close-in debris, fallout moved in divergent directions at higher altitudes. On a map, Simon's march across the country resembles the twisted, gnarled tines of a damaged fork: at ten thousand feet, debris arched northeast toward Salt Lake City and then straight through the Midwest; at eighteen thousand feet, it wandered over northern Arizona and New Mexico before making a sharp swing northward and passing over Pennsylvania; at thirty thousand feet, fallout dipped south, passing over Louisiana and Florida; debris at forty thousand feet adopted a southernly route, before abruptly swinging northeast as it passed over Mississippi and reaching speeds over one hundred knots (115 miles per hour).[33] This upper-level stream traversed the country well before lower-level segments and in a more concentrated mass.[34] And as it barreled over upstate New York, it ran into a thunderstorm, bringing hail, high winds, flooding, and radioactivity.

On the morning of April 27, Dr. Herbert M. Clark, a professor in the chemistry department at Rensselaer Polytechnic Institute (RPI) in Troy, New York, accompanied his students to a lab filled with Geiger counters clicking at an abnormal pace—with the one closest to an outside wall chattering the fastest.[35] Clark and his students soon realized that outside on the pavement, in puddles, and underneath rain gutters, gamma radiation was far beyond natural background levels. "Clark and his students turned the event into a radiochemistry project," Miller notes.[36] A detonation thirty-six hours earlier in Nevada, Clark surmised, had to be the source. He phoned the AEC's Health and Safety Laboratory in New York City but was met with disbelief.[37] Nonetheless, his students detected radioactivity on asphalt roof shingles, on vegetation, in household tap water, and in local reservoirs.[38] "Typical readings were twenty to a hundred times normal, with hot spots up to ten times higher than that," Ernest Sternglass remarks.[39] Radioactivity was detected not just in Troy but in Albany and much of upstate New York, southern Vermont, and parts of Massachusetts. Samples of tap water in Albany registered radiation activity 2,630 higher than normal.[40] Both the count and intensity of gamma radiation were surprisingly high but were not deemed by the AEC and the New York State Department of Health to be hazardous, and no actions were taken to inform the public or engage in cleanup efforts.[41]

AEC officials were slow to react to the Troy-Albany rainout, discounted the necessity of cleanup efforts, and were reluctant to admit the potential dose that was imposed on area residents. An aerial radiological survey was conducted five days after Clark's first phone call and seven days after detonation. As fallout moved eastward over Nevada, roadblocks were set up; vehicles registering 7 millirads (0.007 rads) or higher were subjected to mandatory decontamination. Yet much-higher readings were detected at Troy-Albany, and no remediation efforts were attempted.[42] Clark's research showed how difficult it would probably be. He and his students found that radioactive rain is absorbed by porous surfaces and requires repeated scrubbing to achieve successively lower readings.[43] AEC officials insisted that the likely dose to residents was 100 millirads (0.1 rads), but during a classified meeting of the AEC commissioners on May 13, 1953, the figure was 2,000 millirads (2 rads).[44]

The AEC conducted follow-up studies, but they were classified. Clark also continued to study the issue, under contract with the AEC. He published in *Science* in 1954, depicting the effort to assess the intensity and extent of radioactivity in the wake of Simon.[45] A coauthored article in an obscure academic outlet, the *Journal of the American Water Works Association*, was particularly telling, as it documented that radioactivity in reservoirs serving Troy residents continued to spike upward due to rainout of subsequent shots in the Upshot-Knothole series.[46] In 1982, a study by the Defense Nuclear Agency estimated the maximum dose of the Troy-Albany rainout to be 2 rem. The average resident would normally receive 14 rems of radiation in a lifetime. In a hypothetical scenario in which all five hundred thousand residents of the Albany area were exposed to 2 rems, it would produce an expected one hundred additional fatal cases of cancer.[47]

A one-foot-square sticky sheet of acetate film at ninety-five weather stations nationwide and twenty-six outside the United States constituted the AEC's system for monitoring radioactive debris beyond the NTS region. This was the primary data source for estimating the radioactive burden confronting millions of Americans in the 1950s and in reanalysis in subsequent years. To the degree that these data are unreliable, then, the radiation imposed on the public is underestimated. And there are numerous reasons to suspect that the commission's gummed-film network was just that: a flawed measure of redeposition.[48]

Measurement of fallout within a two-hundred- to five-hundred-mile radius of the NTS depended on stationary gummed-film stations and mobile monitoring teams collecting samples of debris as fallout passed overhead. Beyond five hundred miles, the gummed-film stations were the sole ground-based means of collecting data, while aerial surveys trailed fallout overhead.[49] As Lulejian stated in a classified report, aerial surveys detected large-scale contamination but proved unreliable in assessing contamination on the ground within areas less than five square miles.[50] They were useful for constructing a broad picture but not a fine-tuned assessment. Moreover, it was not aerial surveying that detected rainout over Troy and Albany but a chemistry class at RPI.

A gummed-film station was located at Albany, but on the morning of April 27, when Clark and his students were greeted by clicking Geiger counters, the data had not yet been collected. Moreover, aerial surveys did not provide real-time information. Samples were removed from the aircraft and shipped to laboratories for analysis.[51] This took time and was not conducive to a flexible response to events as they occurred.[52] Moreover, data from the gummed-film stations consisted of a beta radiation count measured in disintegration rate units, a measure of activity, requiring conversion to gamma radiation roentgen equivalent units.[53] This also took time.

Merril Eisenbud was director of the AEC's Health and Safety Laboratory (HASL) of the New York Operations Office (NYOO) and proposed establishment of a nationwide fallout-monitoring system after detection of radioactive snow in upstate New York in January 1951—a consequence of the first detonation conducted in Nevada days earlier.[54] Akin to what happened at RPI, a Geiger counter alerted employees of a Kodak manufacturing facility in Rochester of radioactivity drifting down with snow. Prior to Eisenbud's proposal in 1951, there were no plans to establish a fallout-monitoring system beyond two hundred miles from Nevada. In contacting AEC officials to discuss the Rochester incident, Eisenbud explains, "Much to our surprise, a call to Washington informed us that virtually no monitoring was being conducted in the U.S."[55] The gummed-film network, however, was originally intended to protect not the general public but the film industry. Aerial-sampling aircraft tracked where debris seemed to be heading and forwarded information to film manufacturers so they could take steps to avoid economic losses.[56] AEC officials feared that failure to do so would lead to negative press and civil lawsuits. Residents downwind were not privy to what the film manufacturers routinely received: timely information to engage in protective measures should they decide to act on such information.

In June 1954, Robert J. List of the US Weather Bureau authored a report examining radioactive debris traversing the country during the Upshot-Knothole series. List observed, "An integration of the total fallout from this series as of June 14, 1953, indicated roughly that 9.3% of the total fission product beta activity produced fell in the test site area. 2.8% was accounted for by the gummed film network in the United States, and 13.1% in the rest of the world, for a total of 25.2%. It is concluded that the gummed film network fails to detect a substantial proportion of the deposited debris."[57] That is, the AEC's nationwide gummed-film platforms were remarkably ineffective.

Despite evidence that the gummed-film network was flawed, AEC and military officials referred to these data in arguing that neither Upshot-Knothole nor any other series imposed on the public a hazardous level of radiation near or far from Nevada. A nationwide monitoring system that routinely did not account for a bulk of radioactive debris served as the empirical cornerstone of the AEC's efforts to solicit quiescence from the US public. List's report was classified, however, and neither the public nor experts unaffiliated with the AEC were privy to this conclusion. Moreover, List confided that rain made collection of debris on gummed film particularly inefficient, as "a large part of the activity is lost in water that splatters or runs off the gummed film stands."[58]

List then turned to a troubling question: What if rainout occurred not twenty-two hundred miles from Nevada but much closer? What if it occurred, say, in Kansas? He suggested that this would present a very significant public health hazard. List turned to yet another troubling question: Were there instances of rainout prior to the Troy-Albany episode that went undetected? Parsing the meteorological and fallout data, "Several such situations were found. They are: New York and New England on November 1, 1951, November 3, 1951, and April 7, 1953; Nebraska on May 26, 1952; and Wyoming on May 8, 1952. The latter case appears to be the one most likely to have produced intense local fallout."[59]

The number and spacing of gummed-film stations around the country were also a concern. There were six stations within 150 miles of Troy-Albany, but if there had not been one at Albany, these six stations would have vastly underestimated the radioactivity raining out over Troy and Albany. This indicated, List remarked, the potential for "a small area of very intense fallout" to escape detection.[60] This includes dry debris and the occurrence of rainout.

List's report described the fluidity, vibrancy, and recalcitrance of nature. In commenting on the Troy-Albany hot spot, Emory Jessee argues, "The complex atmospheric mechanisms that contributed to these hot spots presented a direct challenge to the AEC's assumptions about the passivity and dilution capacity of the atmosphere and, by extension, their ability to control the boundaries of the NTS."[61] Indeed, Simon's impact extended well beyond the downwind communities. It was one of the dirtiest detonations ever conducted at the NTS with regard to cesium-137 deposition in the continental US, population-weighted exposure to cesium-137, and population-weighted exposure to radioactive debris overall.[62] It underscored the inadequacies of the AEC's monitoring of fallout nationwide.

Dirty Harry

On May 8, 1953, Lieutenant Colonel R .P. Campbell authored a memo to Brigadier General Kenneth Fields, director of the Division of Military

Application. Campbell noted that the maximum level of radiation exposure for the public was 3.9 roentgens, but Bunkerville, Nevada, with a population of 250 people, already exceeded that due to fallout from Nancy, and Riverside Cabins, Nevada, with a population of 14 people, also exceeded that level due to Simon. Campbell did not argue that the Upshot-Knothole series should be suspended given these instances of overexposure but stressed the hazards of high-yield tower detonations: "Fallout in a populated area such as Las Vegas is quite possible. Winds for Shot No. 2 shifted 30° to the south from predicted azimuth. Another 15° would have caused fallout to hit Las Vegas." Campbell was referring to shot Nancy, and he suggested that AEC officials consider a maximum yield for tower shots—twenty-five kilotons. "The alternatives will otherwise almost certainly be ultimate over-exposure of nearby populations and conceivably the enforced closure of the Nevada Proving Ground."[63] Beyond Campbell's prescient concerns, the memo raises the question of how the 3.9 roentgen threshold was to be enforced. Would a series be suspended if one individual exceeded the threshold, a handful of people, the largest town in the downwind region?

Nancy came to earth where ranchers were tending their sheep. Simon rained down on upstate New York. And then came Harry. A direct hit on the largest town just downwind was afoot. Splitting the atom is not synonymous with dominion over the nonhuman world but with provocation of it, along with the occurrence of rebound effects. Nature exhibits independent force, fluidity, efficacy, and change. Nature has agency.

The Test Organization in Nevada included Alvin Graves as scientific test director; his deputy, John C. Clark; and Carroll L. Tyler, serving as test manager. An Advisory Panel assisted Tyler in parsing the meteorological data before each detonation and making recommendations on whether to proceed. The Advisory Panel included members of the armed services, the Public Health Service, the US Weather Bureau, Los Alamos, and the University of California Radiation Laboratory.[64] The Advisory Panel and the test site manager (Tyler) took note of those towns that were close to exceeding the established threshold due to prior shots in a series, in order to choose conditions under which they were less likely to receive further radioactivity. Despite the winds shifting hither and yon, as with Nancy ("Shot No. 2"), the established threshold was an operational constraint that the Test Organization took seriously, but it was not an absolute limit. No series in Nevada was ever suspended due to offsite fallout exceeding the maximum threshold that pertained at the time.

Tuesday morning, May 19, was overcast and the winds fickle, but the momentum to detonate had been building for some time. There were distinguished guests from afar, and residents of downwind towns assembled onsite to watch the spectacle. Harry was scheduled for May 2, but radioactivity in the shot area, a remnant of Simon, made it hazardous to assemble the many experimental devices.[65] Harry was rescheduled for May 16. Thirty-seven members of Congress assembled onsite, but the shot was delayed—"in two 24-hour increments"—due to unfavorable winds. "The congressional contingent dwindled to twenty-three as impatience mounted," Fradkin notes.[66] Harry was rescheduled for Monday, May 18, but the meteorological forecast on Sunday observed that the winds were blustery and heading toward Las Vegas. "The Advisory Panel discussed the fallout situation, and decided anything was better than the winds forecast." Harry was rescheduled for Tuesday, May 19. The meteorological report on Monday showed prevailing winds moving away from Las Vegas but at higher speeds, at every altitude, than detected on Sunday. High winds aloft were a problem, as fallout arrived before radioactive decay could occur and in a more concentrated mass. Further, the meteorological report on Monday was unambiguous: "The 10,000' trajectory will go straight east."[67] Straight east 135 miles lay St. George, Utah.

On the evening of May 18, the device was raised to the top of the tower as a succession of weather balloons ascended into the night sky to assess the speed and direction of winds at different altitudes, and the data indicated a parade of forces aloft. "Weather conditions were far from ideal. A low-pressure trough was positioned off the West Coast and was pushing a series of fronts to the east," Fradkin explains.[68] There was always tension between expedient weapons testing and public safety, and AEC officials were losing their balance in this regard. In orchestrating an atomic spectacle, they risked losing sight of their fiduciary obligations to residents downwind. The Advisory Panel meeting on May 18 depicts a dizzying scenario:

> Machta said the forecast being/was the best at the moment. There is a 50 percent likelihood of a 10° shift in the winds.

Dunning said the question to consider was whether we are willing to accept 10 R on a population of 1000. Inside the area there might well be
a higher small area, over 10 R. Clark said it was not fair to expect 10
R—the 10 R dose assumes people will be outside 24 hour per day. . . .

The Advisory Panel discussed the possibility of fallout in the populated
areas. Andrews did not like the situation, as it stood. Placak said he
would be in favor of shooting, if one accepted the *calculated risk.*

Morgan said the 48-hour forecast was not too optimistic. Winds would
probably be favorable but he could not predict what cloud cover
would be.

Fackler said we should approach this cautiously. The Laboratory has
some results they want to obtain, and they will not be able to get
them if the cloud level is at 36,000' . . .

Mr. Tyler said the fallout picture looked better than it had for several
days. He was inclined to be somewhat less pessimistic than the others on the sampling, and felt the samples could be obtained. For the
present we will go ahead, and watch the weather during the night.[69]

Harry roared to life at 5:05 a.m.—at thirty-two kilotons—and soon
debris dispersed offsite in a variety of directions: upper segments moved
southeast toward El Paso, Texas, while lower segments tracked northeast
toward Ogden, Utah, even as another stream of dust and debris flowed
eastward. "Heavy dry fallout" from the stem at the eighteen-thousand-
and ten-thousand-foot trajectories was detected at ground-level in Colorado and Kansas, and precipitation over the upper Midwest brought
down debris on May 21 as another stream crossed near Chicago.[70] It was
a lower-level segment moving eastward, however, that was the most immediate concern, and a hot spot was again detected north of Glendale,
Nevada.[71] Roadblocks were quickly set up, and cars were checked for
radioactivity.

Frank Butrico was the Public Health Service monitor in St. George,
Utah, on the morning of May 19. Around 7:50 a.m., he received his first
radio instructions: set up a roadblock outside town.[72] With the assistance of local law enforcement, he stopped cars heading south toward
Mesquite and Bunkerville, Nevada, although he had no idea that fallout
was moving in his direction.[73] Before long, there were more than two
hundred cars backed up, with their occupants impatiently waiting.[74]

Figure 5.2. St. George Boulevard, 1953. (Courtesy of the Washington County Historical Society)

AEC officials only knew where ground-level fallout was when someone detected it, but by then, it was too late to do much about it. "The warning system was akin to shutting the barn door after the animals had escaped," Fradkin suggests.[75] Butrico got in his car, and by 8:30 a.m., he was logging radiation readings at various places in town. Soon these readings would push the needle as far right as it would go.[76] Figure 5.2 is a photo of St. George Boulevard in 1953. The population was around forty-five hundred people.

Table 5.2 outlines the timeline of Harry's arrival in St. George as articulated by Judge Bruce S. Jenkins in his memorandum opinion in *Allen et al.* and by an allegedly falsified AEC report. Jenkins's timeline is based on AEC documents that were ordered declassified by congressional subpoena, civil litigation, as well as the testimony of firsthand witnesses.[77] Indeed, by the late 1970s, reports, memos, and documents that had been hidden from public view for decades were unearthed. One is (in)famous: "Report on the Sequence of Events Occurring in St. George, Utah, as a Result of the Detonation of Shot IX" is dated May 30, 1953, and it was (supposedly) authored by Frank Butrico.[78] The typed report recounts Butrico's actions in St. George on May 19 and when he undertook

TABLE 5.2. Timeline of Shot Harry in St. George, Utah

Time	Jenkins timeline (*Allen et al. v. United States*)	Time	Allegedly falsified monitor's report
8:50 a.m.	Butrico begins to detect increasing radioactivity.	8:45 a.m.	Butrico begins to detect increasing radioactivity.
9:15–9:30 a.m.	Butrico detects 350+ milliroentgens at various places. He telephones William Johnson at Mercury.	8:45–9:15 a.m.	Butrico detects 350+ milliroentgens at various places. He telephones William Johnson at Mercury.
9:45 a.m.	Butrico again telephones Johnson for additional instructions and is told to issue a shelter-in-place announcement.	9:25 a.m.	Butrico again telephones Johnson for additional instructions and is told to issue a shelter-in-place announcement.
10:15 a.m.	The first radio announcement advising residents to shelter in place is issued.	9:25–9:40 a.m.	The first radio announcement advising residents to shelter in place is issued; the population is under cover

key steps leading to the AEC's shelter-in-place directive. Upon viewing this report, twenty-seven years later, Butrico declared that parts were falsified—in particular, the time of the first radio bulletin instructing residents to take cover.[79] And he did not recognize the signature that was purportedly his own. Butrico repeated these assertions under oath in *Allen et al.* This allegedly falsified report is depicted in the right-hand column of table 5.2.[80] Further, the communications log of the mobile offsite monitors, dated May 24, 1953, was also available by the late 1970s, and it is clearly noted that, according to a phone call from Butrico, the radio broadcast instructing St. George residents to shelter in place was first made at *10:15 a.m.*[81] The allegedly falsified document stated otherwise. In turn, the differing timelines outlined in table 5.2 demonstrate what residents of St. George experienced and how sleight-of-hand can paper over very real material-biophysical dynamics whose implications are often recognizable only over time. And time matters. It certainly mattered for St. George residents on the morning May 19.

Harry moved quickly. Fallout descended on Alamo, Nevada, roughly 52 miles downwind in two hours, averaging 26 miles per hour (mph), and onward to Bunkerville, Nevada, a distance of 109 miles, in less than two and half hours, clocking in at 43 mph, and proceeded to St. George,

135 miles from the NTS, in about three hours and forty-five minutes—an average speed of 36 mph.[82] Such pacing made offsite monitoring difficult and is evidence of the blustery conditions in the hours after detonation. As at Trinity, rad-safe teams chased radioactive debris marching haphazardly across the landscape.

There is agreement regarding when Harry arrived in St. George, as outlined in table 5.2. At 8:50 a.m., Butrico began to detect radioactivity. A narrow, long, and dusty cloud, against an overcast sky, was moving over the town.[83] But this is where the timelines diverge. Jenkins surmised that Butrico detected 350-plus milliroentgens between 9:15 and 9:30 a.m. At approximately 9:30 a.m., Butrico telephoned William S. Johnson, the rad-safe operations officer stationed at Mercury, and he was told to wait for further instructions. The allegedly fraudulent document suggests that this period ranges from 8:45 to 9:15 a.m., shaving fifteen minutes off the timeline.

Butrico received the shelter-in-place directive around 9:45 a.m., but the allegedly fraudulent document states that this occurred at 9:25 a.m., a difference of twenty minutes. As Jenkins highlights in his judicial opinion, after Butrico received the directive, he met with the mayor of St. George, who phoned a radio station in Cedar City—and thirty minutes passed before the first radio announcement was broadcast. The allegedly fraudulent document depicts a different scene: Butrico received the shelter-in-place directive at 9:25 a.m.—rather than 9:45 a.m.—and the report states, "At 0940 the bulk of the population in the city of St. George was under cover. The effectiveness of the operation was amazing."[84] The allegedly fraudulent report asserts that residents were under cover thirty-five minutes prior to 10:15, when Jenkins determined that the first radio announcement was broadcast. This thirty-five minutes corresponded with the peak period of fallout. Reanalysis by Virgil Quinn and colleagues suggests that the cloud persisted over St. George for approximately two hours, or roughly 8:45–10:45 a.m., and the shelter-in-place directive was rescinded around noon.[85]

The allegedly fraudulent report suggests that most residents were under cover less than an hour from the time Butrico first detected abnormal readings and were spared the brunt of the fallout. Jenkins's timeline reveals that residents began sheltering in place only after, at minimum, one hour of hard fallout (9:15–10:15 a.m.) and just before

Harry left town. At separate meetings conducted by the DOE in 1980, in preparation for *Allen et al.*, Butrico and Johnson both conceded that the shelter-in-place directive spared few residents from radioactivity.[86] Butrico indicated that he drove around town after the radio broadcast and that not everyone received the shelter-in-place directive: "My first venture up one street was quite obvious that the school hadn't gotten the message because the children were out there in recess."[87] The radio bulletin was first issued around 10:15, but it is not clear when people took cover or how many took cover at all. Johnson confided in 1980, "The period of time that one should remain indoors is most effective if it is during cloud passage. . . . I think we have to admit that we got the people in St. George indoors too late in this case."[88] Smaller communities adjacent to St. George did not have an offsite monitor present, nor were residents instructed to remain indoors despite also being dusted with fallout.[89]

Per the allegedly fraudulent report, Butrico was asked to issue a shelter-in-place directive, Butrico contacted the sheriff (he stated that it was actually the mayor), this person contacted the radio station, the radio station relayed the directive, and the "bulk of the population" was under cover within fifteen minutes. If this is what happened, it probably constitutes the most seamless example of crisis management the world has ever witnessed. Per Jenkins, on Butrico's third phone call to Johnson, around 9:45, Butrico was told to arrange for the announcement of a shelter-in-place directive, he contacted the mayor of St. George, the mayor told him that there was a radio station in Cedar City (fifty miles away), the mayor then called the radio station, and the first announcement was broadcast at 10:15, or approximately thirty minutes after Johnson relayed the order.

Stewart Udall, one of the plaintiff's attorneys in *Allen et al.*, argues that Gordon M. Dunning emerged as the "quarterback of the AEC's damage-control effort" in the wake of shot Harry, and it is a role Dunning repeated time and again.[90] In 1968, Dunning presented a paper titled "Protective and Remedial Measures Taken Following Three Incidents of Fallout" at an academic conference in Switzerland. In it, Dunning recounted Harry's appearance in St. George fifteen years prior. His analysis was modeled on the allegedly fraudulent report depicted in table 5.2, which he attributed to Frank Butrico. "The St. George, Utah incident in 1953 shows the favorable results from a program of education of local officials and the public and the close cooperation with the local authorities," Dunning declared.[91] That is, Dunning utilized the allegedly fraudulent report to demonstrate the remarkable success of the AEC's effort to ensure that St. George residents took cover in a timely manner. It was meant to serve as an example from which others could learn the fine details of crisis management amid the flux and chaos of an untoward event. Udall countered in 1994, "The absence of any plan to protect the civilians produced a safety fiasco for the inhabitants of St. George. The appropriate time to warn people to take cover was *before* the fallout cloud arrived, but Butrico was not told the cloud was headed for St. George until it was over the city."[92]

Under oath in *Allen et al.*, Butrico was pressed to recount what steps AEC officials took in the hours after Harry swept through town:

Q: Did you take readings on your own body?

A: Yes, indoors. So it was obvious that it was off me and not around me.

Q: Do you remember what those readings were?

A: Approximately what they were outside just for brief moments, but then they dropped considerably. And I'd anticipate I was going to go ahead and shower as much as I could and get some shampoo and get some of that stuff out of my hair. . . .

Q: Who were you talking to?

A: This was to Mr. Johnson at Mercury (NTS)

Q: Did he tell you to decontaminate yourself?

A: Well, we were just having this kind of conversation that I was pretty hot, and that I was going to do this, and this was a good idea.[93]

Butrico explained that he was advised to purchase new clothes and discard those he was wearing. He testified,

Q: Did you take a shower?

A: Yes, a number of them that afternoon.

Q: Did Mr. Johnson tell you that you should tell other people in St. George to decontaminate themselves?

A: No. That subject was not brought up.

Q: But the [people of St. George] who were out of doors between 8:50 and 10:15 when the radio message came would have gotten the same exposures that you did?

A: Yes. That's a reasonable assumption.

Q: And they were never advised to shower and discard their clothes as were you?

A: That was never brought up, nor was it suggested.[94]

Offsite monitors had a list of things to accomplish with each detonation: collecting data on passage of the cloud, measuring levels of airborne activity, measuring levels of residual radioactivity on the ground, charting the duration of fallout, and accounting for the size and shape of fallout passing overhead—and all from the boundary of the NTS to two hundred miles out.[95] At the DOE workshop in 1980, William Johnson declared that as Simon passed by the monitors did not anticipate setting up roadblocks, scanning vehicles, and decontaminating those vehicles that were deemed too radioactive. It was "laid on us without any warning," Johnson claimed. "If you can imagine, this is on two U.S. highways on a Saturday. The main highway between Las Vegas and Salt Lake City was one of them."[96]

This task was again abruptly assigned to offsite monitors on the morning of May 19. Among the monitors, including Butrico in St. George, hundreds of cars were stopped, and many were directed into an adjacent town to be washed. And there were only thirty-seven mobile offsite monitors tasked with responding to Harry's turbulent march downwind.[97] In a classified after-action report submitted by Lieutenant Colonel Tom D. Collison to the test director at the NTS, Collison attested, "Offsite monitors were required to devote considerable time to the roadblocks established in the vicinity of Mesquite and St. George. This interfered to some extent with the performance of their monitoring duties in this area."[98] The offsite communications log, akin to a postfight debriefing, related the experience of a monitor working in Mesquite, Nevada: "It is physically impossible to carry out the assigned duties of offsite monitoring and run a roadblock operation at the same time!"[99]

The rad-safe teams were outmatched by Harry. They probably missed hot spots within the towns downwind. The data that were collected, in turn, may not be representative of the full burden of gamma radiation that residents encountered.

Looking for Radioiodine?

Frank Butrico arrived back at Mercury but agreed to return to St. George to ease concerns that residents might have. "From there on, frankly, it was a PR thing: Just be there, console people, talk to them about what happened, and that probably there was no cause for concern." Moreover, the atmosphere was tense: "In fact, when I got back to Mercury, it was quite obvious in the first critique with the top-level people that the primary concern was whether this was going to affect the testing, the further testing, and less of it as to what was really happening and what could be done about it."[100] Before returning to St. George, however, Butrico was given another assignment in addition to the "PR thing." He was asked to obtain milk samples. There is controversy over what AEC officials were looking for. A clear answer would indicate whether they suspected that radioiodine was a hazard long before they were alerted to this fact by a mathematician working in the AEC named Harold Knapp, in 1962.

On May 21, milk samples were collected in an indirect manner so that residents would not know it was occurring. Rather than obtaining fresh milk from a local dairy or herd, Butrico purchased bottled and packaged milk from retail stores in St. George. This was decided by Butrico and a senior-level PHS officer at Mercury. Regarding the covert means employed, "Offsite personnel in the St. George area decided that extensive inquiry into such details would indicate the concern of the test program's rad-safe group with the possibility of milk contamination and alarm an already-worried community."[101]

The report based on analysis of the "samples" does not mention radioiodine but "fission-product contamination."[102] At a DOE workshop in Las Vegas in 1980, however, William Johnson contended that at the time of shot Harry, there was no analytical test to assess the presence of radioiodine in milk: "So the first thing we had to do was to develop a laboratory procedure for the analysis of milk for radioiodine." He insisted that a chemist from Los Alamos was dispatched to Nevada to develop such a procedure.[103] In a strange twist, however, when the workshop reconvened the next morning, Johnson made a "clarifying statement" as the meeting began: the milk samples collected after shot Harry were not analyzed for radioiodine, as he had stated the day before. "Because we did not isolate iodine or in any way identify iodine. The results were really fresh fission products in milk, rather than iodine in milk."[104]

Two years after shot Harry, Gordon Dunning published an article in the *Scientific Monthly* describing the uptake of radioiodine in grazing animals around the country due to detonations in Nevada and the South Pacific, and he remarked that people accumulate radioactive iodine, as well. "Radioactive iodine present in fallout material may find its way into the body through ingestion or inhalation and will concentrate in the thyroid gland." This point is unambiguous. Nonetheless, he insisted that radioiodine detected in people was "far below that needed to produce any detectable effects."[105] Arguably, AEC officials understood that radioiodine was a potential hazard by 1955, if not earlier, but did not recognize *how* significant a threat it presented. Why? In 1964, Dunning argued, "Direct measurements of iodine 131 in milk were not made around the Nevada Test Site during earlier times of testing since it was the consensus of scientists within and outside the AEC and government at that time that the limiting factor was the potential external whole body exposure."[106] This is akin to arguing, "We didn't believe radioiodine was a problem because we didn't believe radioiodine was a problem." The AEC and its Division of Biology and Medicine were responsible for ensuring the safety of detonations in Nevada. The AEC was one of the principal institutional bodies tasked with conducting research on the menagerie of radionuclides raining down on the public.

The US Department of Agriculture understood that radioiodine was a hazard long before 1962.[107] In a pamphlet titled "Defense against Radioactive Fallout on the Farm," published in 1957, farmers were advised, "Get your dairy cattle under cover first. If they eat fallout, or drink it in water, some of the radioactive material will be in their manure and urine, and some will be in their milk." The pamphlet continued, "Some of the radioactive chemical elements in the fallout, such as radioactive strontium and radioactive iodine, can cause serious internal radiation damage if taken into the body in sufficiently large amounts."[108]

Whatever the AEC was looking for, it had little impact on operational conduct. There was no systematic sampling of milk in the downwind towns in the 1950s.[109]

The Perils of Data Manipulation

The AEC commissioners met in Washington, DC, on May 21, two days after shot Harry, and were advised that radioactivity in St. George was as high as 6 roentgens, but the commission stressed publicly that the average dose for the Upshot-Knothole series was less than the 3.9 guideline over thirteen weeks.[110] A conceit lay embedded in the commission's statements. By averaging discrepant data points, it was possible to derive a single figure below the established threshold.[111] The data could be averaged for a given detonation or even an entire series, without also indicating the range of high and low values. Through strategic averaging, it was possible to hide data points above the threshold.

Averaging of data does not mean that every individual was exposed to the mean figure, as individual exposures could be higher or lower, but AEC officials typically maintained that everyone—every individual—in a given town was not exposed to more than the average, a conclusion not supported by the data. And when AEC officials declared that a community had not exceed the established threshold, as derived from the averaging of data, this really indicated that the *average resident* had not exceeded the exposure threshold.

Averaging of data was not the only issue. In the early years of testing, when the AEC did release data on gamma radiation, it reported the "infinity dose," or the unadjusted but averaged readings. Then the commission began calculating the "effective biological dose" to account for sheltering effects and biological repair. This typically lowered reported levels of external gamma radiation by more than 50 percent.[112] "Sheltering" includes seeking cover in houses, automobiles, workplaces, or schools. It is understood that the body can repair minor radiation damage, depending on the rate at which it is experienced, as a given dose applied quickly is more damaging than the same dose administered over a longer period. To the degree that people are shielded from direct contact with gamma radiation, exposure will be reduced, but research on the reductive effects of various building materials was scarce before the mid-1950s.[113] Inferences regarding biological repair were particularly tenuous.

Consideration of sheltering effects and biological repair was applicable to the many civil defense activities ongoing in the 1950s, but deployment of reduction factors in describing the gamma radiation burden imposed downwind was a more contestable practice. In the absence of concerted research underpinning the adjustment protocol, the adjustment protocol may function as a sleight-of-hand rather than scientifically informed corrections. And the representation became the reality for AEC and military officials and was depicted as such for the public. It was strategic ignorance, or effort employed in order not to know something that it is advantageous not to know.[114] Jenkins attested in *Allen et al.*, "Even in dealing with 'hard' external gamma radiation, dose estimation for the offsite residents amounted at best to an educated guess. Hastily taken, surface gamma measurements were generalized into smooth isodose lines on a fallout map and then 'adjusted' downward by a factor of 2 or more to account for *assumptions* made about the attenuating effects of housing materials, automobiles, clothing, topography, distance and other factors."[115]

Film badges were one strategy for confronting uneven deposition of external gamma radiation. A sampling of individuals wearing film badges, clipped to one's clothing, is a more valid and reliable protocol than fixed monitoring stations or the capricious behavior of mobile offsite monitors. Onsite NTS personnel routinely wore them, but downwind residents were never issued film badges in a systematic manner, as the AEC preferred to distribute them selectively as testing progressed.[116] This was arguably due to the expectation that issuing the badges more widely would elicit public anxiety. An experiment in 1957 suggested that such concerns were unjustified. Film badges were issued to residents of Alamo, Nevada, a town of about three hundred people. A subsequent report observed the willingness of residents to cooperate in the effort.[117] Further, there were no blood tests or urine samples collected from the offsite public, as occurred at the national laboratories and for onsite personnel at the NTS.[118]

Enchanted by the Atomic Spectacular

By the beginning of 1953, the *New York Times* had published over one hundred articles addressing some aspect of testing in Nevada, but none challenged the AEC's assertions as to the safety of testing, articulated the protocol for managing offsite radioactivity, or raised questions regarding the commission's reticence to release information about fallout location, intensity, and duration. No experts outside the commission were consulted regarding the public health risks of testing, nor was the point of view of any downwind resident sought in *Times* coverage over the period 1951–1953. Coverage was dominated by the authoritative and reassuring statements of AEC officials.

Thousands of dead sheep were potentially a tipping point reshaping news coverage and calling into question predominant interpretive frameworks. The *Times* limited its coverage to the commission's insistence that the sheep did not die of radioactivity.[119] There were no interviews with the ranchers whose animals were afflicted. If not dead sheep, then Harry certainly constituted a critical discourse moment. There was coverage in the *Times* of the shelter-in-place order at St. George, but the event was not discussed in detail.[120] In a follow-up story five days later, Gladwin Hill stridently reiterated the commission's official position: any radiation encountered was well below that which could cause injury. Further, he discounted the idea that fallout could be dangerous. His story in the *Times* began, "Is there any danger to the public from the Nevada atomic tests? The Atomic Energy Commission, which had drafted more than two years of such experiments, still was being plied with this question on the eve of the final detonation in the spring test series at its proving ground here—the scheduled trial tomorrow of the Army's atomic cannon. The commission's categorical answer was no."[121]

Between 1951 and 1953, the *Times* devoted substantial attention to speculation about an "atomic cannon." The first, and only, atomic cannon shot by the US occurred soon after Harry. Shot Grable was an artillery shell with a fifteen-kiloton yield, launched seven miles from an eighty-five-ton artillery emplacement.[122] Figure 5.3 depicts journalists at the NTS observing shot Grable. It was the crowning spectacle of the series, as portrayed in *Times* coverage. That Grable occurred six days after Harry arguably served to distract *Times* reporters from the problems surrounding Harry. Gladwin Hill published a front-page story with vivid narrative and photographs depicting the Grable detonation.[123] "Dirty" Harry, as it came to be known in southern Utah, was overshadowed by the newest twist on the atomic spectacular.

Upshot-Knothole ended on June 4 with a dramatic sixty-one-kiloton airdrop detonated above the desert. In a front-page story, Hill remarked,

Figure 5.3. Journalists document shot Grable, May 25, 1953.
(Courtesy of the Nevada National Security Site)

"What probably was the greatest atomic detonation ever unleashed on this continent—on the order of double the blast that devastated Hiroshima, Japan, in 1945—was set off in a great and protracted holocaust of cosmic incandescence high over the Atomic Energy Commission's desert proving ground just before dawn today. . . . Acres of Joshua trees, cactus and sagebrush five miles and more away from 'ground zero,' immediately under the blast, burst into flame in a great desert 'forest fire.'"[124]

Diverging editorial commentary between the Salt Lake City newspapers and the *Las Vegas Review-Journal* arose in the days after shot Harry. The Salt Lake City newspapers expressed dismay over the shelter-in-place order at St. George and skepticism regarding the AEC's assurances of safety. The *Deseret News* highlighted the uncertainties: "Human life is too precious to be risked in experimentation and guesswork." The *Salt Lake Tribune* argued that detonations were a nuisance and should be conducted elsewhere—"preferably on some uninhabited island far out to sea." Three days later, the *Las Vegas Review-Journal* published a scathing response reiterating the safety of testing and taking aim at Douglas Stringfellow, a congressional representative from Utah who expressed concern for the welfare of his constituents. The *Review-Journal* mocked the idea that a politician from Utah should have an opinion on what occurs in Nevada. "The fact that atomic clouds float over other areas than the state of Nevada is a vagary of the weather." Further, the *Review-Journal* questioned the patriotism of those who doubted the AEC's assurances of safety: "We like the AEC. We welcome them to Nevada for their tests because we, as patriotic Americans, believe we are contributing something, in our small way, to the protection of the land we love."[125]

The US West is steeped in state-sanctioned environmental violence, enacting a toll on vulnerable segments of the population. Others, of course, benefit. The embrace of the AEC in Las Vegas undercut recognition of the tribulations of residents elsewhere in Nevada, Utah, and Arizona who were more directly affected by testing. Fallout moved in one direction and federal spending and tourist dollars in another. What Las Vegans did not know is how easily things could have been different. As Campbell asserted in a classified memo, an additional fifteen-degree swing in the prevailing winds and shot Nancy would have landed on the Strip.[126] This probably would have dampened enthusiasm for the commission's atomic spectacular.

That the public is not uniformly affected is key to the sustained enactment of environmental violence by powerful actors in society. And being able to point out the benefits, at least to some, is a key rhetorical strategy for obscuring the costs borne by others. When the implications of an activity have uneven impacts, it is easier to leverage segments of the public against each other—particularly when media outlets such as the *Times* are busy commenting on the color of the most recent mushroom cloud.

And while residents of St. George and surrounding communities were reeling from the Upshot-Knothole series, tourists in Las Vegas were drinking atomic cocktails at casino watch parties, with the mushroom clouds visible from downtown. "Indeed, it was not long before the mushroom cloud was vying with the showgirl for top billing along the Las Vegas Strip," A. Costandina Titus recounts. The atomic hairdo was created, the Sands Hotel sponsored a "Miss Atomic Bomb Contest," and the Chamber of Commerce distributed a map illustrating the best roadside vantage points.[127] Las Vegas epitomizes self-promotion, and the bomb fit that bill, so long as others were bearing the costs. As the AEC was orchestrating an atomic spectacle in the desert, Las Vegans were devising their own community theater.

Nevada Testing in Jeopardy?

The Test Organization in Nevada thought it handled Harry well, but the AEC commissioners in Washington, DC, were worried.[128] Simon rained out over upstate New York, Harry rolled over the largest town just downwind, and there were reports of dead sheep elsewhere in southern Utah. Signals were arising that suggested containment of fallout were overly optimistic. However, these biophysical-material surprises were not interpreted as impending hazards to public health but as public-relations issues. The AEC commissioners feared that testing in Nevada was in jeopardy.

On May 13, 1953 John C. Bugher, the director of the Division of Biology and Medicine, briefed the commissioners on shot Simon. Bugher confided that upstate New York experienced a potential dose of 2 roentgens. This was surprisingly high so far from Nevada but within the 3.9 roentgen threshold, Bugher stressed. The commissioners quizzed him on the criteria for determining whether to detonate in Nevada versus the South Pacific: "The Chairman observed that apparently it had simply been understood that relatively large yield devices would be detonated outside the continental U.S. However, no firm criteria for deciding such issues had been established," the transcript notes.[129] The December 1950 memo that President Truman signed creating the NTS called for "relatively low order detonations."[130] More than three years later, what constituted "relatively low order" remained ambiguous.

The AEC commissioners met again on May 21, and shot Harry was the focus of attention. Bugher described the shelter in place at

St. George, two days earlier, and stated that gamma radiation was as high as 6 roentgens, but it was unlikely that anyone exceeded the 3.9 roentgen threshold, given the precautions undertaken that day. He explained that residents were advised to remain indoors from 9:00 a.m. until noon, an erroneous timeline. The meeting notes read, "Mr. Smyth and Mr. Zuckert commented that the absence of any serious fall-out or rain-out on populated areas, although due largely to careful planning of the shots, was due in some part to good fortune." That AEC commissioners Henry D. Smyth and Eugene M. Zuckert did not define the Troy-Albany or St. George incidents as serious from a public health standpoint suggests a degree of detachment from issues arising beyond DC. "Mr. Dean asked that the concern of the Commission be conveyed to the test organization and that everything be done to avoid another fall-out over St. George."[131]

The next day, May 22, the commissioners were apprised of a letter from Dr. Lyle Borst of Salt Lake City. Borst was formerly employed at the commission's Brookhaven Laboratory and currently a professor at the University of Utah. He was familiar with radiation. And it was not his first letter to the commission. Borst detected fallout in Salt Lake City from shot Easy, a twelve-kiloton tower detonation on May 7, 1952, and he wrote to express his concern.[132] In a letter dated May 16, 1953, he again voiced apprehension after detecting radioactivity on his front lawn and on his children. "When I find contamination on my children the equivalent of any contamination I have ever received in eighteen years of nuclear work, I cannot consider it inconsequential nor trivial." In a point that Jenkins also emphasized in his memorandum opinion in *Allen et al. v. United States*, Borst questioned a single standard of gamma radiation for the general population when some people are more vulnerable than others. He contended that while he was working at Brookhaven, pregnant women were routinely removed from areas where they might encounter radiation: "Yet women who happen to live near the test site are given no such consideration." Borst insisted, "I was appalled to learn from Dr. Bugher that the AEC has no policy of warning residents in an area to keep their children indoors at predicted levels appreciably below the evacuation level of 3 roentgens. I see no justification nor excuse for the lack of such warnings."[133]

Borst's concern about the lack of information afforded the public and the contrast between the standards at national laboratories and the looser approach downwind of Nevada foreshadow Jenkins's argument in his memorandum opinion, thirty-one years later: "The notion that far greater releases of radioactive materials than at the other national laboratories somehow justifies far less monitoring than undertaken at those laboratories defies reason and logic, falling well beyond any notion of reasonable care under the circumstances."[134]

The *Deseret News* published a story in which Borst declared that he would not allow his children to play outside on days when fallout traveled the country. "A-Cloud Dangers in S.L. Studied: University Nuclear Expert Finds High Rays Concentration" outlined Borst's contention that radioactivity may have a cumulative effect, such that even small doses, experienced successively, can add up to significant health effects. This, of course, was at odds with the AEC's insistence on a threshold up to which ill effects do not occur. Borst also lamented that the public was never told the levels of fallout in Salt Lake City.[135] The *Deseret News* ran an editorial the following day titled "The Safest Way," reiterating many of Borst's concerns. The editorial proposed that whenever fallout moved over Salt Lake City, it was an opportunity to enact federal civil defense activities, "a full dress rehearsal for the public," as if it were enemy bombardment. The objective was twofold: "people would learn what to do in case of actual enemy attack and at the same time they would be giving themselves the fullest possible protection from whatever real or potential dangers the nuclear radiation may carry."[136] It was noteworthy, but AEC officials were far too fearful of the public to embrace such an idea.

Ralph J. Hafen, a student attending the University of Utah, critiqued the AEC in the Cedar City newspaper. He urged Utahns to consider the potential long-term effects of radioactivity and insisted that the burden of proof rested with the AEC to illustrate that fallout was not hazardous. Moreover, he argued that testing should cease until the AEC did, in fact, fulfill its obligation in this regard.[137] Howard Ball explains, "By the end of the 1953 test series, the radiation issue was clearly identified by the media, especially the *Deseret News*, as an issue of major concern for the downwind citizens."[138]

Harry also elicited the attention of Utah politicians, who were responding to anxious letters and phone calls. US Representative Douglas R. Stringfellow and Senator Arthur V. Watkins expressed concern for the welfare of their constituents. Just four days after Harry, in a letter to Gordon Dean, chairman of the AEC, Watkins stressed, "It seems to me, however, that it is expecting rather much to ask them to submit to repeated experiences like the one they went through on May 19th."[139] Stringfellow was invited onsite to watch the Grable detonation and came away an enthusiastic convert. That evening, he met with residents in St. George and assured them that "every possible precaution was being taken."[140] Stringfellow promptly retracted his previous criticism of the AEC in the press and, in doing so, recited a biblical admonition: "Know the truth and the truth shall make ye free."[141]

Nonetheless, in the wake of Upshot-Knothole, concern over fallout was more pronounced within the AEC, even as its public pronouncements maintained a reassuring tone. "The public and private dialogues went on at different levels, and their content bore little relationship to each other, though the subject matter was the same," Fradkin notes.[142]

On June 10, the AEC commissioners were told that sheep grazing within fifty miles of the NTS had later died, while others appeared to have beta burns in their nostrils and along their backs. Gordon Dunning assured them that researchers would have a report compiled within weeks, even as the commissioners again lamented the public-relations implications. It was determined that the AEC would embark on a concerted public education program in response.[143] On July 15, the commissioners again discussed the sheep problem, and John Bugher admitted that the situation was "not so clear." He suggested that the sheep probably died from eating toxic plants, but further research was needed to determine whether radioactivity was also a factor.[144]

Everything began and circled back to Los Alamos, and the circuit proceeded through many nodes of the atomic state, around the country, in the interim. The Test Organization in Nevada was close geographically and operationally to Los Alamos and under pressure to keep up the pace of detonations.[145] It was vital to closing the loop between design, production, and stockpile. The armed services applied pressure through the Division of Military Application in shaping operations in Nevada. And in important respects, the AEC commissioners in DC existed outside this insular circuit. "Except for major policy decisions, such as whether to develop a thermonuclear weapon, Los Alamos—not the personnel at the AEC headquarters in Washington—formulated specific policy," Fradkin observes.[146] And although both Truman and Eisenhower first approved a test series before it began, they were minimally involved in decisions regarding how weapons were tested. "Executive control over atomic affairs was tenuous."[147] The AEC commissioners were political appointees and concerned with public opinion as well as continued congressional funding, but the people in the field directly confronted the imperatives for which Los Alamos was founded in the first place. The AEC commissioners often appeared disconnected from both the forces bearing down on operations at the NTS and the specifics of policy implementation.

In the wake of the Upshot-Knothole series, the AEC commissioners concluded that another committee should be established to, again, examine the suitability of operations in Nevada and to make recommendations as to appropriate course corrections. This committee was led by Carroll Tyler and began its study in the summer of 1953, just months after the first committee released its report in May. This time, there would be no more detonations in Nevada until the second committee—known as the Tyler Committee—released its report.[148] The AEC commissioners were beginning to apply their own pressure on the Test Organization in Nevada.

Sensitive Dependence on Initial Conditions in Nevada

Testing relied on meteorological forecasting at the expense of robust management of offsite fallout. And when radioactivity became a substantive threat, the AEC was not prepared to respond in an effective manner, nor were downwind residents educated regarding simple decontamination procedures such as those in which Frank Butrico engaged in the wake of Harry. "The shortage of public information about fallout compounds itself: operational personnel decided that inadequate public education justified a failure to warn," Jenkins argues in his opinion in *Allen et al.*[149]

The fluidity of fallout challenged the AEC's insistence that testing was a scripted, rational endeavor. This conceit rested on the assumption that it was possible to identify the prevailing winds and leverage this knowledge in a practical manner. Set off in the predicted direction, radioactive particles would dilute and diffuse in the atmosphere, with hazard evaporating into the ether. However, fallout undercut the commission's pretense of certainty and assumptions of a fundamental distinction between society and nature. Indeed, detonations revealed that nature was not a vast repository but a conduit in which the activities of a cloistered technocratic order were imposed on others in an indiscriminate manner. Practical engagement with nonhuman entities, processes, and things often occurs before knowledge accrues, rather than knowledge preceding and shaping the contours of entanglement.[150] And this clashed with the commission's public persona. Andrew Pickering argues, "Knowledge is not some free-standing entity in its own right; it should be understood as threaded through practice, performance, in a world which cannot be itself reduced to knowledge. I am inclined to say that performance is the ground from which knowledge emerges and to which it returns."[151]

Pickering refers to "performative excess" in describing nonhuman movement, efficacy, and force set in motion through human activity but escaping containment.[152] Differential power and privilege are often predicated on manipulating nonhuman materiality but then denying the unbounded implications. The contemporary environmental justice movement is a testament to the confluence of residual materiality and marginalized segments of the population, unbounded assemblages in which others are enrolled without their knowledge or consent.

The ultimate objective of sociotechnical innovation is to routinize the capture of nonhuman materiality and confine the dance of human and nonhuman performativity to within acceptable boundaries. The goal is to construct islands of stability reflecting cultural assumptions of duality and mastery. "Cartesian dualism must not be mistaken for a universal description of how the world generally is. It is, rather, a way of being in the world that the world sometimes tolerates," Pickering insists.[153]

Dirty Harry was emblematic of learning through practical engagement. It was an expression of performative engagement with nature and *sociotechnical instability* in the desert. Nausea and beta burns on skin were very real indications of instability, but the effort to control nature often evolves into the imperative to control marginalized segments of the population, as the unintended consequences reverberate beyond what is intended. And Harry underscored the tension between discourse and the biophysical and material effects of nature that is a central dilemma of modernity. Biophysical and material effects call into question prior discursive statements as well as the conduct of powerful organizations. From the moment Harry sprung to life, AEC and military officials, assisted by the PHS, were busy responding to the unexpected twists and turns as fallout moved in different directions, at different altitudes, and at a different tempo. As at Trinity, this hardly approximated the rigorous conduct of physics in the laboratory. In response, AEC officials classified information and released misinformation to symbolically erase the very real effects of fallout. This is exemplified in the fraudulent report describing Harry's timeline in St. George but characterizes the commission's conduct as a whole.

Western cultures have long assumed that the biophysical-material world is set apart from the human, is eminently knowable, and is amenable to rearrangement.[154] Nature is envisioned as mechanical and uniform, and the goal of the applied sciences is to decipher its fixed laws and relations for the betterment of humanity. Human mastery is presupposed.[155]

In this sense, AEC officials approached the weather as they did the decoding of uranium: a task to be broken down into discrete, logical procedures formalized and replicated over time. Time and again, however, their best efforts were met with the nonlinear behavior of nature. Divergent and capricious winds aloft, varied topography, and precipitation hundreds of miles from ground zero proved intractable. Even as atmospheric testing contributed to visualization of ecological processes, it also demonstrated emergent dynamics that were resistant to mechanistic and reductionist explanations. At the height of the Cold War, the armed services funded research on weather modification.[156] The objective was not simply prediction but control, in order either to boost domestic agricultural production or to be detrimental to an enemy. However, such research revealed that it is one thing to provoke and disturb but another to govern primal nature.

In 1961, the meteorologist Edward Lorenz stumbled on a counterintuitive result while experimenting with a computer program designed to model weather patterns. It sparked an insight relevant to the AEC's difficulties in anticipating, let alone controlling, the vagaries of fallout. Enter data on the initial conditions, and—based on knowledge of the relevant physical laws—they should produce a predictable, deterministic outcome illustrating enduring weather patterns. "Given a particular starting point, the weather would unfold exactly the same each time. Given a slightly different starting point, the weather should unfold in a slightly different way," James Gleick explains. In viewing the computer output, however, Lorenz saw lines initially running in tandem but soon veering off in wild, undulating movements, and the further in time, the greater the divergence. It was a startling realization. Despite profound differences in outcome, the initial starting points were so small as to be impossible to measure. Lorenz discovered that within complex systems, "sensitive dependence on initial conditions" leads to widely divergent outcomes, and he coined the term "Butterfly Effect" to describe this phenomenon.[157] Change as minuscule as a butterfly flapping its wings can reverberate through the system and contribute to large-scale alterations. If something as anomalous as a butterfly in Brazil can alter conditions to create a tornado in Texas, then long-term weather forecasting is impossible, as is any effort to control the weather. Moreover, Lorenz argued that the world is replete with other examples of complex systems.

Chaos and complexity theories reveal nonlinearity; threshold effects, in which change occurs rapidly; feedback, such that the effect exacerbates or constrains the initial causal influence; bifurcation, or irreversible systemic shifts; and the domino effect, wherein change in one system disrupts other systemic patterns. And it highlights the difficulties of sustaining the pretense of Cartesian dualism. Sensitive dependence on initial conditions is a stake through the heart of pretensions of prediction, let alone the interactive stability of human intention and nonhuman materiality. As with Einstein's iconic equation, the Butterfly Effect is simple in its formulation but profound in its implications. Both are ontological. They point to a performative relationship between society and the nonhuman world but particularly in an era when the power of humanity to manipulate is so vast. The Butterfly Effect also illustrates severe epistemological conundrums such that the presupposition of the eventual end of knowledge is tragically naïve. Rather than unfolding in a Newtonian manner, many aspects of the nonhuman world display chaotic novelty. "Nonlinearity means the act of playing the game has a way of changing the rules. . . . That twisted changeability makes nonlinearity hard to calculate, but it also creates rich kinds of behavior that never occur in linear systems," Gleick notes.[158] "Rich kinds of behavior" is not what AEC officials were anticipating.

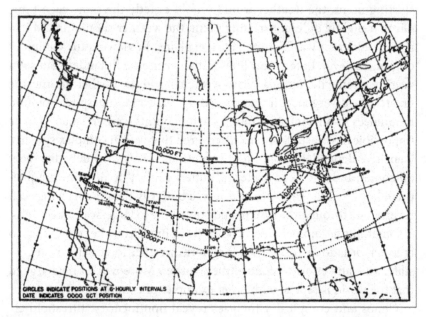

Figure 5.4. Shot Simon's trajectory across the United States. (Robert J. List, "The Transport of Atomic Debris from Operation Upshot-Knothole," US Weather Bureau, June 25, 1954)

Figure 5.4 depicts shot Simon's trajectory across the country—a sensitive dependence on initial conditions for which the Test Organization in Nevada was not prepared. And this contributed to a deepening material-discursive bind that AEC officials confronted with more robust public relations. This held at bay an increasingly restless public, for a time, but did little to change the very real implications of radioactive fallout.

Harry remains a point of reference around which the downwinders pin their frustrations and grievances, a symbol of impersonal forces imposing on their lives.[159] Harry left behind enduring questions and only limited and hard-won approximations of institutional accountability. And the data raise troubling considerations. Cumulative collective external gamma radiation exposure over the period 1951–1958 is estimated at 85,000 person-R, and of this, the Upshot-Knothole series is responsible for 40,000, or nearly one-half. Harry alone accounts for 30,000 person-R, or 75 percent of the Upshot-Knothole series and nearly one-third of the overall total between 1951 and 1958.[160] Harry's geographic reach was far more extensive than previous shots, as the 1 roentgen infinity dose reached as far as 250 miles from ground zero.[161] St. George accumulated the highest community-population estimate of external gamma exposure, at 18,000 person-R during the era of open-air detonations, in large part due to Harry. St. George accounts for one-fifth of the cumulative external gamma radiation exposure between 1951 and 1958. Other towns such as Hurricane, Washington, La Verkin, and Santa Clara in Utah, as well as Ely, McGill, and Ruth in Nevada, also bore a disproportionate burden.[162] Harry deposited more radioiodine around the country than any other open-air detonation ever conducted by the AEC.[163] Nancy, Simon, and Harry account for 78 percent of the fallout detected nationwide on gummed-film filter stations during the Upshot-Knothole series.[164]

The Upshot-Knothole series accounted for about 23 percent of the cesium-137 deposited around the country due to testing in Nevada, but the Plumbbob series of 1957 deposited approximately 35 percent of the total.[165] Upshot-Knothole included the dirtiest shots of the atmospheric era, but Plumbbob was the longest series ever conducted in Nevada, a remarkable twenty-four detonations, and had the highest total explosive yield of any series.[166] The imposition of radioactive fallout on the US public garnered more attention after the Upshot-Knothole series, but testing continued; and while records were broken in 1953, so too were many during the Plumbbob series of 1957.

6

Dead Sheep and the Fluidity of Fallout

The herds found refuge from the snow grazing on the public lands of the Great Basin in Nevada, lumbering back to Cedar City, Utah, from March to April, in time for the arrival of newborn lambs and shearing of wool, and then departed for cool, rich mountain pastures. It was a geographic and biological cycle repeated year in and year out, except they returned from the winter range in 1953 with a malady trailing alongside.

An interorganizational investigation coordinated by the AEC Division of Biology and Medicine considered three scenarios: (1) Radioactivity was a direct cause. (2) Radioactivity was a contributing factor aggravating poor range conditions and the stresses of trailing and pregnancy. (3) Radioactivity played no role in the deaths of about forty-four hundred sheep during the Upshot-Knothole series. The commission settled on the least plausible explanation: "Considering all of the information and data available at this time, it is apparent that the peculiar lesions observed in the sheep around Cedar City in the spring of 1953 and the abnormal losses suffered by the several sheepmen cannot be accounted for by radiation or attributed to the atomic tests conducted at the Nevada Proving Grounds." In presenting this conclusion to the ranchers in January 1954, AEC officials did not offer a definitive alternative explanation.[1] In a press release, however, it was implied that malnutrition was the problem, and a report to Congress summarizing AEC activities over the period January–June 1953—released while the investigation was in its early stages—stated, "Coincident with the 1953 test series, sheep and cattle grazing in Southern Nevada and Utah died. Malnutrition was a major factor in the deaths."[2]

The ranchers were surprised.[3] With their decades of experience raising sheep in the hardscrabble valleys and mountain ranges of southern Nevada and Utah, the commission's explanation did not ring true to them. Having a more personal knowledge of malnutrition in sheep, the ranchers nonetheless confronted an account by experts and officials operating under the rational-legal auspices of an organization charged with maintaining atomic military development.

The conclusion that radioactivity played no role in the sheep deaths was based on willful and strategic ignorance. In this sense, the investigation successively came to know less rather than more over seven chaotic months. Knowledge is based on gathering data, evidence, and experience to better understand some aspect of the world, but willful and strategic ignorance seeks to forestall a more expansive understanding.

As outlined in table 6.1, *provisional ignorance* refers to a deficit of knowledge that ostensibly can be overcome through concerted effort.[4] It is thus assumed to be temporary. In contrast, *willful ignorance* is the refusal to foster a better understanding or recognize the validity of new knowledge that may have disadvantageous implications. It involves deliberate rather than inadvertent neglect. "In a nutshell, willful ignorance can be seen as ignorance that is due to one's own will rather than external barriers," Jan Wieland explains.[5] It is an expression of biased decision-making processes. Willful ignorance is useful, but it is not without culpability. Alexander Sarch notes, "Courts routinely allow willful ignorance to substitute for knowledge. The willfully ignorant may be punished as if they had the knowledge required for the crime."[6] There are, in turn, two basic requirements for establishing willful ignorance in court: (1) there is a high probability that the defendant is aware of the significance of what they elected to ignore; and (2) there is evidence of a deliberate effort to avoid learning of a given situation or fact. Moreover, willful ignorance is distinct from self-deception. The latter suggests that one truly believes that which is demonstrably false, while the former suggests a degree of choice in remaining unaware of something that may be inconvenient or require substantial operational changes.

AEC officials often displayed willful ignorance during the era of open-air testing in Nevada, and they possessed considerable latitude to do so. They had unprecedented power to restrict and classify information. When there is little outside accountability amid the strict control of information and delayed injurious effects of failure to act responsibly, organizational actors are more likely to lapse into willful ignorance.

Strategic ignorance involves the methodical, coordinated production of uncertainty where it is not reasonably, and empirically, justified. Strategic ignorance is twofold: (1) emphasis on the existing unknowns in a

TABLE 6.1. Typology of Ignorance

Type	Description
Provisional ignorance	A lack or deficit of knowledge; being unaware; an original state in which awareness and/or knowledge has simply not been accrued.
Willful ignorance	The deliberate refusal to foster or recognize the validity of new knowledge that may have disadvantageous implications.
Strategic (manufactured) ignorance	Uncertainty is actively produced where it is not reasonably, and empirically, justified; a more coordinated and methodical enactment than willful ignorance. It is twofold: (1) emphasis on the unknowns in a situation to assert control and deny liability; and (2) the manufacture of unreasonable doubt.
Sanctioned ignorance	Ignorance viewed as justifiable and even essential to promote individual and social welfare.

situation to assert control and deny liability; and (2) the manufacture of unreasonable doubt. Strategic ignorance can be proactive or employed to not know in the present something actors do not want to confront.[7] It can also be defensive or employed to convince others that organizational actors could not have known something in the past in the same manner that it is understood in the present. Linsey McGoey argues that the ability to authoritatively define for others the unknowns in a given situation is crucial to organizational control: "We need to resist the tendency to assume that knowledge is more powerful than ignorance or that key actors have an overarching interest in expanding knowledge." The procedures and conceptual language of experts, in turn, is central to the organizational production of strategic ignorance: "Expert status is often dependent on maintaining a monopoly over what remains difficult or impossible to know empirically," McGoey observes.[8]

"Truth" is slippery, but one way to approximate it—or to verify knowledge—is through systematic, repeated observation.[9] Verifiable knowledge rests on an evidentiary foundation and not simply social consensus or an institutionally preferred interpretation. It is more than speculation. Strategic ignorance is often enacted by attacking, misrepresenting, and otherwise undermining verifiable knowledge accumulated at a given point in time. The tobacco industry perfected the manufacture of unreasonable uncertainty. "Doubt is our product," an infamous leaked memo declared. Indeed, the tobacco companies not only produced an addictive product but spent millions of dollars fostering confusion, even as the empirical evidence increasingly demonstrated the carcinogenic effects of smoking. Their effort included positing "phony facts," cherry-picking data, focusing on anomalies and irrelevant details, buying favorable scientific opinion, engaging in ad hominem attacks, creating fake grassroots campaigns, and, in sum, muddying the water even as the empirical evidence was clear.[10]

Ulrich Beck argues that the Chernobyl disaster is emblematic of manufactured non-knowing: "The nuclear explosion was accompanied by an explosion of non-knowledge, and this inextricable intermeshing of nuclear contamination and non-knowledge constitutes the strange, symptomatic, thoroughly Kafkaesque character of the post-Chernobyl world."[11] Franz Kafka conceived of the modern individual as trapped by the contradictory demands placed on them by overarching bureaucratic structures in society. They are ensnared as they struggle to negotiate impersonal administrative structures that are unresponsive to their distress and whose dictates cannot be avoided.[12] Their immediate concerns—which give their lives meaning—amount to little as they confront rule-bound structures whose procedures are often arbitrary and ambiguous. Kafka's fictional parable is reminiscent of Max Weber's conception of the iron cage of bureaucracy viewed from the perspective of the individual, and strategic ignorance is emblematic of the unintended contradictions of a world populated by large organizations pursuing novel science and technology.[13]

Strategic ignorance is a political project in which organizational actors utilize their authoritative status, available resources, and access to privileged information to obscure inconvenient biophysical and material effects. Radioactivity is particularly subject to the production of ignorance as it is *already* imperceptible to human senses.[14] There is a "double-twist" as its sense of imperceptibility necessitates expert knowledge and technique to make radioactivity visible in the first place, but too often expertise is co-opted to accentuate its invisibility both socially and politically.[15]

Many luminaries of the Scientific Revolution anticipated a point in time when science would triumph over provisional ignorance.[16] The irony of modernity is that uncertainty "is the *product* of more and better science," Beck suggests.[17] Provisional non-knowledge does not diminish in proportion to innovation but in tandem with it, particularly regarding the unintended consequences. It is a double bind: knowledge and uncertainty advance in lockstep. "The world is not a solid continent of facts sprinkled by a few lakes of uncertainties, but a vast ocean of uncertainties speckled by a few islands of calibrated and stabilized forms," Bruno Latour argues.[18] And sometimes power is employed to accentuate uncertainty.

Sanctioned ignorance refers to unawareness that is deemed socially legitimate and, indeed, necessary to promote individual or collective welfare.[19] There are laws protecting personal information and the trade secrets of corporations and ensuring national security. However, AEC officials abused their power to restrict and classify information. Willful and strategic ignorance is particularly pernicious as it was the commission's responsibility, inscribed in the McMahon Act of 1946, not simply to promote weapons development but to ensure public safety. The commission was granted substantial powers of sanctioned ignorance to promote national security—not to hide the hazards of fallout from the public.

In the context of novel and risky sociotechnical development, AEC officials worked to convince the public that there was little provisional ignorance in regard to radioactive fallout and yet employed willful and strategic ignorance when events on the ground threatened to undermine prior assurances of knowledge and control. And the tighter the material-discursive bind, the greater the incentives for willful and strategic ignorance. This balancing act is depicted in figure 6.1. In the risk society, more broadly, organizational actors often project certainty publicly but then circumvent new knowledge that may be disadvantageous and even manufacture ignorance to undermine verifiable claims that are at odds with organizational objectives. It is a dance in which certainty is projected, on the one hand, but uncertainty is promoted, on the other, when it is expedient.

The sheep investigation was characterized by three stages.[20] The first encompassed arrival of AEC officials and representatives in Cedar City, Utah, on June 5 and ended with a meeting of the investigators in Salt Lake City in early August. When local, state, and then AEC officials first inspected the affected herds, the majority of deaths had already

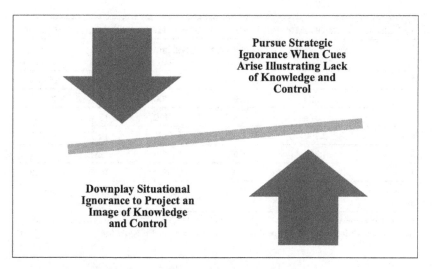

Figure 6.1. Organizational manipulation of certainty and ignorance.

occurred, but it was apparent that radioactivity was present in abundance. Indeed, it was clear that radioactivity was a contributing, if not the sole, factor responsible for the sheep losses.[21] At this point, evidence of radiation injury evoked willful ignorance and a refusal to follow where the evidence might lead. The second stage began with the August meeting and ended with a conference at Los Alamos in October. During this second stage, AEC officials solicited experimental and laboratory research from a broader cadre of investigators who were more tightly aligned to the AEC. These additional participants introduced a more strategic and manufactured ignorance into the investigation. The third stage began with the conference in Los Alamos on October 27 and ended with a meeting in Cedar City in January 1954, where AEC officials assured the ranchers that radiation did not kill their sheep. This third stage was characterized by the heightened employment of manufactured uncertainty to pave the way for the final report.

Stage 1: Preliminary Findings, June 5–August 3, 1953

Table 6.2 summarizes the estimated losses of the six herds downwind from the test site after shot Nancy. Approximately 1,420 ewes died across the six herds, but two-thirds of all deaths, around 3,000 in number, were

TABLE 6.2. Data on Affected Cedar City Sheep Herds and Associated Losses

Herd	No. on winter range	No. ewes lost lambing	No. lambs lost	Trailing deaths	Time of greatest loss
1	1,835	200	600	35	5/15–5/25
2	3,200	300	700	6	5/20
3	1,375	200	400	15	5/5–5/20
4	2,100	120	470	10	4/5–4/15
5	1,700	300	200	unknown	5/1–5/31
6	1,500	300	600	10	5/1–5/20
Total sheep deaths	11,710	1,420	2,970	76	4,466

Source: Data summarized from Monroe A. Holmes, "Compiled Report on Co-operative Field Survey of Sheep Deaths in S.W. Utah (Cedar City)," December 31, 1953, NTA Accession No. NV0020585.

lambs lost at birth or shortly thereafter. There were no abnormal losses reported while trailing east in any of the six herds. This suggests that poor range conditions were not the sole cause of the heavy losses. If malnutrition killed the sheep, it is likely that many deaths would have occurred while trailing eastward, and all but one of the six herds were given supplemental feed on the Nevada range.[22]

The first signs of trouble included blisters and scabbing around the mouth and lesions extending up the muzzle, along the head, and down the back underneath the wool, which "slipped" or pulled off in chunks during shearing.[23] At lambing time, the ewes were lethargic but continued to eat until their deaths, which were unnervingly sudden.[24] Many lambs were stunted at birth or displayed malformations.[25]

AEC officials first learned of sheep losses on June 2, and representatives were dispatched to investigate. By their arrival in Cedar City three days later, local, state, and PHS officials were already onsite. The heaviest losses occurred between mid-April and late May.[26] Some carcasses were hastily buried, but many were hauled away or ground up for chicken feed. That AEC representatives were onsite only after the majority of deaths is noteworthy, as the sheep examined as well as those slaughtered for analysis of wool, internal organs, and bone were assumed to be representative of the affected sheep. Veterinarians who were acting on behalf of the AEC and who first examined the remaining herds nonetheless identified beta burns and high dosimeter readings along the throat, indicative of radioiodine within the thyroid.[27]

Table 6.3 lists the key participants in the investigation. In early June, there was the potential for officials from the state of Utah as well as from various federal agencies each investigating the issue. AEC representatives arriving in Cedar City on June 5 included Deputy Field Manager Joseph Sanders; Robert Thompsett, a veterinary consultant from Los Alamos; and Major Robert Veenstra, from the Naval Radiological Defense

TABLE 6.3. Key Actors in the Sheep Investigation

AEC officials and representatives	John C. Bugher, MD, Director, Division of Biology and Medicine, Washington, DC
	Dr. Gordon Dunning, health physicist, Biophysics Branch, Division of Biology and Medicine, Washington, DC
	Dr. Paul B. Pearson, Chief of the Biology Branch, Division of Biology and Medicine, Washington, DC
	Joseph B. Sanders, Deputy Field Manager, Las Vegas and Los Alamos
	Brig. General Kenneth Fields, Director, Division of Military Application, Washington, DC
	Robert E. Thompsett, veterinarian, Los Alamos, under contract with the AEC
	Maj. Robert H. Veenstra, Naval Radiological Defense Laboratory, California
	Lieutenant Colonel Bernard F. Trum, Veterinary Corps, US Army, University of Tennessee, AEC program at Oak Ridge
	Lieutenant Colonel John H. Rust, Veterinary Corps, US Army, University of Tennessee, AEC program at Oak Ridge
	Leo K. Bustad, veterinarian researcher, Hanford, Washington
	Clarence Lushbaugh, MD, pathologist, Los Alamos Health Research Laboratory
US Bureau of Animal Industry	F. H. Melvin, Salt Lake City office
US Public Health Service	Monroe A. Holmes, veterinarian assigned to the Utah State Department of Health, Salt Lake City
	Arthur Wolff, Environmental Health Section, Cincinnati, Ohio
	William Hadlow, veterinary pathologist, Rocky Mountain Laboratory, Montana
	James G. Terrill Jr., Acting Chief, Radiological Health Branch, Washington, DC
State of Utah	George A. Spendlove, MD, Director, Department of Health
Local/county	A. C. Johnson, veterinarian, Cedar City, Utah
	Stephen L. Brower, Iron County Agricultural Agent, Utah State University Extension Service

Laboratory in San Francisco. F. H. Melvin out of the Salt Lake City office of the Bureau of Animal Industry was present, as were the veterinarians Monroe Holmes, William Hadlow, and Arthur Wolff of the PHS. A. C. Johnson, a local veterinarian, was one of the first to examine the affected herds and proclaimed that in thirty years of practice, he had never seen anything like it.[28] Stephen L. Brower also was present during the initial collection of field data by AEC representatives. Between 1950 and 1954, Brower was the Iron County agricultural agent for Utah State University Extension Services.

"This sheep is hotter than a $2 pistol," the rancher Jack Pace recalled in 1979 before congressional investigators as to what he overheard an AEC veterinarian exclaim.[29] The two veterinarian consultants to the AEC, Thompsett and Veenstra, made a number of observations in early June that were reiterated by the stockmen even decades later. Also testifying before congressional investigators in 1979 was Stephen Brower, by then a professor at Brigham Young University. Brower contended that Veenstra and Thompsett both clearly indicated evidence of radioactive contamination.[30] As a county agent familiar with the ranchers and witness to the grisly scene, Brower constituted a potentially loose thread for AEC officials. In the months to come, he would be excluded, despite repeated requests for information, from the inner dialogue of the investigation, through classification of key documents.[31] This made it difficult for Brower to act as an intermediary between AEC officials and the ranchers. More pointedly, he testified under oath during congressional investigation in 1979 and again in the *Allen et al.* trial and the 1982 *Bulloch et al. v. United States* proceedings that amid the initial observations in Cedar City, Paul Pearson of the AEC Division of Biology and Medicine admitted to him that the commission would, "under no circumstance," be held liable for payment of radiation-injured "animals or humans."[32]

Pearson's statement suggests that AEC officials were intent on perpetuating a cover-up. Parsing the archival materials, however, uncovers something less explicit but nonetheless problematic: AEC officials were searching for empirical evidence on which to hang a preferred conclusion, while rationalizing the dismissal of conflicting empirical observations. Rather than an explicit cover-up, they were searching for some way to conclude that radioactivity was not responsible for the sheep losses. And this necessitated the incremental descent into the organizational production of ignorance.

On June 10, Gordon Dunning advised the AEC commissioners in Washington, DC, that sheep grazing fifty miles from the NTS had beta burns in their nostrils and on their backs, and attention quickly turned to the "public relations problems." Commissioners Thomas E. Murray and Eugene M. Zuckert urged that the public be given the "full facts concerning the reported incidents and an explanation of the precautions taken by the AEC to insure against creation of any hazards to health and safety." The AEC commissioners expressed concern but also resolute trust in those who were directly looking into the issue. Dunning assured the commissioners that reports were forthcoming.[33] By mid-June, Brigadier General Kenneth Fields, director of the AEC Division of Military Application, and John C. Bugher, director of the Division of Biology and Medicine, discussed with the AEC commissioners the need to direct the investigation. Fields expressed concern about a fragmentary interorganizational effort, and these fragments were soon assembled under the AEC's purview.[34] Paul Pearson was assigned responsibility for coordinating the investigation. AEC officials now held tight control over the inquiry.

On June 15, AEC officials and consultants gathered in Las Vegas. The meeting included experts from Utah State University, the University of Nevada, Las Vegas, and AEC-affiliated researchers from the University of Tennessee.[35] The PHS veterinarians, including Holmes, Hadlow, and Wolff, were not in attendance, nor were Veenstra and Thompsett. Those who were present for the June 15 meeting had little firsthand knowledge of the situation. As Barton Hacker observes, "When they met, most of them had yet to see an injured animal from the fallout area."[36] This was crucial, as the June 15 meeting established an empirically unwarranted tone. Willful ignorance was arising. Indeed, it was noted during the meeting that many of the sheep exhibited beta burns, but it was also asserted that levels of radioactivity were too low to account for the other dramatic and acute effects. Despite the fact that results from ongoing laboratory tests were not yet available, it was "generally agreed" that there was "little evidence" that radiation killed the sheep.[37] The report outlining the June 15 meeting does not suggest that those who were assembled discussed radioactivity as a contributing factor. The discussion, rather, hinged on whether gamma radiation was high enough to account for the heavy losses—a question presupposing a direct causative effect and introducing a confusing, erroneous trajectory.

Nonetheless, on June 17 Bugher briefed the AEC commissioners in Washington, DC, and offered a more encouraging assessment than Dunning had one week earlier. Bugher assured the commissioners that radioactivity did not kill the Utah sheep, but its role as a contributing factor could not yet be ruled out.[38]

The June 15 meeting in Las Vegas, however, occurred amid of a volley of reports and memoranda passing between the veterinarians who examined the herds firsthand and were now undertaking laboratory studies, and while AEC officials were discussing malnutrition in Las Vegas, the veterinarians were documenting radiation injury. Examining a section of bone, Robert Veenstra found evidence of gamma radiation in the marrow and declared that the lesions on the Utah sheep were "typical" of what was seen with the Trinity cattle.[39] William Hadlow suggested that radioactivity was probably a contributing factor, and Arthur Wolff of the PHS noted that preliminary results indicated a "surprisingly large concentration" of radioactivity in the thyroid.[40] Robert Thompsett personally examined the Trinity cattle and asserted that the lesions discovered at Cedar City, examined macroscopically and in the laboratory, were beta burns contributing to the death of sheep that were facing the additional stresses of trailing and pregnancy. More ominously for AEC officials, Thompsett concluded, "Again with the sheep losses I am of the opinion that the Atomic Energy Commission has contributed to great losses."[41] Veenstra stated in his report, "In view of the hyperplasia, presence of detectable gamma radiations in the bone marrow, skin lesions, possibility of toxins due to the lesions and the presence of pregnancy in many ewes, it is my opinion that radiation was at least a contributing factor to the losses of these animals."[42] Nevada state officials, moreover, reported no abnormal losses or mysterious ailments afflicting sheep herds throughout the state.[43] If it was an infectious disease, it was only affecting sheep that Cedar City–area ranchers were grazing immediately northeast of the NTS.

Despite the optimism of the June 15 meeting in Las Vegas, the preliminary field observations and laboratory analysis of the veterinarians began trickling in, and they were not as encouraging. Radioactivity appeared, at the very least, to be a contributing factor. In late July, Morse Salisbury, the AEC's Director of Information Services, circulated among the five commissioners in Washington, DC, a counterstatement to be released publicly if an "attack from authoritative sources" arose criticizing the commission's management of open-air testing. The statement stressed the gains in weapons development and insisted that the testing posed "no substantial hazards to the public health and safety." Salisbury's statement admitted "some nominal injury" to livestock grazing nearby the NTS but stressed that the animals were located at "close-in areas" and thus that grazing was a "calculated risk" on the part of the ranchers.[44] If it was not possible to deny that radioactivity contributed to the sheep losses, it was at least plausible to blame the ranchers for grazing their herds too close to the test site.

AEC officials originally anticipated a prompt investigation.[45] By July, however, it appeared that the commission would have to accept responsibility for the sheep losses and compensate the ranchers accordingly. Money was not the problem. Adverse publicity was the problem. And the AEC commissioners in Washington, DC, were fearful of public opinion turning against continued testing in Nevada.[46] What AEC officials needed was some kind of game-changer. "One way to get different results was to employ different investigators—namely, those investigators who were more accountable to the parent organization," Philip Fradkin explains.[47]

Researchers at the central nodes of the atomic industrial complex—Los Alamos, Hanford, and Oak Ridge—were asked to contribute to the investigation. At Los Alamos and Oak Ridge, research focused on experimentally induced beta burns on sheep. Researchers at Hanford analyzed radioiodine in experimental sheep. And the final report released in January relied heavily on these studies. In turn, the PHS veterinarians as well as Veenstra and Thompsett, all present in Cedar City at the beginning, were pushed aside as things progressed. Stephen Brower, the Iron County agricultural agent, moreover, was eager by late summer to get copies of the initial field and laboratory analysis, indicating radiation injury, only to discover that the AEC had classified these studies.[48]

Salisbury's press release admitting AEC liability for the sheep deaths was never submitted to the public, as things were about to take a turn. Indeed, they were about to shift from willful to strategic ignorance.

Good news first arrived from Tennessee. Researchers traveled to Cedar City and determined that the lesions were not due to radiation but most likely malnutrition and/or disease. Stunted, malformed, and dead lambs, moreover, "occur in nature." While there were indications of ingestion of radioiodine, the researchers denied that there was evidence of radiation injury.[49] In late July, Lieutenant Colonel John H. Rust forwarded to Paul Pearson a memorandum listing fourteen reasons why the sheep did not demonstrate radiation injury. Rust could not identify what killed the sheep but suggested that malnutrition was a "major contributory factor."[50] Later in the summer, Lieutenant Colonel Bernard Trum submitted a report based on the autopsy of a seven-year-old ewe from Cedar City, sacrificed in August, and he did not detect abnormal levels of radioactivity in wool, skin, or tissue samples.[51] The researchers from Oak Ridge thus gave AEC officials reason to look beyond the seemingly obvious evidence of radiation effects.

Stage 2: Gathering Different Data, August 3–October 27, 1953

A meeting was scheduled on August 3–4 in Salt Lake City to bring together the various participants in the investigation, but it did not result in unanimous agreement. The veterinarians who first encountered the sheep, Veenstra and Thompsett, were reluctant to dismiss fallout as a contributing factor. Thompsett, the veterinarian consultant from Los Alamos, insisted that the lesions seen at Cedar City were beta burns and suggested that the sheep deaths were due to toxins in the blood as a result of the lesions. Veenstra reported on laboratory analysis indicating that thyroid samples were abnormal and agreed with Thompsett's assertion "regarding the significance of the facial lesions." Hans Kornberg, however, summarized research at Hanford illustrating that the levels of radioiodine necessary to induce immediate observable effects in adult sheep was considerably greater than what the sheep at Cedar City probably encountered. The veterinarian Arthur Wolff of the PHS also questioned whether radiation levels were high enough to induce "immediate biological effects." Bernard Trum from Oak Ridge argued that the facial lesions were not beta burns but did not offer an alternative explanation, and he questioned whether the samples from Cedar City illustrated radiation-induced damage in thyroid tissue. Paul

Pearson of the AEC's Division of Biology and Medicine presented data on the below-average rainfall on the range. Kermit Larson, an AEC-affiliated researcher from the University of California, Los Angeles, claimed that rabbits and rodents collected in southern Nevada did not have observable beta burns. However, Larson also raised an issue that he suggested deserved greater attention: intestinal exposure from "insoluble fission products." It was a crucial consideration lost in the maelstrom to come.[52]

Amid continuing disagreement at the Salt Lake City meeting, it was concluded that further research was needed. Those who were in attendance agreed on three issues:

1. The amount of radioactivity found in the thyroid should not have produced the immediate, observed effects on the basis of correct knowledge.
2. Beta burns should be studied experimentally and compared with the findings in the affected sheep before a decision is reached on the significance of this factor.
3. An effort should be made to determine the incidence of the typical symptoms in areas unaffected by fall-out.[53]

A more immediate next step involved assembling in Cedar City on August 9 to update the ranchers on the investigation. The meeting depicts the contrast between AEC officials, who were uncertain as to their message, and the ranchers, who asked probing questions of the experts—questions that the investigation was veering away from by August. Arguably, AEC officials exhibited hesitation rooted in ongoing processes of willful and strategic ignorance. The ranchers, despite their lack of requisite training, raised pertinent comments that did not elicit meaningful dialogue during the meeting.

Before the assembled ranchers, Paul Pearson floated the possibility of some new disease. "This could be a new disease or an old set of circumstances in disguise."[54] It was an unwarranted assertion. The investigation never uncovered evidence suggesting a newly discovered disease, and investigators dismissed the idea soon after the inquiry began.[55]

AEC officials next raised the issue of a drier than normal year on the range. Rainfall was 50 percent of normal, but the ranchers countered that such conditions were not unusual and that they provided supplemental feed accordingly.[56]

A rancher asked, "Isn't it true that your investigations did show effects of radiation?" An AEC representative conceded that radioactivity was detected but argued that the levels were too low to kill the sheep. Another rancher reminded AEC and military officials that the investigation began only after the worst symptoms abated.[57] This rebuttal highlights the question of whether the bone, wool, and tissue samples obtained in June were *representative* of the radioactive damage inflicted on the herds. The ranchers understood that there was a potential problem obtaining samples from sheep that lived and making inferences regarding those that had died as much as six or seven weeks earlier.

The meeting soon turned to prior statements of AEC representatives—probably Veenstra and Thompsett—upon first encountering the afflicted herds in June:

LIVESTOCKMAN: When you made your tests you mentioned some sheep were hot. What did you mean?

AEC: It is true that some sheep had relatively high values who weren't ill. The [*redacted*] herd has as high an external value on the instruments as any.

[REDACTED]: You said "this sheep isn't as hot," or, "This is a hot one." What did you mean? A dead one was unusually "hot."

SANDERS: Any radiation was hot. It didn't mean anything special. External readings have no consistency.[58]

The response by Joseph Sanders, the commission's deputy field manager for testing in Nevada, was peculiar. Colloquially, "hot" refers to levels of radioactivity beyond conventional expectation. But much of the meeting on August 9 reveals an awkward tacking back and forth between ranchers who have given the issue a good deal of thought and AEC officials who offer only ambiguous and conflicting responses.

The discussion next turned to brief consideration of *internal emitters* as opposed to simply external gamma radiation. A rancher asked, "Couldn't animals acquire some radiation from eating plants that had some radiation?" This question was met with oblique reference to ongoing experimental research. The rancher Douglas Clark then pointed to what should have been a central focus of the investigation: "It isn't a question of did they receive lethal doses. Any radiation could have pushed a poor sheep on to the other side."[59] That is, Clark identified radioactivity as a *contributing factor*. This includes external and internal radiation acting in a synergistic manner with the physical effects of trailing and the stresses of gestation and birth. In the final investigation report, AEC officials stressed that radioactivity levels were too low to account for the sheep deaths: if radiation did not kill them outright, then it did not hurt them. This is, of course, nonsensical. Nonetheless, after the August meeting in Cedar City, the question of radioactivity as a contributing factor was eclipsed by a focus on the direct causal effect and, in the final months, debate over whether the lesions were in fact beta burns at all. If the lesions were not beta burns, then, perhaps, radiation damage was minimal, if not absent entirely.

The August 9 discussion soon circled back to malnutrition. The ranchers had seen malnutrition before. "Usually with malnutrition you can put them on good feed and they respond. There were about 25 days when these sheep didn't respond to anything," a rancher declared. An unidentified AEC official responded, "We don't say that is it malnutrition. We've *never* said this is malnutrition." In response, the ranchers reminded the AEC officials that they provided supplemental feed. An AEC official, unnamed in the meeting transcript, then clarified, "I doubt that all this trouble could be attributed to malnutrition. It may have been a contributing factor. They were poor but I have never felt that it was due to malnutrition."[60]

In contrast to the hedging expressed in Cedar City on August 9, the one thing AEC and military officials were certain of throughout seven convoluted months of investigation was that malnutrition was somehow implicated in the sheep deaths. Below-average range conditions and malnutrition were the cornerstone on which willful and strategic ignorance was constructed. If the spring of 1953 had not been drier than average, it is difficult to envision how AEC officials would have found a way to eventually dismiss the effects of radioactivity.

The discussion next turned to whether herds grazing elsewhere in Nevada suffered the same symptoms as the Cedar City herds, but an AEC official, unnamed in the transcript, offered only a vague answer in response.[61] It is a crucial question reminiscent of classic experimental research design: If other herds did not get dusted with fallout and yet displayed symptoms consistent with the Cedar City sheep, then radioactivity was but a distraction—as the real culprit was something else entirely. If herds grazing elsewhere did not exhibit symptoms consistent with the Cedar City sheep—and there are no other demonstrable differences between the herds—then radioactivity constituted the experimental treatment and the unafflicted herds were the control group.

By August, investigators had communicated with Nevada health and agricultural officials, but none indicated abnormal sheep losses elsewhere in the state.[62] Both Joe Sanders and James G. Terrill Jr., acting chief of the radiological health branch of the PHS, attended the August 9 meeting and thereafter traveled throughout Nevada gathering more information. They spoke to ranchers with herds located north and west of the NTS during Upshot-Knothole, but, again, there was no evidence of abnormal sheep losses or the grisly symptoms observed at Cedar City.[63] Whatever killed the Utah sheep only stalked the rangeland northeast of the test site where fallout from shot Nancy came to earth.

Near the end of the August 9 meeting, a crucial observation was briefly and ambiguously raised. "This is an absolutely new condition. These sheep produced poor lambs that couldn't live and couldn't feed them," a rancher declared.[64] No AEC officials, as outlined in the meeting transcript, pursued this discussion. Embedded within this statement was the most crucial question the investigation should have focused on: What happened to the lambs?

Much of the latter half of the investigation involved assessing the degree to which beta burns on experimental sheep resembled the lesions encountered at Cedar City. This was an important consideration, for if the Utah sheep suffered from beta burns, it indicated levels of radioactivity well beyond that indicated by the limited data gathering after shot Nancy. Unanimous agreement was never attained on this question, but more importantly, it came to eclipse a more fundamental question. Given that approximately 68 percent of the forty-four hundred deaths involved lambs, what happened to the lambs? The experimental research at Los Alamos and Oak Ridge—which shaped the content of the final report—involved adult sheep. There was a mismatch between this experimental research and what primarily occurred at Cedar City.

Atomic testing in Nevada began with several crucial operational expectations, including the assumption that fallout exhibits relatively predictable dispersion and dilution patterns. Methodical meteorological forecasting, in turn, could ensure safe operational conduct. Dead sheep called into question such an assumption. Fallout, absent "rainout," could provoke startling adverse effects. Another foundational assumption was the conviction of an unreasoning and irrational public that could not be trusted to engage in meaningful dialogue and debate. In turn, the cues—whether human or nonhuman—from living and working in the shadow of testing were consistently devalued by AEC officials over the course of the 1950s.

And AEC officials were keeping account of the disposition of the Cedar City ranchers, but it was consistently noted that they were neither acting irrationally nor characterized by panic-ridden behavior.[65] James Terrill summarized his impressions of the ranchers during the August 9 meeting: "Fortunately, the A.E.C. is dealing with a group of men who are apparently in financial circumstances where they can view even large sheep losses fairly objectively. There was no evidence of any attempt by individuals or groups to emphasize damages. They all apparently have a patriotic understanding of the national defense needs of the tests."[66]

It is often assumed that people are unable to understand radioactivity and approach such issues in an irredeemably irrational manner: as radioactivity is tasteless, odorless, and invisible, nonexperts are simply incapable of reasoned thought and behavior about something so esoteric. This is a self-serving point of view propagated by people in positions of organizational and political power in order to exclude the public from key decisions involving risky sociotechnical activities. Radioactivity is not something the layperson can readily assess. They need experts, who are generally situated in large, complex public and corporate organizations, to act in a fiduciary manner in assessing the complexities of a given situation. Rather than an irredeemable fear of radioactivity, perhaps what the public really fears is that people in positions of organizational and political power—whom they are involuntarily dependent on to act in good faith—will fail to act with integrity. Perhaps it is not all things atomic that people fear but those who betray their trust, particularly when evidence of such betrayal may not arise until years into the future.

The August 9 meeting in Cedar City illustrates the advantages of a broader point of view, even if there is scarce evidence that the ranchers shaped the investigation in any meaningful way. Indeed, parsing the meeting minutes raises the question of whether the ranchers were approaching the issue in a rational and methodical manner, whereas AEC officials were mired in a contrived process of coming to know less rather than more as the investigation wore on. Willful and strategic ignorance was building not due to "bounded rationality" or organizational decision-making hampered by a dearth of available information but to undue reliance on information oriented toward justifying a preferred conclusion.

During the August 9 meeting in Cedar City, George A. Spendlove from the Utah Department of Health stressed the importance of the investigation: "We are interested in the health of the sheep and in the health of the people. We are primarily interested in the health of the people but the health of the people has a relationship to the sheep."[67] Examining the reports and memoranda of the investigation, however, shows little evidence that AEC officials assumed that what happened to the sheep was symptomatic of what was happening to the people. Willful and strategic ignorance is arguably easier to lapse into when there is not a perceived harm to others.

That AEC officials did not associate the radiological risks imposed on sheep to the people living downwind was indicative of a geographic and social distance between those who were conducting testing and those who were experiencing the deleterious effects firsthand. This detachment between the people engaging in risky sociotechnical experimentation and the people involuntarily placed at risk rested on the hubris embedded within the AEC's organizational culture. This arrogance was expressed not simply in a fear of the public standing in the way of atomic weapons development but in contempt for the public's involvement beyond that of quiescent spectators. And hubris can impede recognition of signals suggesting that problems are arising. The sinking of the *Titanic* in 1912 still offers important lessons. Jim Meigs argues, "Technology can outpace judgment."[68] An iceberg was only the most proximate cause of a disaster that was ultimately rooted in arrogance.

The rancher Douglas Clark participated in the August meeting in Cedar City but did not live to see the meeting in January 1954, just prior to the release of the final report. "Clark may have been, in an indirect sense, the first human casualty of fallout from the Nevada site," Fradkin suggests.[69] Testifying before congressional investigators in 1979, Stephen Brower recalled,

> I remember going with [Douglas Clark] and some of the veterinarians who were doing some autopsies one day, and Doug raised some questions with the team of scientists, one of whom was a colonel. I don't recall what his name was, but he seemed to be the leading spokesman to kind of press this issue that it couldn't have been radiation. Doug asked him some fairly technical questions about the effects of radiation on internal organs that he'd gotten from other veterinarians. The man, rather than answering the question, called him a dumb sheepman, told him he was stupid; he couldn't understand the answer if it was given to him, and for just 10 or 15 minutes, just kind of berated him rather than answer the question. It was a tough kind of experience for Doug. I remember he left there to go out to his ranch to meet with the loan company to account for what sheep he had left, and within a couple of hours, he was dead from a heart attack. I think that, you know, part of the stress that he experienced at that time was that abuse he received from these officials.[70]

Stage 3: Arriving at the Final Report, October 27, 1953–January 13, 1954

On October 27, investigators gathered in Los Alamos. They examined induced beta burns on experimental sheep and tissue samples to determine whether the lesions documented in Utah were due to radiation. The examination proceeded despite a compelling alternative explanation. The investigators argued one point against another, and as with the meeting in Salt Lake City, undivided opinion was not achieved. It would take some wrenching to attain even the superficial appearance of unanimous agreement. Robert Thompsett and the PHS veterinarians Arthur Wolff and Monroe Holmes were unconvinced that the lesions on the Utah sheep were due to something other than beta radiation. Robert Veenstra was not in attendance at the October 27 meeting. Those who were more closely tied to the parent organization, including Bernard Trum and Clarence Lushbaugh, insisted that there were observable differences and that therefore the Utah sheep did not display radiation injury.[71]

Gordon Dunning, a health physicist with the AEC's Division of Biology and Medicine, was only tangentially involved in the investigation, but he played an exceptionally odd role near the end. Dunning joined the AEC in 1951 and spent twenty years working in a variety of positions. As Stewart Udall observes, his most prominent role at any given time was earnest protector of the commission's public image.[72] Dunning found a way to make problems disappear.

Near the end of the Los Alamos meeting, Dunning drafted a document articulating several conclusions that the investigators could agree on and serving as a framework in moving forward. The "Dunning statements" were out of sync with the empirical evidence, and yet those who were in attendance signed the document. The major conclusions included:

1) The same lesions were noted on sheep in areas of little or no fall-out with approximately the same incidence as in areas of relatively heavy fall-out.

2) The general amounts of fall-out in the areas under question have been determined. These quantities of radiation dosage are not known to be sufficient to produce the lesions noted.

3) Evidence has been gleaned from microscopic examination and comparisons of tissues both from the sheep in fall-out areas and some sheep on which skin beta burns were produced experimentally. All of these data present a preponderance of evidence to support the conclusion that the lesions were not produced by radioactive fall-out.

4) In considering radiation damage to the internal organs, the most critical is the thyroid, due to the uptake of radioactive iodine from the fall-out material. The amount of highest radiation dosage to the thyroid has been calculated to be far below the quantity necessary to produce detectable injury.[73]

Statement 1 is simply incorrect. There was no evidence that sheep grazing in areas with little or no fallout exhibited the peculiar lesions that plagued the Cedar City sheep.[74] Dunning's assertion was an empirically unwarranted effort, arising in the final months of the investigation, to disassociate the Cedar City sheep from fallout.

Statement 2 is fiction. Monitoring was spotty, and many readings were collected near roadways. There were probably numerous hot spots in the areas where the sheep were located. There was a tendency within the AEC, however, to treat the lack of data as the lack of radioactivity. Undue focus on gamma radiation, moreover, was a red herring—a logical fallacy distracting from more essential considerations. The central issue was the lambing ewes nipping at foliage while trailing east and ingesting internal emitters of radiation, which passed to their gestating lambs.

Statement 3 is erroneous. There was never evidence that the burns exhibited by the sheep at Cedar City were due to anything other than beta radiation. Experimental research on adult sheep involving the induction of beta burns and comparison to the Utah sheep was a contrived effort of manufacturing dissenting evidence. This effort, moreover, obscured consideration of the toxicity that probably accumulated in the blood of the ewes (due to beta burns) and passed to their gestating lambs.

Statement 4 is misleading. Radioiodine was a crucial consideration but not necessarily the "most critical," particularly in reference to the lambing ewes. Undue focus on radioiodine obscured consideration of the radioactive assault on the sheep's gastrointestinal system. Further, Dunning asserted that the highest radiation dosage calculated in the thyroid of Cedar City sheep was "far below" that necessary to produce observable injury. Dunning was mistaken, as he failed to appreciate the concentrating effects of radioiodine in the fetal lambs.

The Dunning statements are puzzling. Adding to the mystery, if not the crushing irony, is that while employed by the AEC, Dunning frequently published in academic journals and a consistent theme was developing one's critical thinking skills.[75] He endeavored to educate others on avoiding mistakes in reasoning while extolling the virtues of science to illuminate real-world issues of concern.

Why would well-educated, competent individuals immersed in the data agree to Dunning's statements proffered near the end of the October 27 meeting in Los Alamos? The answer involves sleight-of-hand. Soon after the meeting, Monroe Holmes of the PHS indicated that he signed, even though he did not agree with the conclusions, because he understood that it was a meeting attendance sheet.[76] Arthur Wolff signed because Dunning stated that it was necessary to appease Eugene

Zuckert, one of the five AEC commissioners located in Washington, DC, and to ensure continued funding of testing in Nevada.[77] F. H. Melvin of the US Bureau of Animal Industry also remarked that there was confusion over the intent of the document.[78]

In 1984, Dunning penned a letter to congressional investigators to "correct the record." Dunning insisted that he did not harbor deceitful intent in drafting the statements. Indeed, he declared that he did not participate in developing the conclusions but simply acted as a "scientific secretary," persuading those who were in attendance to summarize their discussion on paper.[79] Just after the October 27, 1953, meeting in Los Alamos, however, Dunning wrote a memorandum addressed to Morse Salisbury, director of the AEC's Division of Information Services, in which he admitted a more active role than he remembered thirty years later. "I doubt if we will ever obtain more positive conclusions from this committee than are contained in the attached statements," Dunning confided.[80] And in a series of memoranda passing between AEC officials in the wake of the October 27 meeting, the Dunning statements were embraced with few reservations.[81]

The Dunning statements muddied the water. This is one strategy underlying strategic ignorance: contribute to confusion and disorder. The muddier the water, the easier it is to arrive at a preferred conclusion that is hardly justified by the empirical evidence.

Within this context of declining visibility, Paul Pearson wrote a first draft of the investigation report. Table 6.4 outlines the major conclusions articulated in this first draft and in the final report. The Dunning statements were less consistent with Pearson's first draft than with the final report released two months later. Indeed, the Dunning statements foreshadowed where the investigation would end up.

The draft report, dated November 4, 1953, is clear: most of the investigators are of the opinion that radiation did not contribute to abnormal losses, but there was a lack of unanimity on this point. The final report is more decisive: radiation was not the cause, and undivided agreement among the investigators is implied. The draft report articulates a more provisional decision in asserting that research does not "justify" the conclusion that radiation contributed to the sheep losses, and the failure to identify an alternative explanation is acknowledged. The final report raises the singular issue of the lesions observed at Cedar City and insists

TABLE 6.4. Comparison of Draft and Final Sheep Investigation Reports

Draft report, November 4, 1953	Final report, January 6, 1954
Lack of consensus:	Consensus implied:
"While the majority of those who made a study of the sheep losses are of the opinion that radiation and the activities incident to the spring test did not contribute to abnormal losses of sheep, there was a lack of unanimity of opinion."	"Considering all of the information and data available at this time, it is apparent that the peculiar lesions observed in the sheep around Cedar City in the spring of 1953 and the abnormal losses suffered by the several sheepmen cannot be accounted for by radiation or attributed to the atomic tests conducted at the Nevada Proving Grounds."
Conclusion is provisional, not final:	Conclusion is scientifically sound:
"While in the opinion of the Division of Biology and Medicine the evidence so far does not justify a present conclusion by us that the spring tests were factors in these losses, the lack of unanimity of scientific opinion, combined also with the lack of positive findings as to the etiology of the losses, necessitates additional work before a final scientific conclusion can be reached."	"Considering all of the data it is apparent that the peculiar lesions observed on sheep around Cedar City in the spring of 1953 cannot be reasonably accounted for by radiation or attributed to the atomic tests conducted by the U.S. Atomic Energy Commission. This opinion is concurred in by the U.S. Public Health Service (HEW) and the Bureau of Animal Industry."
Fallout occurred where the herds were wintering:	Presence and intensity of fallout is downplayed:
"There was fall-out of radioactive material from the March 24 detonation in the area north of the NPG where many sheep were wintered. There was also fall-out around Cedar City from the May 19 detonation. The herds in which heavy losses occurred had been moved to the Cedar City area prior to May 19."	"There was some fall-out of radioactive material from the March 24 detonation in the area where the sheep were north of the Nevada Proving Grounds and again around Cedar City from the May 19 detonation. In these areas it is estimated that the measured and integrated gamma dosage of radiation did not significantly exceed five roentgens."
Sheepmen satisfied with the AEC's handling of the problem:	Comment on sheepmen's attitude deleted
"The sheepmen have on several occasions expressed considerable satisfaction with the manner in which the Commission is handling the problem."	

Source: AEC, "Sheep Losses Adjacent to the Nevada Proving Grounds," November 4, 1953, NTA Accession No. NV0020417; AEC, "Report on Sheep Losses Adjacent to the Nevada Proving Grounds," January 6, 1954, NTA Accession No. NV0020422.

that they cannot "reasonably" be accounted for by radiation. This implies not only that the Cedar City sheep did not die of radiation injury but that evidence of radiation injury was negligible.

From the beginning, AEC officials could not conduct the investigation as an exclusively internal affair. By early June, there were state, county, and other federal agencies examining the herds at Cedar City.

AEC officials needed these outside entities to publicly concur in the final report to assure the ranchers that the investigation was valid.[82] The explicit concurrence of the PHS and Bureau of Animal Industry, noted in the final report, was a significant achievement for the commission but obscured disagreement behind the scenes. Few readers probably noticed the failure of the Departments of Agriculture and Health from the state of Utah to "concur" with the final report. Rather, the Department of Health "reviewed" the final report.

The draft report explicitly acknowledges that the sheep and radioactive material from shots Nancy and Harry intertwined. In contrast, the final report acknowledges that "some" fallout was detected coincident to the herds grazing on the Nevada range and at Cedar City, but the gamma levels did not "significantly" exceed 5 roentgens. Further, the draft report acknowledges the rancher's satisfaction with the commission's handling of the issue. The final report does not. It was, perhaps, too presumptuous to inform the ranchers that the lesions were not due to radioactivity and that the abnormal losses could not be accounted for by activities at the NTS and then declare their "considerable satisfaction" with the commission's investigation.

Between Pearson's draft and the final report were several revised drafts, and within this context, the "Bustad report," dated November 30, was released. Based on analysis ongoing at Hanford since 1950, it provided compelling evidence that fallout could not have killed the Utah sheep.[83] The persuasiveness of the report, however, hinged on the omission of a crucial empirical observation: ewes ingesting radioiodine gave birth to stillborn, stunted, and weak lambs.[84] This is significant, as it mirrors what occurred at Cedar City, and yet the report authored by Leo K. Bustad and colleagues only stated that "a significant reduction in birth weight of lambs occurred."[85] Further, the major conclusions of the Bustad report focused on the effects of radioiodine on adult sheep. Approximately three thousand lambs were stillborn or died soon after birth, and yet the Bustad report did not mention congruent effects among experimental sheep. It is an astonishing omission of data that are directly relevant to the central question at the heart of the investigation: the fate of the lambs. Finally, researchers had data pertinent to what occurred among the Utah herds, but Bustad and colleagues failed to articulate this in their report.

There are many aspects to the AEC's production of strategic ignorance. But the effort to redefine the lesions observed at Cedar City, to
disassociate the Utah sheep from fallout exposure while on the Nevada
range, and the omission of key data by Bustad and colleagues were key
steps paving the way for the content of the final report.

Prior to publication of the final report, Monroe Holmes wrote a
pointed critique documenting problems that remained unaddressed.
The issues Holmes identified were consistent with what the ranchers articulated in August. In a memorandum to George Spendlove, Holmes
declared, (1) there was no conclusive evidence that radiation could not
have been a contributing factor; (2) specimens obtained for further
study were obtained from sheep that may not have been representative
of those that died earlier; (3) there was no mention in the final report
of the effects of radioactivity on gestation and pregnancy; (4) sufficient
evidence had not been presented to conclude that the Cedar City sheep
were not exposed to higher levels of radiation than what the sparse
monitoring suggested; (5) the radiation experiments conducted in the
course of the investigation did not simulate conditions prevailing on
the open range; and (6) the lesions observed at Cedar City had never
occurred before, on the basis of the long-term experiences of local veterinarians and ranchers. Further, Holmes remarked, "In the body of the
report there has been no mention concerning the reason for the stunted
lambs and their deaths, and deaths of ewes during birth or shortly thereafter. The main concern of the report seems to be largely on the external
body lesions and not necessarily the over-all clinical and pathological
picture. No reference has been made of the lack of general pathological condition, although this again may be contributed to the fact that
the sheep which were examined were possibly those which had been the
least affected."[86]

The critique by Holmes did not alter the content of the final report
but probably contributed to the failure of the AEC to obtain the Utah
Department of Health's concurrence. Indeed, Spendlove, director of
the Department of Health, expressed dissatisfaction with the commission's conclusions.

The last hurdle involved presenting the final report before the ranchers. It is clear that AEC and military officials took this seriously, as the
day before, they engaged in a run-through behind closed doors.[87] AEC

officials were in a strange position. They were committed to the conclusion that fallout did not constitute even a contributing factor, and yet there was no plausible scenario in which radiation was not complicit to some degree. Infectious disease and toxic plants were ruled out quickly, and malnutrition may have been a factor but hardly explained the myriad of symptoms, including lesions and scabbing.[88]

During the January 13 meeting in Cedar City, AEC officials and representatives had a more definitive message than the previous August: it was not exactly clear what killed the sheep, but it clearly was not radiation. The questions and concerns articulated by the ranchers, nonetheless, were similar to what they articulated five months earlier—questions and concerns that AEC officials rarely responded to with a straightforward answer. As with the August meeting, AEC officials were not in Cedar City to engage in meaningful dialogue and debate. They were there to ensure that the final report stood as the established definition of the situation. Before the assembled ranchers, AEC and military officials argued that fallout and the sheep did not coincide on the Nevada range, that the peculiar lesions observed were not due to radioactivity, and that gamma radiation levels were too low to cause observable injury. Further, they suggested, it could be some new disease; levels of radiation were so small that they could not cause as much damage as a drop in temperature could; you can get more radiation from an X-ray than the sheep received; and it is important to keep in mind that radiation has many beneficial uses. At the January 13 meeting, AEC officials did not stress malnutrition, but it was suggested in the final report and in an AEC press release accompanying its publication.[89] Ultimately, the mystery of identifying what did kill the sheep was never unraveled by the AEC. If there were some new disease, it did not reappear as the herds returned to Cedar City in the spring of 1954 or in any other spring thereafter.[90]

"Kafkaesque" refers to situations in which the individual is enmeshed in an illogically complex bureaucratic maze that they cannot negotiate or evade, disorienting imperatives and procedures that seem only to exist for the perpetuation of the bureaucracy itself.[91] In Franz Kafka's estimation, there was little one could do in the face of such forces: "In the struggle between yourself and the world, hold the world's coat."[92] Rather than enlightenment and liberation, then, the progress of modernity could be distinctly unnerving. Sheep showing unmistakable signs of radiation injury yet not being affected by radiation and not demonstrating malnutrition yet probably dying of malnourishment constitute the Kafkaesque bind that many downwind residents confronted: a strange and surreal encounter with technical expertise in pursuit of the inexplicable and absurd.

Dead and dying sheep signaled the fluidity of fallout, and the ranchers were not convinced that radioactivity played no role in the die-off. "The AEC's final report was accepted in all parts of the country except in Cedar City, Utah, where a tiny band of livestock men, ignorant about radioactive fallout but knowledgeable about the behavior of sheep, were certain this announcement was not accurate," Udall notes.[93] The ranchers possessed expert knowledge not of radioactivity but of sheep, weather, sage, and death, acquired over decades, and they would not leave the issue to rest.

PART III

The Long Road to Taming the Atomic State

7

Respite, Reconfiguration, and Operation Teapot, 1954–1955

Few things embody paradox as does the Great Salt Lake. It has no outlet to the sea and is briny, salty, and stale. It lies within eyeshot of the towering symbols of Mormonism, although Mormon culture traditionally has little use for nature lacking in practical, instrumental ends. And yet here they found refuge from religious persecution in the Midwest. "Water in the desert that no one can drink. It is the liquid lie of the West," Terry Tempest Williams remarks.[1]

The AEC's outdoor proving grounds lay at the western edge of Mormon Country. The gathering of Zion emigrated to a valley framed by the Great Salt Lake, the towering Wasatch Mountains to the east, and the westward expanse of the Great Basin. This is the place, it was hoped, that would be unwanted as others ventured forth to Oregon and California. "They staked it with farms and villages, outposts that worked outward from Salt Lake City and in the course of little more than a generation occupied virtually every acre of arable ground for hundreds of miles," Wallace Stegner explains.[2] Friendly, industrious, but imbued with a cloistered worldview, Mormonism has long conflated God, country, and militarism. It is characterized by a keen awareness of those who are in and outside the fold, and yet Mormons long for recognition and acceptance. Mormon theology stipulates that America is ordained by God for the restoration of the gospel of Christ; it is the new promised land. AEC officials could hardly have found a population more willing to grant a benefit of the doubt. Compliance is paramount in Mormonism. Independent thought is not taught in Sunday school. False consciousness is not discussed in Sacrament Meetings.

Williams documents parallel changes as the Great Salt Lake rose to ever-higher levels, engulfing freshwater marshes, as her mother battled ovarian cancer, the relentless march of uncontainable progression. Born of the "clan of one-breasted women," Williams ponders the meaning of land, family, and time and wonders aloud over that which is normally kept to oneself in Mormon culture: "The price of obedience has become too high. The fear and inability to question authority that ultimately killed rural communities in Utah during atmospheric testing of atomic weapons is the same fear I saw in my mother's body. Sheep. Dead sheep. The evidence is buried. . . . Tolerating blind obedience in the name of patriotism or religion ultimately takes our lives."[3]

As militarism expanded in complexity and scale, there arose a need for sacrificial spaces. Viewed from an economic, instrumental mindset, deserts are wastelands. Traci Voyles adapts the idea of a *wasteland discourse* in underscoring marginalization enacted through "wastelanding," as a verb. Wasteland is the thing so constituted; wastelanding is the process of sustaining a particular definition of the situation with material and biophysical consequence.[4] That which the desert is taken for becomes a self-fulfilling prophecy. Slow environmental violence is the enduring consequence.

Testing in Nevada embodied the turning of colonial impulses inward due to the frictions of geographic distance. Native American settlements were at risk, but so too was Mormon Country, a population typically exempt from racialized environmental violence.[5] Rural, white, Mormon towns primarily lay in the path of the prevailing winds, and controversy surrounding atmospheric testing rests in part on the failure to uphold the conventions of colonial devaluation, whereby a demonstrable Other was at risk. The atomic-industrial complex intruded on the lives and land of scattered Mormon outposts that had intruded on Native American tribes throughout the Great Basin. All Americans were at risk, but some confronted a disproportionate burden.

Internal Review #2: An Escalation of Commitment

Fallout dispersing in an uneven manner, the influence of different radionuclides in the food chain, their impact on the human body by age and gender—all were considerations that AEC officials did not fully untangle. Commission officials professed knowledge and control, and yet detonating atop a tower was a volatile arrangement. Suspended alluvial soil raised the potential for concentrated fallout, and the vagaries of wind direction, speed, shear, and topography ensured that the aftermath was fluid. Tower shots were characterized by human-material interactive instability, and what emerged was greater risk in the downwind communities. Tower shots were necessary, however, to obtain diagnostic data and to facilitate troop maneuvers and effects testing.

Fallout spurred creation of another internal committee tasked with reviewing operational conduct. The Committee to Study Nevada Proving Grounds (Tyler Committee) was established in July 1953, composed of AEC officials and military officers, and met periodically in August and September. This was concurrent with investigation into the sheep deaths. Indeed, the Tyler Committee report was released just weeks after the AEC announced that the sheep deaths could not be accounted for by radioactivity. Although the Tyler Committee suggested more thorough pre-shot and post-shot surveys of livestock grazing offsite, there is little evidence that the committee made the connection between sheep and people in formulating its assessment and recommendations.[6]

Nonetheless, Harry rattled nerves within the AEC, and the Tyler Committee cogently declared in a classified internal report what the commission would not admit publicly: predictions regarding fallout trajectory and intensity had been "insufficiently accurate." On average, wind direction deviated from prediction by thirty degrees and wind speed an average of fifteen knots.[7] A study appended to the Tyler Committee report by Colonel B. G. Holzman and Lester Machta of the US Weather Bureau noted that the "average directional error" was forty-five degrees at ten thousand feet due to "terrain effects."[8] The committee recommended that the final weather briefing be issued just one hour before detonation to improve post-shot prediction.[9]

In recounting the rationale for establishing the NTS, the report asserted, "The consensus of official opinion was that there was adequate—not absolute—assurance of human and property safety, and that the degree of risk involved was acceptable in view of the national necessity." The Tyler Committee recognized that testing in Nevada saved time and money amid a looming Soviet threat and thus that the public must accept some degree of risk. "Adequate" was imprecise, however. "Acceptable" was vague, as public opinion was never solicited before the creation of a continental test site. But the report also conceded, "The consensus was that Upshot-Knothole exposure levels in the more extreme cases were above the levels the public should be expected to accept."[10]

The committee unanimously recommended continued testing in Nevada and underscored the economic investment that would be "largely lost" if an alternative site were deemed necessary. Investment was estimated at around $20 million.[11] The report outlined recommendations for improving operations, and many were adopted with Operation Teapot. One was to raise the height of shot towers from three hundred to five hundred feet. Beginning with Operation Plumbbob in 1957, AEC officials experimented with seven-hundred-foot towers and lofting devices skyward by balloon, tethered to the ground.

The Tyler Committee argued that the AEC had not made a concerted effort to inform downwind residents about test dates and times, fallout direction and intensity, or practical steps they could take to minimize exposure to radioactivity. It was clear, too, that the expectation of unreasoning public opinion was one reason why. Preventative measures were not draconian. "Exposure may be materially reduced by remaining inside structures or vehicles, or by removing dust particles through changing clothes, bathing, combing the hair, shaking, etc.," the report counseled.[12] Withholding information was based on the conviction that the public could not be trusted to respond in a deliberative, even-handed manner. But lack of information ensured that downwind residents were not prepared to respond to fallout in a prompt and effective manner. Failure to inform became an inability to respond, which became another incentive for the suppression of information.

The Momentum of Systemic Imperatives

Nevada was a downstream node in a larger system spread across the country, and what happened in the desert was influenced by systemic needs and objectives upstream. "Systems nestle hierarchically like a Russian Easter egg into a pattern of system and subsystems," Thomas Hughes explains.[13] While other parts of the system were embroiled with the materiality of nature within industrial production facilities and in the laboratory, large-scale provocation occurred in Nevada and the Pacific Proving Grounds. And although the pace of testing and average yield of devices significantly expanded over time, the Tyler Committee stressed that Los Alamos officials were frustrated: "LASL does not yet feel that the rate of testing is as rapid as the generation of new ideas would really warrant."[14]

The Tyler Committee suggested that increasing complexity at the NTS demonstrated the efficient use of fissionable material. For each shot, the more experiments and exercises the more achieved, given the expenditures involved; and the more detonations per series, the lower the monetary cost per shot. The committee thus did not envision the pace and competing uses of the NTS as necessarily complicating management of fallout. Nonetheless, the committee recommended "a definition and interpretation of the full measure of the radiation hazard (or exposure) which the publics in various geographic regions are asked to accept, or may experience in any degree of emergency." It was a notable recommendation, but even the Tyler Committee did not have a comprehensive conception of the hazards. The report reasoned that radiation exposure is due to five mechanisms: passage of the cloud overhead, ground-level fallout, inhalation of aerosol, fallout on the skin, and fallout in public water supplies. There was no mention of pasture, cows, and fresh milk. There was no discussion of the differing susceptibility of women and children to gamma radiation or internal emitters such as strontium-90 or radioiodine. The Tyler Committee declared, "There has been no instance of injury or death of people, and those most intimately aware of radiation's biological effects have concluded that there is no reason to assume that fallout has at any time really endangered people in the sense of immediate or of lingering effects."[15]

In the appendix to the Tyler Committee's report were separate studies forming the basis for its assessment and recommendations, and one provided a more nuanced view than articulated in the preceding quote. Captain Howard L. Andrews of the National Institute of Health conceded that the permissible exposure standard did not include a "very large factor of safety":

> With radiation (except for beta ray burns) clinical evidence of moderate overexposure will be delayed for many years. Increased incidence of radiation-induced cancer may require 10–15 years to become evident, and several generations will be needed before adverse genetic effects appear. . . . A second disturbing consideration arises from the fact that rapidly growing tissue is more radiation sensitive than slower

metabolizing tissue. The permissible figure of 0.3 r/week was established for adult workers. The amount by which this value should be reduced for children is unknown but a factor of 10 has been suggested. Because of these considerations it seems necessary to adopt a rather conservative attitude toward the involuntary exposure of general populations. An error on the radical side will not be immediately apparent but *the chickens will inevitably come home to roost at some later date.*[16]

Reduction by a factor of ten would have put the AEC out of business in Nevada. In important respects, AEC and military officials opted to let the chickens come home to roost at a later point in time.

The problems encountered during Upshot-Knothole did elicit reflection and operational changes. Rigidities in belief and perception were indeed called into question. Altered definitions of the situation and operational changes, however, were not as substantive as they could have been. But neither were AEC and military officials unaware of their obligation to the public. As Hughes notes, "As systems mature, they develop style and momentum."[17] "Style" refers to organizational culture and patterned ways of accomplishing particular tasks. "Momentum" is born of prior investment of time, money, and resources toward a particular course of action. Testing in Nevada exhibited momentum, and it was easier and cheaper to make operational changes than to pull up stakes and start over elsewhere. The commission would not abandon testing in Nevada.

Castle Bravo as a Frame Break

There was no testing in Nevada in 1954, but the Bravo detonation in March of that year at Bikini Atoll in the Marshall Islands resonated in unexpected ways throughout US society. Indeed, it constituted a *frame break*, or shift of interpretive understanding.[18] It was the first of six detonations in the Castle series, which totaled over forty-eight megatons (forty-eight thousand kilotons) of explosive yield.[19] Bravo was a thermonuclear or fusion device with a fission trigger. Its anticipated yield was six megatons, but it approached fifteen megatons while coinciding with a shift in the prevailing winds, a yield one thousand times larger than the bomb dropped over Hiroshima. It pulled up coral reefs and sand, forming a cloud of debris engulfing atolls eastward while ascending to

114,000 feet. The base of the mushroom cloud hovered at 60,000 feet.[20] Twenty-eight American servicemen and 236 Marshallese were exposed to gamma radiation and internal emitters as long-lived radionuclides dispersed downwind.[21] A Japanese fishing trawler named *Fukuryu Maru,* or "Lucky Dragon," was eighty to ninety miles east of Bikini and outside the commission's demarcated hazard zone when white ash began to fall. Two weeks hence, the twenty-three sailors returned to port suffering from radiation poisoning. All were hospitalized, and one died.

The Japanese government sponsored an expedition to measure radioactivity in the waters off the Marshall Islands and detected surprising levels of radioactivity, and irradiated fish caught far from Bikini appeared at Japanese ports and markets.[22] Bravo contaminated approximately twenty thousand square miles while injecting radioactive debris into the stratosphere.[23] "Slowly it would become evident that, while this weapon had been tested in the Marshall Islands, its detonation was a global event," Robert Jacobs argues.[24]

Bravo was the largest US atmospheric detonation ever conducted, and the unanticipated yield, in tandem with a shift in the prevailing winds, put the AEC on the defensive. And from the start, officials worked to downplay the incident and cast doubt on the extent of the effects. Their statements appeared inconsistent with the empirical data.[25] Largely because of Bravo, thermonuclear testing elicited increasing public scrutiny as the 1950s wore on, due to long-lived radionuclides with half-lives measured in decades. Of these, strontium-90 was the most feared. With the advent of atomic military technology, and thermonuclear devices in particular, science and technology closed the distance between the state and the individual. The technology was a hazard not only in war but in peacetime, and after Bravo, there were questions whether the AEC, as the institutional body with oversight on behalf of the American people, was fulfilling its fiduciary duty.[26] Ralph Lapp insisted at the time, "The Atomic Energy Commission, like Macbeth, is being haunted by the ghost of things which will not die. Bomb makers have learned how to convert mass into a fearful output of energy. The have developed the fine art of splitting and fusing atoms, but as if in protest the split atom dies slowly. The ghosts of fission, that is the split atoms of uranium, must seem like veritable specters to AEC officials."[27]

Radionuclides served as tracers, illustrating connections at a local, regional, and global scale, and Bravo spurred the AEC and the armed services to increase funding in the earth and life sciences.[28] "Like radiation medicine administered to a patient to make the internal system visible to doctors, the movement of radionuclides through the ecosystem revealed a systemic interconnectedness that had been previously invisible," Jacobs explains.[29]

Contemporary representations of the complexity of the planet are derived from Cold War military planning and serve as a baseline from which change is measured. The physics and engineering of increasingly destructive devices and their more efficient delivery was paramount during the Cold War, but so too was atmospheric, geomorphological, and oceanic research, as it was anticipated that conflict with the Soviet Union would be a total war, not simply with regard to civilian casualties but unlimited in scope and strategy. Nothing would be off-limits— including "environmental warfare." This encompasses aerial dispersal of radioactive wastes over enemy territory, production of biological and chemical weapons, employment of pests designed to destroy an enemy's food supply, and defoliants such as Agent Orange.[30] "Without nuclear weapons tests, much less would be known about the atmosphere than what is understood today," Paul Edwards observes.[31]

Facts, models, and data do not speak for themselves but must be assembled and presented to the public, the scientific community, and people in positions of political power, but even as the AEC and the armed services funded innovative research, officials typically downplayed new knowledge when it called into question public assurances of minimal threat from weapons testing. Over the course of the 1950s, AEC officials frequently were beset by discursive momentum, in which prior statements were contradicted by subsequent empirical observation. When the empirical data suggest that socionatural entanglement is not how it has been depicted, then prevailing discursive constructions must be amended or rhetorical strategies employed to cast doubt on the data or interpretation. Often, AEC officials chose the latter.

"You Are in a Very Real Sense, Active Participants in the Nation's Atomic Test Program"

Upshot-Knothole and Castle Bravo were the impetus for an extensive public-relations campaign prior to Operation Teapot. And as testing resumed in the spring of 1955, the AEC distributed a revised pamphlet accentuating some aspects of testing while concealing dimensions that might evoke concern. "Atomic Test Effects in the Nevada Test Site Region" began with a personal appeal from NTS test manager James E. Reeves, illustrating the paternalistic role adopted by AEC officials: "You are in a very real sense active participants in the nation's atomic test program. You have been close observers of the tests which have contributed greatly to building the defenses of our own country and of the free world. . . . Some of you have been inconvenienced by our operations. At times some of you have been exposed to potential risk from flash, blast, or fall-out. You have accepted the inconvenience or the risk without fuss, without alarm, and without panic. Your coop- eration has helped achieve an unusual record of safety." Reeves further counseled, "In a world in which free people have no atomic monopoly, we must keep our atomic strength at peak level. Time is a key factor in this task and Nevada tests help us 'buy' precious time."[32] Time scarcity is invoked not simply to expedite accomplishment of a given objec- tive but as a mechanism of social control.[33] When time is construed as scarce, critical reflection and inquiry appear unreasonable. That time is of the essence was a refrain often reiterated in AEC public-relations materials and was not necessarily an overstatement. It does not explain the shortcomings in AEC operational conduct, however. There was lit- tle reason to infer from the reaction of downwind residents that a more open posture would jeopardize continental testing. To assert the imme- diacy of a problem or task, moreover, is distinct from actively stoking fear in the public. Over the course of the 1950s, the AEC crossed this line repeatedly, as it employed the Soviet threat as a rhetorical tactic to gain political leverage, increase defense spending, and expand the military-industrial complex.[34]

The 1955 pamphlet was more informal in tone and less technical in its overview of fallout and the means employed to measure radioactiv- ity than the 1953 version was. Indeed, it was less informative and more casual than the version published in 1957, as well. And despite the con- clusion articulated in the Tyler Committee report, asserting that AEC officials should issue "prompt, frank and comprehensive warnings and reports," the revised pamphlet contained only a superficial discussion in this regard: "There will, however, always remain a possibility of offsite effects from flash, blast, and radioactive fall-out. The potential exposure of the public will be low and it can be reduced still further by continued public cooperation."[35] Cooperation can mean different things, but here it refers to quiescence. Agitation and concern only jeopardize public safety as well as national security. The downwinders are "active participants" whose primary role entails inactivity. "Your best action is not to be worried about fall-out."[36]

The Tyler Committee asserted that the public must accept some degree of risk due to the demands of the arms race, but the 1955 pamphlet steadfastly asserted that the risk was minimal. Further, the committee criticized the AEC's reticence to encourage downwind residents to engage in simple preventative measures to minimize exposure.[37] The 1955 pamphlet addressed such concerns but in a manner that fell far short of the committee's recommendation, which advised the AEC to release information in a timely manner, allowing downwind residents to exercise some degree of discretion over their exposure. The 1955 pamphlet advised, "If you are in an area exposed to fall-out, you will be so advised by our radiation monitors. . . . If you have been outdoors during the fall-out you may be advised to bathe, wash your hair, dust your clothes, shake off your shoes, etc."[38]

And while the 1953 pamphlet included photos of lab technicians engaging in sophisticated procedures, the version distributed in tandem with Operation Teapot and the Plumbbob series of 1957 included "Disney-like" drawings that risked trivializing fallout.[39] Figure 7.1 is a typical illustration. It depicts a cowboy on a horse watching an ascending mushroom cloud. Neither appear particularly alarmed.

Many people residing downwind owned Geiger counters, which they used routinely to locate uranium deposits, but fallout made prospecting problematic, as radioactivity dispersed across the landscape. The AEC struggled to preempt such visceral, material effects. Figure 7.2 shows an alarmed prospector whose Geiger counter is registering radioactivity. The 1955 pamphlet counseled, "We can expect many reports that 'Geiger counters were going crazy here today.' Reports like this may worry people unnecessarily. Don't let them bother you. . . . If the fall-out is heavy enough to be of any significance, our monitors will be in the area and will tell you what is happening."[40]

Figure 7.1. Cowboy downwind of a detonation. (AEC, "Atomic Test Effects in the Nevada Test Site Region," March 7, 1955)

Dirty Harry on Film

The AEC's public-relations effort in preparation for Operation Teapot included a movie, twenty-five minutes in length, titled *Atomic Tests in Nevada: The Story of AEC's Continental Proving Ground*.[41] Many scenes were shot in St. George with residents, and the film premiered there in the spring of 1955. The film begins with a milkman, gas station attendant, and police officer performing their respective duties in the predawn darkness. The town is quiet. There are few cars on the street. Suddenly, the sky is aglow. The milkman, gas station attendant, and police officer do not appear fazed in the least. The narrator interrupts: "That great flash in the western sky! An atomic bomb at the Nevada Test Site 140 miles to the west. But it's old stuff to St. George. Routine. They've seen a lot of them ever since 1951. Nothing to get excited about anymore."

The film depicts a radio bulletin announcing a shelter-in-place directive due to Harry. Mothers and young children are at home, a homemaker

Figure 7.2. A bewildered prospector detecting radioactive fallout. (AEC, "Atomic Test Effects in the Nevada Test Site Region," March 7, 1955)

is peeling potatoes, and men are at work when they turn their attention to a nearby radio:

Ladies and gentlemen, we interrupt this program to bring you important news. Word has just been received from the Atomic Energy Commission that due to a change in wind direction, the residue from this morning's atomic detonation is drifting in the direction of St. George. It is suggested that everyone remain indoors for one hour or until further notice. There is no danger. This is simply routine Atomic Energy Commission safety procedure. To prevent unnecessary exposure to radiation, it is better to take cover during this period. Parents need not be alarmed about their children at school. No recesses outdoors will be permitted. Please stay indoors and advise your friends and neighbors who may not hear this announcement to do likewise.

Focus turns to a discussion of testing in Nevada as a matter of "national survival." Newspaper headlines announcing Soviet advances in atomic weapons development fill the screen. "We have no choice," the narrator intones. "To ensure our defense, we have to keep our atomic strength at top level by testing new ideas and principles and applying these principles in new weapons." The film cuts to scenes of the NTS: "set in some of the loneliest acres the world has ever seen." Scientists and technicians are shown working with sophisticated devices; military officers engage in spirited debate around a table strewn with reports; the atolls of the Marshall Islands are portrayed as the narrator explains that high-yield devices are tested there and low-yield shots in Nevada. The logistical challenges, time, and money involved in testing five thousand miles from the West Coast are stressed, while, the narrator cautions, "atomic progress waits."

Problems that cannot be addressed in the laboratory, the narrator advises, await the "outdoor workshop" in Nevada. There are scenes of diagnostic and experimental devices encircling ground zero, troops digging foxholes, a house, a concrete and steel bridge, and cars of various makes and models standing ready for effects testing to aid civil defense preparations. The work of planning for and conducting an open-air detonation is highlighted. Towers are built, electrical lines are strewn over the terrain, weather maps are sketched, and ascending mushroom clouds are viewed from a distance. "Without a continental test site, we would now find ourselves years behind our present atomic development. This is why the Nevada Test Site is absolutely essential as a backyard workshop," the narrator clarifies.

An overview of natural, background radiation reveals that it is not a new phenomenon, and the question of interference with the weather is scrutinized. "Following extensive studies, no relationship between atomic detonations and weather has been established. An atomic bomb is puny compared to the forces of nature and is completely lost in the vast oceans of the sky," the narrator assures.

The film returns to a reenactment of the morning Harry swept through southern Utah. "What, then, was the story at St. George? Here are the facts." The shelter-in-place directive is depicted as prompt, and the reaction of residents is orderly and relaxed. Everyone took cover in time to avoid fallout. The film describes events that did not happen for an audience that did not enact them to ease their concerns over fallout, which—as portrayed in the film—they scarcely possess. The narrator explains, "The word was spread throughout the community, and the citizens calmly took cover. One hour and fifty minutes later, at 11:25 a.m., the all-clear announcement was broadcast. Actually, when the invisible cloud had passed, the total amount of radiation deposited on St. George was far from hazardous. Then, you may ask, why were the people asked to stay indoors? A very simple reason: the Atomic Energy Commission doesn't take chances on safety."

Operation Teapot, 1955

Preparing for Operation Teapot, the DOD scheduled a ten-kiloton surface detonation, but AEC officials rejected the proposal due to the risk of heavy on- and offsite fallout.[42] They did not want a repeat of shot Harry, and Operation Teapot included both a stepped-up public-relations campaign and a renewed emphasis on minimizing offsite fallout. However, these efforts ran in tandem with an increasing sense among many AEC and military leaders that the public must accept a degree of risk for the US to keep pace with the Soviets.

Beginning with Teapot, the PHS assumed primary responsibility for offsite radiological monitoring, under the direction of the AEC. During Upshot-Knothole, only twelve offsite monitors were involved; for Teapot, the number was increased to sixty-six, and for the first time, they lived in communities downwind for the duration of the series.[43] Residents, like radioactive fallout, were something to be managed and contained.[44] Monitors were instructed to integrate themselves into the community and respond to concerns in an "informal and down-to-earth basis." PHS personnel arrived seven to ten days before the opening detonation, in February, to allow for a "continuous educational program." The monitor in Glendale, Nevada, taught Sunday school, while the monitor in Alamo, Nevada, plastered the ceiling of a hotel room. They were advised to sit on the boards of civic organizations, speak at public schools, and reach out to local leaders who could then spread the commission's message to others. "Every opportunity to reach the public through talks and film showings was accepted."[45] Moreover, prominent residents were invited onsite to witness a detonation. This included mayors, town council members, county commissioners, school principals, local editors and journalists, science teachers, and even Mormon bishops presiding over local wards.[46] The commission also invited physicians from southern Nevada and Utah to a conference in St. George to discuss the public health aspects of the upcoming test series.[47]

To reduce offsite fallout, the commission experimented with soil stabilization techniques around towers, such as oil or water to reduce updraft, and final weather briefings one hour from shot time to keep abreast of shifting winds. Consistent with recommendations of the National Committee on Radiation Protection, the offsite threshold was changed from 3.9 roentgens of gamma radiation in a thirteen-week period to 3.9 in a year, and guidelines detailing when residents would be evacuated or asked to shelter in place were formulated.[48]

The effort was well scripted, but as Teapot began, the AEC commissioners in Washington, DC, learned of a letter from Senator Clinton P. Anderson, a Democrat from New Mexico and chairman of the JCAE, to AEC chairman Lewis Strauss. It caused a stir. Anderson was born in South Dakota, but suffering from tuberculosis at just twenty-two years old, he moved to Albuquerque, New Mexico, for treatment, his doctor advising that he had but six months to live. He served in the US House of Representatives, was secretary of agriculture in the Truman administration, and he was a US senator from 1948 to 1972. He sought a position on the JCAE because of the AEC's influence within his home state of New Mexico.[49]

In the letter to Strauss, Anderson pondered whether only "very small yield devices" were suitable for testing in Nevada. Anderson had been at the NTS in 1953 to observe a detonation that had been repeatedly rescheduled due to the weather, and now he was questioning whether continental testing was economically feasible and expedient. The idea was greeted with trepidation by the AEC commissioners at their meeting on February 23, 1955, just days after the start of Operation Teapot. They feared that testing in Nevada was, once again, in jeopardy:

COMMISSIONER [WILLARD] LIBBY: I am pretty disturbed by this.
I noted, Lewis, that you had cooled off about the Nevada Site.

CHAIRMAN [LEWIS] STRAUSS: I had been cool long before. My cool-
ness started in the spring of 1953, but I have never discussed this with
Anderson. This is spontaneous.

COMMISSIONER LIBBY: I think this will set the weapons program
back a lot to go to the Pacific.[50]

What Anderson did not know was that several AEC commissioners
were having their own doubts after the Upshot-Knothole series. Nonetheless, further discussion centered on the conviction that testing in
Nevada could not be abandoned anytime soon. It was essential to the
weapons development program. Americans must, of necessity, bear the
burden:

COMMISSIONER LIBBY: People have got to learn to live with the facts
of life, and part of the facts of life are fallout.

CHAIRMAN STRAUSS: It is certainly all right they say if you don't live
next door to it.

GENERAL KENNETH D. NICHOLS: Or live under it.[51]

Examining a chart, the commissioners discussed the direction fallout
must travel to bypass towns downwind and the difficulties that shifts in
the prevailing winds created for ensuring that it did, indeed, do so. And
the issue of hot spots was raised:

CHAIRMAN STRAUSS: South of those two places is a very narrow
corridor where if the wind shifts ten degrees in either direction, then
they are in trouble again. Of course, they really never paid much
attention to that before.

GENERAL NICHOLS: Those charts are the same as when they explained them to the Commission at the time we went over the
criteria.

CHAIRMAN STRAUSS: At that time they did not have on them what
they describe as hot spots. What are they, Bill?

COMMISSIONER LIBBY: A fluke of the weather or winds causing local
precipitation.[52]

A weather anomaly causing "local precipitation" is not the definition of a hot spot. Four years after testing began on the continent, the AEC commissioners appeared at a loss to identify the most crucial dimensions of open-air detonations. Moreover, Lewis Strauss seemed to suggest that prior to the Upshot-Knothole series, the Test Organization in Nevada devoted little attention to what shot Harry would force them to consider: a direct hit on a town downwind.

After a two-year respite, detonations resumed in February 1955 and continued until mid-May. As shown in table 7.1, Teapot consisted of three airdrops, a subsurface burst, and ten tower detonations, the largest codenamed Turk. All the devices tested during the Teapot series, except Turk, were tactical weapons designed for various battlefield tasks. The first detonation, Wasp, was a one-kiloton airdrop detonated at eight hundred feet above the desert. It was followed by Moth, a two-kiloton tower shot and a prototype of an atomic missile warhead; Tesla was next and was a seven-kiloton tower shot.[53] Despite the relatively low yield, Tesla produced a hot spot in an unpopulated area northeast of the test site.[54] And then came Turk. It was a robust forty-three-kiloton shot and the first atop a five-hundred-foot tower, to reduce the updraft of alluvial soil. "It would stand above the desert only 55 feet shorter than the Washington Monument," Richard Miller explains. For the first time, moreover, ground zero was coated with asphalt. As Turk migrated offsite, winds moving in different directions at variant altitudes twisted debris in spirals. "The cloud was spreading in every direction except south," Miller observes.[55] A 30 roentgen reading was detected beyond the test site boundaries but in an unpopulated area.[56]

Hornet was next. It was a four-kiloton tower shot, and the device was small, the size of a suitcase. Bee followed and consisted of an eight-kiloton tower detonation that unexpectedly moved southeast toward Las Vegas. It was the first time fallout moved over the city. Gamma radiation was low but the population high, around fifty thousand residents. Even though Turk was a "monster," Bee produced the highest population exposure readings of the series.[57] It exemplified the fluidity of fallout and the fact that explosive yield was crucial but not the only factor at play. And these other factors were maddeningly unpredictable. Shot ESS was next, a one-kiloton subsurface shot buried sixty-seven feet deep. It was

TABLE 7.1. Operation Teapot Detonations

Code name	Type	Yield (kilotons)
Wasp	airdrop	1
Moth	tower	2
Tesla	tower	7
Turk	tower	43
Hornet	tower	4
Bee	tower	8
ESS	crater	1
Apple-1	tower	14
Wasp Prime	airdrop	3
HA (high altitude)	airdrop	3
Post	tower	2
MET (military effects test)	tower	22
Apple-2	tower	29
Zucchini	tower	28
Total yield		166

Source: DOE, *United States Nuclear Tests: July 1945 through September 1992* (Las Vegas: Nevada Operations Office, 2000).

another experiment in miniaturization and about the size of a suitcase and thus easily transportable.

Apple-1, a fourteen-kiloton tower shot, and Wasp Prime, a three-kiloton airdrop, were both conducted on March 29. It was the first time multiple shots were fired on the same day and demonstrated that over the preceding two years the weapons laboratories had accumulated a backlog of experiments they wanted to test. Apple-1 generated fallout over Alamo, Nevada, but within the established threshold. The next shot, HA (high altitude), was a three-kiloton device detonated five miles above the desert and the first atomic antiaircraft experiment conducted in Nevada. Post soon followed, a two-kiloton tower shot that lingered. "Before it slowly drifted to the east, Post stayed around long enough to drop a significant amount of fallout over the entire test site," Miller notes.[58] MET (military effects test), a twenty-two-kiloton shot atop a four-hundred-foot tower, was accompanied by a variety of experiments, but perhaps the most notable was assessment of radioactivity on the uniforms worn by Chinese and Russian soldiers.[59] MET generated remarkably high gamma readings north of the test but in an unpopulated area, and it then hovered over the continent for an unusually long period, as long as twenty days.[60]

Apple-2 on May 5 was the most extravagant shot of the series, and it was accompanied by the largest civil defense program ever conducted in Nevada. It was an open shot, and the media and members of the public were in attendance. Operation Cue entailed construction of homes at differing distances from ground zero, complete with food, appliances, and mannequins dressed in J. C. Penney clothing. It was named "Survival Town, U.S.A.," and its destruction was televised live to a national audience. As with Operation Doorstep in 1953, it was an opportunity for the FCDA to conduct its own menagerie of experiments and instill a healthy dose of nuclear fear in Americans. The goal was to assess the prospects of life after a nuclear exchange and motivate the public to take civil defense seriously without invoking dread. It depicted the destruction of white, middle-class America, acted out amid the starkness of the Nevada desert. Tanfer Emin Tunc suggests, "Survival town simultaneously symbolizes many aspects of the era at once: the family as the literal and figurative battleground of the Cold War, the nuclear anxiety that permeated every facet of American life, atomic style and kitsch and, above all, consumer culture."[61]

Figure 7.3 depicts a mannequin mother and children at home just before an atomic detonation. "Conceptually, the argument was that at any moment of the day—while enjoying one's breakfast for example—the bomb could drop," Joseph Masco observes.[62]

One experiment placed file cabinets full of documents at differing distances from ground zero, another placed bottled and canned beer and soft drinks, and another involved strategic placement of travel trailers in the shadow of ground zero; amidst a nuclear exchange, some survivors would want to hit the road.[63] The file cabinets survived with their contents intact, as did most of the beer and soda. Beverages closest to ground zero were not overly radioactive but had a funny taste. "Immediate taste tests indicated that the beverages, both beer and soft drinks, were still of commercial quality, although there was evidence of a slight flavor change in some of the products exposed at 1270 ft from GZ [ground zero]," research indicated.[64]

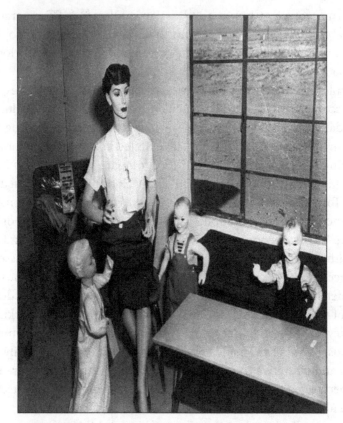

Figure 7.3. Mannequin mother and children awaiting an atomic detonation during Operation Doorstep in March 1953. (Courtesy of the National Archives and Records Administration, College Park, MD)

Food was a focus of Operation Cue, but so too was a psychological and social reprogramming of Americans amid spiraling militarism and the threat of nuclear holocaust. In fostering a new state-citizen contract, moreover, wives and mothers were a key focus.[65] As depicted in Operation Cue, they bore primary responsibility in preparing the family and community for life in a postapocalyptic landscape.

Operation Teapot ended with Zucchini, a twenty-eight-kiloton tower shot that, as with Bee, generated a comparatively high collective population exposure estimate. Nonetheless, Teapot was deemed a success in regard to minimizing offsite fallout, particularly compared to Operation

Upshot-Knothole. And yet collective estimated population exposure to gamma radiation was notably higher during Teapot than during the Tumbler-Snapper series of 1952.[66]

An AEC report lamented that public-relations efforts during Teapot garnered considerable goodwill, with one notable exception: a newspaper editor in Tonopah, Nevada, who "maintained a highly critical attitude towards test activities."[67] Robert A. Crandall was the editor of the *Tonopah Times-Bonanza*. In contrast to other local newspapers in the downwind communities, Crandall was a frequent critic of the AEC, and his weekly columns often earned him a phone call or visit from AEC and military officials.[68] Pointed editorials in the *Deseret News*, out of Salt Lake City, also arose during Operation Teapot. In an editorial titled "Don't Hurry Back," the *Deseret News* commented, "But if the AEC decides to conduct subsequent tests in the ocean or Antarctica or somewhere else, they'll find no complaints here."[69] The AEC made significant operational changes in 1955, but the memory of Dirty Harry echoed downwind.

8

Bulloch et al. v. United States

Deception and Dirty Science, 1956

In January 1954, the AEC released its report concluding that radioactivity played no role in the mysterious loss of over four thousand sheep. The report did not offer a definitive alternative explanation but promised that research would continue. In this regard, the commission funded a multi-year project titled "The Effect of the Level of Nutrition on the Pathology and Productivity of Range Sheep."[1] It was coordinated through Utah State University, and it focused on malnutrition. The ranchers displayed a lack of interest in the study.[2] Indeed, there was a degree of absurdity in funding a research program to shed light on the symptoms displayed at Cedar City but refusing to examine radiological injury.[3] If the project was intended to demonstrate the commission's goodwill and integrity, it probably promoted neither.[4]

A court of law is an arena where truth claims are presented, counter-arguments and evidence are offered, and a determination is rendered as to the definition of the situation that is accorded institutional support. The plaintiff asserts they have been harmed by the defendant and seeks to rectify the damage inflicted. The procedures dictating the conduct of participants and rules of evidence lay the foundation for adjudicating between the veracity of competing truth claims, but this rests on the integrity of all parties involved.

When AEC and military officials presented their conclusion to the Cedar City ranchers, they discouraged them from filing lawsuits.[5] It did not work. In February 1955, just as testing resumed in Nevada, lawsuits were filed in federal district court in Salt Lake City, and *Bulloch et al. v. United States* was the test case.[6] The ranchers sought financial compensation for their losses, but they had to show that the government could be sued in this instance and, if so, to establish by a preponderance of the evidence that fallout was the sole or a contributing factor.

Preparing for trial, the AEC had a problem: an astute judge would surely detect the logical inconsistencies necessary to conclude that radioactivity played no role in the sheep deaths of 1953. *Bulloch*, in turn, necessitated an even more substantial production of strategic ignorance. Moreover, it would not be possible to convince a federal judge of the validity of the investigation without sacrificing the public standing of key participants. In particular, the problem was the dissenting opinions of Robert Thompsett of Los Alamos, Major Robert Veenstra of the Naval Radiological Defense Laboratory in California, and Monroe Holmes of the Public Health Service. All three veterinarians identified radioactive injury in the Utah sheep, and all three had to be pressured to conform, once and for all, with the commission's self-serving narrative before the trial began.

"The first move the AEC made was to take soundings among the offsite population," Philip Fradkin explains.[7] Major Grant Kuhn visited ranches along an easterly arc from Tonopah, Nevada, to Cedar City, Utah, between March 9 and 22, 1955. Kuhn was a veterinarian attached to the AEC, and he documented lesions on sheep that had not yet fully healed over the preceding two years, as well as lingering evidence of beta burns on horses.[8] Holmes also uncovered similar ailments in the spring and summer of 1955.[9] Whatever plagued the herds left lasting evidence of its occurrence.

From March 31 to April 20, 1955, Kuhn embarked on another tour downwind alongside US Army Lieutenant Colonel Bernard Trum, who was serving as the technical adviser to the government attorneys preparing for *Bulloch*. They found evidence of lingering effects on sheep as well as distrust among ranchers who felt they had been lied to by the AEC.[10] Trum's primary concern, however, was not the ranchers but the veterinarians who had participated in the investigation and might be deposed or called to testify.[11] In correspondence dated March 25, 1955, Trum wrote to Veenstra and pushed him to defend any doubts he might have about the commission's official version of events. Trum inquired, "I got the distinct feeling that you felt that there was a chance that radiation could have caused the death of some of the sheep at least." Trum then outlined recent research at Oak Ridge, the facility with which he was affiliated, seemingly casting doubt on the radiological origin of the injuries displayed by the Utah sheep. Trum quizzed the veterinarian, "I've been wondering if you might not have changed your mind about these things? If you haven't changed your mind, I'd like to know what you are basing your opinion on, for I shouldn't like to go into this thing divided within our own Corps if we can avoid it."[12]

Before Veenstra could respond to Trum's correspondence of March 25, Trum and Kuhn visited him at his laboratory in San Francisco on April 2. Veenstra indicated he was of the opinion that radioactivity was a contributing factor and that he was not convinced research conducted by Clarence Lushbaugh at Los Alamos, examining beta burns on skin, and Leo K. Bustad at Hanford, analyzing radioiodine accumulation, proved otherwise. Trum noted in his subsequent report a concern expressed by Veenstra that, in retrospect, was a crucial consideration: "*The accumulated effects of the total fission product spectrum was not adequately considered.*" Fifteen months after the AEC issued its final report, one of the principal experts involved in the investigation still would not concede to the tidy conclusion presented to the public as unassailable fact. But Trum's report suggested that Veenstra expressed a willingness to reevaluate his position, given ongoing research conducted by the commission.[13]

After the meeting with Kuhn and Trum on April 2, Veenstra wrote a telling letter, dated April 7 and addressed to Trum—*a letter that was never sent.* He reiterated his concern that Bustad's experiments only examined thyroid damage when this was probably not the only radiation-induced injury. Moreover, the experiments were performed on healthy sheep, whereas those that died were subject to the stresses and strains of the range. In reference to Lushbaugh's experiments, Veenstra questioned the relevance of studying experimentally induced beta burns with regard to the Utah sheep: "We feel that our position has not been materially changed: basically we are still of the opinion that radiation could have contributed to the deaths of the animals. Although it is of course difficult to decide what degree of damage would have followed from a given exposure of this type, we are aware of the additional studies that have been carried out and we are convinced of their validity, but not completely of their relevance."[14]

In a strange turn of events—also on April 7, 1955—Veenstra forwarded a handwritten note to Trum indicating that, if called on, he would downplay the idea of radiological injury.[15]

On April 12, Trum and Kuhn met with George Spendlove and Monroe Holmes in Salt Lake City. According to Trum's report, despite Holmes's identification of radioactivity in his prior reports, he indicated that he had "not taken a stand." Moreover, Trum remarked, "He will disqualify himself as an expert witness in radiation and therefore will not attempt to evaluate the role of radiation in sheep losses in 1953. He will merely state what he saw and that sheep died of causes unknown."[16]

"Trum then went to work on Thompsett," Fradkin observes. The Los Alamos veterinarian was in a vulnerable position because his small animal hospital relied on contracts with the AEC.[17] Trum sent Thompsett a prewritten "model letter" aligned with the commission's final report but at odds with Thompsett's insistence that fallout was a contributing factor.[18] Thompsett signed the letter.

Trum's efforts to persuade Veenstra, Holmes, and Thompsett to fall in line with the AEC narrative was remarkably successful. Veenstra and Holmes agreed to disqualify themselves as expert witnesses, and Thompsett abandoned his position that fallout was a contributing factor. Moreover, Trum came up with another explanation to add to the commission's list of improbable alternatives to radioactivity: cold weather. May 1953 registered below-average temperatures in the Cedar City area, and he hypothesized that this may have contributed to the deaths of the lambs.[19]

Trum authored a report delivered to Llewellyn O. Thomas, a Justice Department attorney in Salt Lake City, dated May 12, 1955, reporting the good news: Thompsett and Veenstra had the opportunity to "reevaluate the situation." Trum declared, "They will definitely state that they do not have evidence that radiation injury *was* either the cause or contributed to the death of the sheep." That this conclusion was illogical was perhaps lost amid the willingness to believe in a preferred institutional narrative. "Therefore, contrary to rumored opinion, the syndrome and deaths of sheep can be accounted for without introducing a radiation factor."[20]

Trum assured Thomas that the sheepmen were not "officially warned" of danger while on the range in 1953, "since there was no danger." He noted that the slippage of wool was common among sheep grazing on a nutritionally poor range but did not mention that the ranchers provided supplemental feed.[21] He suggested that the lesions along the mouth and bridge of the nose, which Thompsett and Veenstra identified as typical of beta radiation, were demonstrated by Lushbaugh at Los Alamos in 1953 not to constitute evidence of radiation injury.[22] Trum did not recount the variance of opinion in the fall of 1953 in regard to Lushbaugh's experimental results, particularly Thompsett's refusal to concede that the Utah sheep did not display beta burns. And Trum assured Thomas that the lesions were "considered by many to be the result of photosensitization." He did not include a citation in support of this conclusion, and it is not clear who the "many" may have been. The Army lieutenant colonel referenced research at the University of Utah and Oak Ridge, suggesting that samples taken from the Utah sheep did not exhibit undue levels of radioactivity. He contended, "Therefore it can be correctly assumed that the sheep in the Cedar City area were exposed to fall-out after the major loses had occurred."[23] This was a strikingly indefensible conclusion. Trum asserted that the sheep did not encounter fallout on the range after shot Nancy in March but only due to Harry in May.

Trum posited a hypothetical scenario: assume that the Utah sheep were exposed to 5 roentgens of gamma radioactivity. "According to medical testimony which can be developed, a physician would not hesitate to give this much radiation to a sick person without fear of harm."[24] That is, he suggested that since a sick person may be administered 5 roentgens for therapeutic purposes, this indicated that even if the sheep *were* exposed to this level of gamma radiation, it would not have been a problem. Few investigators postulated that gamma radiation was solely the problem, however. Trum then recounted research at the University of Tennessee suggesting that it took over 400 roentgens to kill an adult sheep. This was irrelevant. Most deaths comprised pregnant ewes and newborn lambs.

Trum then reiterated Bustad's research at Hanford illustrating that adult sheep fed I-131 at a single point in time did not die unless given very high doses, suggesting that it was not radioiodine that killed the Utah sheep. Given that two-thirds of the sheep deaths at Cedar City were fetal and newborn lambs, this was irrelevant. Given that the other one-third were pregnant ewes, such results were only relevant if one assumed that radioiodine was the sole cause of death. It is not clear that any of the investigators suspected that radioiodine alone was to blame for the death of the ewes absent the rigors of trailing while pregnant—and Bustad's research did not reproduce such conditions. "However, it might be said that the internal irradiation from accumulated pick-up may have been great." As an aside, he hit the nail on the head. Accumulation of internal emitters in the gastrointestinal tract was not assessed in a substantive manner amid the chaos. Trum asserted that if a "formidable" amount of internal irradiation occurred in the Utah sheep, "there would have to be hematological and pathological signs of radiation injuries."[25] There was such evidence. Trum then reiterated the cold weather angle, noting below-average temperatures in Cedar City concurrent with the die-off.[26] The effect of below-average temperatures on newborn lambs was never evaluated empirically. It was more conjecture, making the water a little muddier.

Holmes, Thompsett, and Veenstra were in Cedar City in early June 1953 and examined the herds firsthand. They each expressed little doubt that radioactivity was, at the very least, a contributing factor in the die-off. Thompsett was of the opinion that the AEC contributed to "great losses," and, having seen radiation-induced lesions on cattle downwind of Trinity, suggested that the lesions were "absolutely typical" of what he had observed. Moreover, he noted that the ailments were not found on a herd leaving the Nevada range before commencement of the Upshot-Knothole series.[27] Veenstra also indicated that the lesions at Cedar City were "typical" of those seen on cattle after the Trinity detonation and that radiation was "at least" a contributing factor to the heavy losses. He further remarked that radiation injury was obvious both from field observation and laboratory analysis.[28] Holmes, for his part, did not see evidence of "contagion or toxic plants" and reiterated that it was primarily the younger, pregnant ewes that died.[29] This is consistent with Thompsett's hypothesis: radiation injury, trailing long distances, and pregnancy were a lethal combination.[30] In Holmes's report dated November 9, 1953, he concluded that the lesions produced experimentally at Los Alamos by Lushbaugh's team—which reoriented the investigation back toward malnutrition—were "similar" to those he had documented while in southern Utah.[31]

The reports authored by the veterinarians were a problem, but it was possible to convince Veenstra and Holmes to disqualify themselves as expert witnesses and Thompsett to change his opinion. Indeed, government attorneys worked to discredit the veterinarians' qualifications and experience in depositions and in the courtroom. The AEC was closing ranks, and some reputations would be sacrificed.

The Commission Goes to Court

Bulloch et al. v. United States went to trial in September 1956 before Judge A. Sherman Christensen. The ranchers sought compensation for their losses, but the government refused to negotiate. *Bulloch* was never about money but about procuring a definition of the situation that would be ascribed with the weight of civil law and that stipulated that fallout did not kill sheep downwind of the NTS. And the government had unlimited assets to devote to the cause, whereas Dan S. Bushnell, representing the plaintiffs, was a "country lawyer" with limited resources. It was David v. Goliath.[32]

Robert Thompsett did not testify at trial but was deposed by government lawyers in March 1956. When Thompsett was asked by Bushnell if he would be available to testify, he was reticent and stressed the difficulties of leaving behind his veterinarian practice to travel out of state.[33] Monroe Holmes was deposed in September 1956, just prior to trial. The lawyer for the government, Llewellyn O. Thomas, quizzed Holmes as to his training in diagnosing radiation injury. "I do not consider myself an expert in radiological damage," Holmes responded.[34] Before the court, Bushnell pondered the irony of the government questioning the qualifications of veterinarians whose expertise they originally solicited: "Mr. Finn said yesterday he's never been more proud of the witnesses presented. Dr. Pearson said that they called in the best experts in the field to make this examination. They called Dr. Veenstra, they called in Dr. Thompsett, and they called in Dr. Holmes. Yet, now the ones of which they are proud are those which support their conclusion, which they would like this Court to come to, and have attempted to discredit these men which were called in at their own investigation as men qualified to make this investigation."[35]

Moreover, at trial, the government floated numerous alternative explanations, including the cold-weather angle, but offered no compelling evidence. Judge Christensen pushed John J. Finn, representing the government, to account for the sheep deaths absent the influence of radioactivity:

THE COURT: What do you say about the explanation—Well, let me ask this: Don't you suppose that the evidence indicates very substantial and abnormal losses for this particular year?

FINN: I would say yes, your Honor.

THE COURT: And your view is that it was a mere coincidence that those losses occurred following the detonations on the 17th and the 24th of March?

FINN: I wouldn't want to characterize it by "mere coincidence," but it was a coincidence, yes, sir.

THE COURT: Yes. And—

FINN: I don't want your Honor to think I am so hard hearted as not to appreciate the unfortunate situation of these sheepmen, these sheepmen ran into. But I think it was, as I explained, a combination of causes, which had nothing to do with radiation.

THE COURT: Is there any evidence here to indicate that that combination of causes was confined to this particular area where radioactive fallout occurred?

FINN: There isn't. There isn't any, your Honor. But there, on the other hand, is some evidence that in other areas where there was no radio, or no possibility of radioactive fallout, the same conditions were found in sheep.[36]

The same conditions were never found in sheep herds that were not wintering northeast of the test site after shot Nancy. Investigators consistently came up empty in this regard.

Moreover, Christensen pushed Finn to admit that the government did not know the exact location of the herds after shot Nancy. This is significant because attorneys for the government argued that the levels of gamma radioactivity were simply too low to cause injury, but it was not clear that the monitors monitored where the sheep were grazing.

FINN: Under the evidence that's in the record, your Honor, all that I think we could contend for was that all, prior to the series itself, that is, prior to March 17, 1953, we had an idea as to where the sheep generally were. But we did not, I don't believe, know.

THE COURT: They weren't under any controlled conditions, I mean?

FINN: No, sir.

THE COURT: At the time of the shot, they could have moved from one
   area to another?
FINN: Correct.[37]

Nonetheless, on October 2, 1956, Christensen announced his decision: the evidence did not support the conclusion that radioactivity was
the cause of or substantively contributed to the injuries exhibited by the
sheep. Christensen contended that witnesses for the government included the most well-regarded experts in the country, and he concluded
that the levels of radioactivity to which the sheep were exposed could
not account for the grisly symptoms. However, the judge conceded that
he was troubled by the lack of evidence pointing to an alternative explanation: "I am not satisfied on the exact cause of these excessive losses.
That part has bothered me a good deal."[38]

The AEC assured the public that it possessed the knowledge and
expertise to conduct testing in a manner that was not detrimental to
human health, agriculture, or livestock production, and now a court of
law upheld the commission's insistence that an unprecedented die-off
was in no manner linked to fallout. It was just a coincidence. It was not
cause and effect. There were no lessons to be drawn connecting the animals and the people, the lambs and the children.

To initiate a chain reaction, it is necessary to possess a critical mass
of fissionable material. In a similar manner, in order for an entrenched
definition of the situation to be disrupted, there must be a critical
social mass sufficient to sustain a chain reaction of altered opinions
and assumptions. "We can define this *social critical mass* as the smallest
number of people who are cognizant of a particular problem and who,
under the right social and cultural circumstances, can facilitate the con-
tinual exchange of information about that problem," Sarah Fox argues.[39]
Like the successive collisions of neutrons and protons at the heart of an
atomic device, the dominant interpretive framework must be disrupted
in a cascading manner. The ruling against the plaintiff in *Bulloch*, however, short-circuited the critical social mass sufficient to fragment the
AEC's preferred narrative. Dead sheep could have provoked a reassess-
ment within the AEC of assumptions regarding the efficacy of internal
emitters of radiation. If the sheep were a canary among the sagebrush,

what did their demise portend for the downwinders, particularly the children?

David v. Goliath II

By the 1970s, some residents downwind had been sketching maps with the names, ages, dates, and the cancers befalling friends, family, and acquaintances.[40] These maps seemed to depict patterns that were not a part of the ebb and flow of daily life but at odds with it, something amiss not due simply to happenstance but to some outside force. Governor Scott Matheson of Utah took an interest in the issue. Despite being a Democrat in a staunchly Republican state, Matheson was elected in 1976 and pressured federal officials to release documents that had been hidden for decades. In 1979, Congress conducted hearings in Salt Lake City, Las Vegas, and Washington, DC, chaired by Senator Edward M. Kennedy.

The AEC no longer existed. Congress dissolved the commission in 1974, and one reason was the tension between promotion and oversight of nuclear technology vested within the same agency.[41] The Nuclear Regulatory Commission was given the role of overseeing nuclear technology, while the DOE was tasked with research and promotion. Nevertheless, by 1979, old ghosts came back to haunt the memory of the commission. One was Dan Bushnell. Another was Steve Brower. Bushnell began looking for documents that were newly available that he had not seen twenty years earlier and that might have changed the questions he asked, the witnesses he called, and the strategies he employed at trial. Brower was the Iron County agricultural agent from 1950 to 1954 and witnessed the interactions among the ranchers, AEC officials and military officers, and those who were brought in to contribute to the sheep investigation. Brower testified before congressional investigators in 1979 and penned a trenchant letter to Governor Matheson articulating his opinion of the 1956 *Bulloch* trial: "When we finally did get this case into court, none of the technical data from experts was allowed to be used and of course the Iron County sheepmen could not make a case that would stand up in court without this data. The original, preplanned strategy of the AEC worked because they had the indiscriminate power to intimidate, to withhold, screen, change, and classify any and all information, reports, and data."[42]

Documents were coming to light that described events that did not garner widespread attention two decades earlier, and the allegations were striking. The ranchers petitioned A. Sherman Christensen, still on the bench, asking to set aside the original judgment on the basis of AEC reports that were newly available due to litigation and congressional inquiry.[43] After four days of testimony, Christensen issued an unprecedented rebuke, insisting that government lawyers perpetrated "fraud upon the court" in 1956. He vacated the original judgment and ordered a new trial. "Fraud upon the court" is a legal term describing intentional conduct that inhibits the rendering of an impartial judgment. Christensen identified four instances: (1) the Bustad report omitted results that indicated the death of fetal and newborn lambs; (2) the government presented misleading information at trial in reference to offsite measurement of radioactivity; (3) witnesses were pressured to testify in a manner consistent with the government's case; and (4) government lawyers gave misleading and evasive answers to pretrial interrogatories. In 1983, a three-judge panel of the US Court of Appeals for the Tenth Circuit rejected Christensen's finding of fraud upon the court. Two years later, the full court reaffirmed this ruling by a vote of five to two.[44] This ruling was not justified based on the evidence at hand, however. Duplicity and malfeasance are not always glaring and overt but are often nuanced and subtle, and the majority opinion of the Tenth Circuit court ignored the incriminating details.

Leo K. Bustad and colleagues at Hanford, Washington, began radioiodine experiments on sheep in 1950 and were asked to compare ongoing research to the die-off in southern Utah. The Bustad report was dated November 30, 1953, and claimed, "The Utah sheep showed no evidence of the radiation damage observed in the experimentally treated sheep."[45] This conclusion paved the way for the final report absolving radioactivity, released two months later. There was a crucial empirical result that was not discussed in the Bustad report, however, but in an obscure study cited on the last page. This citation contained a game-changer: Bustad and colleagues had evidence that radioiodine administered to pregnant ewes produced dire effects on their newborn lambs. It was an incomprehensible omission. Indeed, in their report dated November 30, Bustad

TABLE 8.1. Key Sections of the Bustad Report of 1953

a. "Although details of the experiments are available elsewhere (3, 4) a brief summary of the methods applied together with a *summary of experimental observations of immediate comparative applicability* is included in this report" (2; emphasis added).

b. "The lambs born to ewes fed 240 uc/day during the latter half of gestation of the first lambing season were normal in size. *However, in subsequent lambing seasons in this group a significant reduction in birth weight of lambs occurred*" (4; emphasis added).

c. "The pathologic examination of tissues obtained from flocks adjacent to the Nevada Proving Grounds showed no evidence of significant abnormal findings in the thyroid gland, bone marrow or any organ other than the liver" (42).

d. "Premature lambing described in some Utah sheep flocks was not observed in the experimental sheep at Hanford although some test animals were fed up to 1800 uc/day for 420 days" (43).

e. "The large number of deaths observed in Utah sheep was not observed in any acute or subacute Hanford experimental sheep fed up to 280,000 uc in a single dose. Deaths observed in experimental sheep did not occur until over five months following the initiation of I^{131} feeding and was attributed to systemic disease complicating the chronic hypothyroid state" (43).

f. "Tissue sections from newborn Utah lambs were not available for comparison. *However, no increase in deaths in newborn experimental animals was observed in any group in which there was absence of definite evidence of thyroid damage*" (43; emphasis added).

g. L. K. Bustad, D. E. Warner, and H. A. Kornberg, "Toxicity of I^{131} in Sheep. V. General (Low-Level Chronic Effects)," Biology Research—Annual Report 1952, Document HW-28636, 136–147 (1953) (UNCLASSIFIED).

and colleagues should have stated, "But we did kill newborn lambs by feeding their mothers radioiodine."

Sections of the Bustad report are presented in table 8.1. The report, as outlined in table 8.1a, was a "summary of experimental observations of immediate comparative applicability." This section lists two citations— not referred to again in the text. The citation outlined in table 8.1g., "Toxicity of I^{131} in Sheep. V. General (Low-level Chronic Effects)," is crucial. One would barely deduce it from the title, but in time, this came to be known as the "fetal lamb report." Government lawyers had this report but did not present it at trial in 1956.[46]

A section of the Bustad report, just three paragraphs in length, commented on weight changes among the animals. As reiterated in table 8.1b, "However, in subsequent lambing seasons in this group a significant reduction in birth weight of lambs occurred." The report did not expand on this finding. A significant reduction in birth weight is one thing, but weak and soon-to-die newborn lambs is quite another. The latter was of "immediate comparative applicability," but Bustad and colleagues cryptically referred simply to a reduction in birth weight. Information on the dead lambs was contained in the source cited in table 8.1g.

One year after *Bulloch* went to trial, in 1956, the research team at Hanford published in a peer-reviewed academic journal and outlined the impact of radioiodine on pregnant ewes and their offspring. Peer-reviewed academic journals are another arena in which truth claims are asserted, and within this article lay data that were not discussed in the Bustad report or at trial in 1956. As early as 1952, the Hanford experiments revealed that radioiodine contributed to weak, stunted, and dead lambs. The effect was observable even at a moderate dosage level. The 1957 article proclaimed, "Thyroid glands were obtained from 4 offspring of the ewes fed 1,800 uc. per day, 2 being stillborn, one having lived for 15 hours, and one for 3 days. The pathologic effects were severe despite the fact that the fetal exposure to I^{131} occurred only during the last trimester of pregnancy."[47] Bustad and colleagues illustrated that ewes that were fed radioiodine produced stillborn or soon-to-perish lambs. The Cedar City ranchers would have recognized this grim pattern. If this information had been presented at trial, it is unlikely that the AEC would have persuaded a federal court judge that radioactivity and the sheep die-off were merely a coincidence.

The Bustad report further concluded that the Utah sheep did not demonstrate substantial "abnormal findings" in the thyroid or bone marrow, as shown in table 8.1c. Throughout the initial investigation, there was concern that investigators arrived on scene only after the die-off occurred. This raised questions as to whether the bone and tissue samples obtained were representative of the animals that died weeks earlier. It makes no sense to examine apples and then make inferences to oranges. Indeed, as articulated in table 8.1f, Bustad and colleagues noted, "Tissue sections from newborn Utah lambs were not available for comparison." That is, the tissue samples examined were from adults. Given that two-thirds of the deaths were newborn lambs—and the Hanford researchers did not have tissue samples from newborn lambs—the lack of abnormal findings, as stipulated in table 8.1c, should not have come as a surprise.

The section of the report shown in table 8.1d proclaimed the failure of experimental results to detect the abnormal incidence of premature lambing. The ranchers at Cedar City did not encounter premature lambing but weak, stunted, and dead lambs at birth or soon after—the kinds of things buried in the source referenced in table 8.1g and discussed in the 1957 journal article but not mentioned in the Bustad report itself.

The section of the report shown in table 8.1e claimed that researchers at Hanford did not find evidence of acute death among sheep that were fed a large amount of radioiodine in a single dose. The crucial question was not whether a single dose of radioiodine can kill adult sheep but what happens to ewes that are subject to ingestion of radioiodine during the stresses and strains of trailing long distances while pregnant, in tandem with a variety of other fission products working their way through the gastrointestinal tract. Given that two-thirds of the deaths involved fetal and newborn lambs, what of their offspring?

There is an additional nuanced but important catch to the Bustad report. It is presented in table 8.1f. Bustad and colleagues insisted that the experimental sheep exhibited hypothyroidism but the Utah sheep did not. They further argued, "no increase in deaths in newborn experimental animals was observed in any group in which there was absence of definite evidence of thyroid damage." That is, Bustad and colleagues maintained that the Utah sheep did not exhibit hypothyroidism or damage to the thyroid—an assertion that is clearly contestable, if not impossible to defend—and that among those experimental sheep that did *not* exhibit "definite evidence of thyroid damage," there was no increase in deaths of newborn lambs. A key question is what constituted "definite evidence," particularly when the samples obtained at Cedar City were probably not comparable to the sheep that died as much as six weeks earlier. In the report, Bustad and colleagues should have then mentioned that *lambs born to ewes that showed thyroid damage dropped dead like flies.* This information was buried in the fetal lamb report referenced on the last page (shown in table 8.1g).

The question confronting the Tenth Circuit court in 1985 was whether the Bustad report suppressed information. The Tenth Circuit stated, "Thus, the issue on review is not whether plaintiffs could have found the fetal lamb information if they had been more diligent but, rather, whether the government sought to diminish the likelihood of their finding the information." The court then refused to render an opinion on whether government lawyers "sought to diminish the likelihood" that Bushnell would uncover the fetal lamb report but stressed that he should have been more meticulous in unearthing information prior to trial (principally the source cited in table 8.1g). The majority argued, "Regardless, it is not our function to determine whether or not the government intended to suppress the fetal lamb report. Bulloch I was tried to the bench." The "issue on review" was whether government lawyers intentionally sought to "diminish the likelihood" that Bushnell would access the fetal lamb report, but the majority opinion of the Tenth Circuit court then refused to take a stance in this regard. Further, the court argued that the Bustad report was a summary of experiments and conclusions and not a comprehensive accounting, as contained in the sources cited on the last page (principally the study highlighted in table 8.1g).[48] However, the Bustad report clearly proclaimed that it was a "summary of experimental observations of immediate comparative applicability," as outlined in table 8.1a. It is unclear how the majority opinion of the Tenth Circuit court disregarded the idea of "immediate comparative applicability."

Some people's reputations were sacrificed because of their affiliation with the AEC, while others benefited. Leo K. Bustad went on to become a distinguished professor at Washington State University, dean of the College of Veterinary Medicine from 1973 to 1984, and a speaker at high schools, community colleges, and universities around the world. So revered is Dr. Bustad that the Veterinary Science Building at Washington State is named in his honor. He wrote books examining the human-animal bond and published a book of speeches addressing subjects as varied as compassion, the responsibilities of a scholar, and empathy toward the nonhuman world. In 1976, he delivered a speech addressing the problems plaguing the country. "I believe history teaches us that the foundation of a republic's power is not restricted to its armies, but rests on the integrity of its institutions and its people," Bustad counseled.[49]

The majority opinion of the Tenth Circuit court in 1985 also rejected the contention that government lawyers provided misleading information at trial in 1956 in reference to measurement of offsite radiation. Before Judge Christensen, three years earlier, Bushnell argued that at trial in 1956, the government neglected to admit the existence of hot spots. The Tenth Circuit did not equate this with fraud upon the court.[50] Consideration of hot spots is significant, as Christensen, in ruling against the ranchers in 1956, relied heavily on evidence presented by the government suggesting that gamma radiation levels were simply too low to account for the sheep deaths. However, this assertion neglected to consider potential hot spots.

By 1956, AEC and military officials clearly had knowledge of the potential for intense radioactivity surrounded by lower overall readings. And it began at Trinity. Prior to 1953, AEC reports noted the existence of unequal distribution of fallout downwind of tower detonations.[51] And Upshot-Knothole only reaffirmed the troubling occurrence of hot spots.[52] The Tyler Committee, assembled in July 1953, clearly had a conception of the vagaries of fallout: "Predictions of fallout location have been insufficiently accurate, both because of weather changes from forecast and because of methods. The primary concern of the forecaster is to determine *where in the general fallout area the hot spot will occur.*"[53] Moreover, a classified report by Lieutenant Colonel N. M. Lulejian, dated November 1953, remarked that after shot Nancy, areas offsite were not "adequately covered." This was "because there are no roads in the region where the close in fall-out occurred."[54] Rad-safe monitors need roads, but sheep do not. Not only did AEC and military officials have a conception of unpredictable hot spots by 1956, but they could have reasonably assumed that the Cedar City herds encountered areas of high-intensity radioactivity that were not accounted for by monitors whose movements were bounded by existing roadways.

An "interrogatory" is a list of questions that one party submits to another in gathering evidence prior to trial. As detailed in table 8.2, government lawyers contended that there was no evidence that the ranchers and their sheep were in an area of intense fallout in the wake of shot Nancy or shot Harry. The final investigation report and follow-up studies had been furnished to the plaintiffs, and no one involved with the AEC's investigation now disagreed with the final investigation report or were of the opinion that radioactivity was a possible cause of injury. During the investigation, there were no sheep found to be "hot." Further, the sheep died due to a combination of factors including poor range conditions, mismanagement, and weather. They may have eaten poisonous plants, too.

During closing arguments in 1956, an attorney for the government, John J. Finn, conceded that AEC officials did not know the exact location of the herds in the aftermath of shot Nancy.[55] And yet, in the pretrial interrogatory and at trial in 1956, government lawyers insisted that the herds were not located in an area of heavy fallout.

TABLE 8.2. Answers Offered by Government Lawyers to the Plaintiff's Interrogatories

Interrogatory 10: Were any men or sheep known to be in an area of relatively heavy fallout?
Answer: No.

Interrogatory 28: Did anyone involved in the investigation disagree with the reports?
Answer: We are not aware of anyone who is involved in the Commission's investigation of the alleged sheep losses who *now* disagrees with the report issued by the Atomic Energy Commission.

Interrogatory 29: Did anyone conclude that radio-active fallout was a possible or probably a cause of the injury to the sheep?
Answer: We do not know of anyone connected with the Atomic Energy Commission's investigation on the alleged sheep losses who has *now* concluded that radioactive fallout was a possible or probable cause of the injury to the sheep.

Interrogatory 31: Were any sheep examined considered to be "hot" or evidencing abnormal radioactivity?
Answer: No.

Interrogatory 33: What explanation is given for the symptoms observed on the sheep from said investigation?
Answer: A combination of factors, including malnutrition, poor management, and adverse weather conditions. The animals probably ate poisonous plants due to poor range conditions, extreme drought on the range, and low temperatures.

Source: Dan Bushnell, "Untrue and Evasive Answers to Interrogatories, Excerpt from Answers to Interrogatories," submitted to the court in *Bulloch et al. v. United States*, 95 F.R.D. 123 (D. Utah 1982).

That none of the investigators *now* disagreed with the final investigation report or were *now* of the opinion that radioactivity was a contributing factor are the most problematic statements. The distinction between past and present tense was not disentangled by Bushnell at trial in 1956—suggesting that he did not have a conception of the machinations employed by Bernard Trum and others in persuading Thompsett, Veenstra, and Holmes to diminish their role in the investigation and the validity of their own research. The answers given in reference to interrogatories 28 and 29 were less than forthcoming.

The ranchers repeatedly stressed that the first investigators to examine the affected herds were taken aback by the frenzied clicking of a Geiger counter and repeatedly invoked the adjective "hot" in describing the situation. This was expressed when the ranchers met with AEC and military officials in Cedar City in August 1953; again in January 1954, when they were informed that radioactivity did not kill their sheep; and in 1979, as the ranchers spoke before congressional investigators.[56] In sum, the answer to interrogatory 31 is false. This is relevant, as are the other misleading answers outlined in table 8.2, as it is evidence of fraud upon the court that the majority opinion of the Tenth Circuit court overlooked.

The majority opinion of the Tenth Circuit further noted that at trial in 1956, alternative explanations for the sheep deaths "were established by the evidence." "In Bulloch I the conclusion of the experts was that the sheep and lambs had died as a result of unprecedented cold weather during the lambing and shearing, together with malnutrition, also to generally poor range conditions and to common diseases which caused the skin lesions and deaths. These conditions were established by the evidence." None of the above was "established by the evidence" at trial in 1956. That cold weather in combination with malnutrition and "common diseases" accounted for the sheep deaths was an absurd conclusion in 1956 and worse still by 1985. When Christensen announced his decision in favor of the government at trial in 1956, he lamented that the government had not advanced a compelling alternative explanation to radioactivity: "I am not satisfied on the exact cause of these excessive losses. That part has bothered me a good deal."[57]

Why, then, would Christensen rule for the government in 1956? During closing arguments, Christensen stated, "Now, then, actually the case pretty well turns on whether there was substantial radioactive damage." By his own admission, Christensen gave considerable weight to the testimony of government witnesses in this regard. Moreover, before the court in 1956, Bushnell refused to concede that witnesses testifying on behalf of the AEC were not acting in good faith. "I don't know that I can say that they have knowingly concealed. I believe they have been greatly influenced by their closeness to the subject matter."[58] In retrospect, this is where Christensen and Bushnell did not exercise due diligence: in giving the benefit of the doubt to AEC and military officials testifying before the court.

Stewart Udall has referred to the majority decision of the Tenth Circuit Court of Appeals in 1985 as a "grotesque episode of American jurisprudence." Udall represented the state of Arizona in the House of Representatives (1954–1961), served as secretary of the interior (1961–1969), and was involved in civil litigation regarding atomic testing in Nevada as well as Navajo uranium miners. Udall had a unique perspective born of working in the federal government as well as finding himself at odds with the government later in his career. He further remarked that the Tenth Circuit's 1985 ruling belongs in the "Legal Hall of Irrational Opinions."[59] Indeed, David v. Goliath seems tepid compared to *Bulloch et al. v. United States*.

A "26-Year Oversight"

In 1954, the AEC concluded, "On the basis of information now available, it is evident that radioactivity from atomic tests was not responsible for deaths and illness among sheep in areas adjacent to the Nevada Proving Grounds."[60] This was a nearly impossible definition of the situation to sustain, but the commission did just that for over two decades. Atomic detonations were the spectacle, but dead sheep were a grim and unruly counterspectacle that few people witnessed, as national news media yielded to the commission's preferred story line, as did local, state, and federal political representatives. Nonetheless, residents downwind had been telling and retelling stories of life amid fallout clouds, and over time, social critical mass sufficient to disrupt the commission's self-serving narrative arose.[61] Congressional hearings are another arena in society in which truth claims are asserted and contested, and in 1979, hearings focusing on open-air atomic testing in Nevada included a tale of thousands of sheep perishing under mysterious circumstances. Despite having worked for the AEC, one person who was surprised by what he heard was Dr. Harold A. Knapp. He was so taken aback that he began sketching a series of mathematical calculations.

"The latter day nemesis of the AEC is Harold Knapp," R. Jeffrey Smith explains.[62] Knapp earned a PhD in mathematics from the Massachusetts Institute of Technology, joined the AEC in 1955, and began working in the Fallout Studies Branch, within the Division of Biology and Medicine, in 1960. He was a skilled mathematician, but he was not employed by the AEC for long. His job entailed estimating the public health hazard of fallout distributed across the country. Such research was in response to pressure from the JCAE.

After testifying before congressional investigators in April 1979 regarding radioiodine, Knapp overheard testimony regarding the sheep deaths of 1953 and submitted his own analysis to congressional investigators the following August. Knapp pointed to a "26-year oversight."[63] He argued that external gamma radiation on the Nevada range was probably much higher than AEC estimates, and he stressed the occurrence of hot spots that were not accounted for by scattered readings. Seated before congressional investigators, he suggested that radioactivity dispersed across the landscape like invisible snowdrifts.[64] Just as the vagaries of wind and topography shape uneven accumulation of snow, so too the depth of radioactivity ebbs and flows. Lumbering through the sagebrush, dry washes, and canyons of the Great Basin amid invisible snowdrifts of radioactivity, the sheep emerged as a material, biophysical sign of what the eye could not see.

Knapp argued that the investigation overlooked three considerations: (1) no experiments were conducted to assess the amount of fallout ingested by sheep in the course of a day's grazing; (2) no experiments were conducted to feed sheep "real fallout" or all of the various fission products at one time; and (3) the investigation failed to sample and analyze the stomach and intestines of sheep that died. "There were a number of things that should have been asked, but the most important one was: How much did the sheep eat and how much fallout does it take to kill a sheep?"[65]

This question had been asked. Indeed, it is reminiscent of queries posed by the ranchers as they met with AEC and military officials in August 1953 to discuss the ongoing investigation. The ranchers approached the issue from a possibilistic perspective, while AEC and military officials stressed the improbability of radioactivity. One rancher inquired, "Couldn't animals acquire radiation from eating plants that had some radiation?" According to the meeting transcript, they never received a direct answer in response.[66]

Knapp explained to congressional investigators that he began by envisioning how much an average-sized sheep eats per day, the expected gamma radiation present, and the proportion of fission products that was probably ingested: "If you know the external gamma dose rate in roentgens per hour or rads per hour, at any given time after a nuclear detonation, then there is a straightforward, although somewhat complex, way of determining how much fallout is present per square meter or per square mile and how much of each of the radioactive nuclides is there."[67]

Strontium-90 sought out the bones, cesium-137 went to the muscles, iodine-131 accumulated in the thyroid, and a "hundred or so" other radionuclides progressed through the stomach and intestines.[68] And while the thyroid probably received the highest dose, the gastrointestinal tract was the most sensitive. Detailing Knapp's calculations, Smith clarifies, "His conclusion was that sheep exposed to only 4 rads of external gamma radiation might have gotten a dose of 1500 to 6000 rads in their gastrointestinal tracts. He estimated fetal lambs may have received a thyroid dose of 20,000 to 40,000 rads."[69] In turn, Knapp argued that acute effects were possible with the levels of gamma radiation measured on the range, let alone if the herds encountered hot spots.[70] Moreover, while Knapp was preparing his report for congressional investigators, he noticed that Bustad and colleagues at Hanford published data in 1957 that were not included in their report to the AEC in 1953 or at trial in 1956. This discovery prompted the ranchers to file a petition before Judge Christensen in 1982 asking to set aside the original judgment in *Bulloch*.[71]

In crucial respects, AEC and military officials in 1953 focused on the wrong question. External gamma radiation levels were an important consideration but not the most important consideration. *The most important question was about the myriad of fission products the animals ingested as they moved eastward, grazing and browsing, one mile after another.*

Knapp's 1979 report provided compelling evidence that ingestion of fission products by the ewes was, at the very least, a contributing factor, in tandem with the physical exertion of trailing, pregnancy, and birth. The radioactive burden confronted by the lambs was probably the sole cause of their deaths. "Unborn lambs may receive a thyroid dose from the isotopes of iodine 10 times as great as that to the thyroid of their mother," Knapp explained. Moreover, unborn lambs were subjected to gamma irradiation from fission products in the gastrointestinal tract of their mother.[72] The ranchers were right. Fallout killed their sheep.

In August 1979, congressional investigators quizzed Knapp on whether it was reasonable to expect AEC officials to recognize the sensitivity of the gastrointestinal tract to radioactivity at the time of the investigation in 1953. Knapp testified that he was alerted to such a consideration reading the testimony of Gordon Dunning that was submitted to a congressional hearing in 1957, illustrating the sensitivity of the gastrointestinal tract and including an equation for estimating the probable burden. Two decades later, Knapp found the equation to be relevant to the issue at hand and incorporated it into the report he prepared for congressional investigators.[73]

Congressional investigators also pushed Harold Knapp to account for a question that he could not answer: Why did AEC officials and affiliated researchers not recognize the significance of the gastrointestinal tract in general and the effects of radioiodine on lambs in particular? "I would certainly say there was no excessive zeal in trying to find all the possible ways in which a sheep could have been killed by fallout from the tests," Knapp remarked.[74]

As organizational actors push forward their instrumentally rational plans, they often cannot know in advance the negative consequences unleashed. At other times, they simply do not want to know, beforehand or after the fact. The calculative, methodical behavior of organizational actors has its limits, sometimes self-imposed. Either way, not knowing is a liability in terms of detecting incubating problems. If AEC officials had been willing to follow the evidence, it could have contributed to a better understanding of the hazards posed by internal emitters of radiation. "First, the sheep died. The sheep were surrogates for humans. They should have been regarded as an early warning so that precautions could have been taken for the people," Fradkin observes.[75]

Dead and dying sheep constituted an acute indication of the empirical reality of fallout, but the willful and strategic ignorance and, in time, heightened deception and malfeasance obscured the social salience of this event. What the AEC accomplished was not the elimination of risk but erasure of the recognition of the risk to public health. That something is not real with regard to its presence in public opinion and perception does not mean, of course, that it is not real.

The public typically engages in possibilistic rather than probabilistic thinking. The public considers a broader range of potential anomalies and contingencies than do experts, who focus more on the probable. The public wonders about all that might go wrong and the potential consequences. Experts focus on what is most likely to go wrong. With novel technology, the probable may be unclear, as there is little experience from which to make methodical calculations. What is probable when there are no historical reference points to consult in assessing the risks of a given technology? Possibilistic thinking often seems irrational to experts who are accustomed to rigorous quantification and formal risk-assessment procedures, but as the complexity, scale, and interconnectedness of sociotechnical systems increases, there is a grain of wisdom to more expansive, consequentialist thought.[76] Indeed, the Cedar City ranchers arguably displayed a more insightful and reasoned assessment of the fate of their animals than did AEC and military officials.

Science is a powerful mechanism for establishing claims to truth. Biased science is a powerful aid in constructing and sustaining a preferred definition of the situation. Expert knowledge imparts status and demands deference, and the capacity of organizational actors to co-opt these characteristics is key to the production of strategic ignorance and the short-circuiting of democratic deliberation and accountability.

To understand the perspective of AEC and military officials during the years of open-air testing, it is instructive to recount the testimony of Charles Dunham, director of the Division of Biology and Medicine, before the JCAE in 1957. The division that Dunham headed was tasked with assessing the public health hazards of atmospheric detonations. Seated before the JCAE, however, he was more concerned that they would be unduly influenced by critics of the AEC who did not understand that technological development entails risk. To eliminate risk, Dunham stressed, is to eliminate technological innovation: "Why should we have any increase in our exposure to radioactivity? Let us stop a moment and consider one of the basic facts of life, namely, to attain any specific objective almost invariably involves the surrender of something else. Reluctance to face this issue is commonly described as wanting to eat one's cake and have it too."[77] Dunham did not address the question of who should be accorded a say in such trade-offs. In a democratic society, should the people be allowed a say in such matters when it presents a risk to public health? Or should the technocracy be in charge? What are the liabilities if authority for such decisions is vested solely in the technocratic system itself?

9

Operation Plumbbob

Accelerated Testing in the Shadow of a Moratorium, 1957

On October 26, 1956, Judge Christensen's ruling absolving the AEC from harm in *Bulloch et al. v. United States* was released to the public. Just two days earlier, a boy from the Warm Springs area, north of the NTS, died in a hospital far from home in Reno, Nevada. Martin Bordoli came home from school feeling tired and running a fever, and he died less than a year from the initial diagnosis: stem cell leukemia.[1] His nickname was "Butch," and at just seven years of age, he may have been the first direct casualty of testing in Nevada.[2]

In 1957, Martin's mother circulated a petition with the signature of seventy-five neighbors. At times apologetic in tone, the petition urges the AEC to suspend detonations or that "some equally positive action be taken to safeguard" area families: "We believe further that it is both un-democratic and un-American to subject one group of citizens to hazards which others are not called upon to face, particularly when the adverse effects may be reflected in future generations yet unborn. We are not excit-able or imaginative people, most of us coming from rugged ranch families, but neither are we without deep feeling for each other and our children."[3]

Martha Bordoli also penned a letter to Lewis Strauss, chairman of the AEC. Strauss's letter in response misspelled her last name, address-ing her as "Mrs. Bordoh," and admonishes her to take heed of President Truman's advice: "Let us keep our sense of proportion on the matter of radioactive fallout." "Of course, we want to keep the fallout from our tests to an absolute minimum, and we are learning to do just that, but the dangers that might occur from fallout in our tests involve a *small sacrifice* when compared to the infinite greater evil of the use of nuclear bombs in war," Strauss urged.[4]

Accusations of alarmism are a tactic employed by organizational ac-tors to curtail public critique, and although the Bordoli family were not contacted by AEC personnel prior to Butch's death, the petition earned them a visit from a lieutenant colonel and a letter from Senator George W. Malone of Nevada urging her not to fall prey to a "minority group of scientists" spreading "'scare' stories" about fallout, as they were probably "Communist inspired."[5]

Martin Bordoli was, perhaps, one small sacrifice, but there were others. Childhood leukemia is a signature effect of ionizing radiation. As Captain Howard Andrews attested in a supplemental paper appended to the Tyler Committee report of 1954, "The chickens will eventually come home to roost."[6]

In the summer of 1977, Gordon Eliot White, a reporter for Salt Lake City's *Deseret News* but based in Washington, DC, discovered an unpublished National Cancer Institute report outlining leukemia rates in Utah between 1950 and 1969. He combined this with data on fallout obtained from a source at the Pentagon. Laying one atop the other, something stood out: "When he put the cancer figures together with the charts of the downwind fallout patterns from the Nevada tests, he saw that the heaviest concentrations of fallout consistently dropped over those five counties that now showed a higher than normal rate of leukemia," Patterson and Russell observe.[7]

White then wrote an article pointing out that the Utah counties of Garfield, Kane, Iron, San Juan, and Washington had rates of leukemia above the state average.[8] "Deaths High in Utah Fallout Area" caught the attention of Joseph L. Lyon, a cancer researcher at the University of Utah who was skeptical that fallout had a demonstrable impact within the state.[9] Two years later, however, Lyon and colleagues' research in the *New England Journal of Medicine* was strikingly consistent with White's reporting. Lyon and colleagues analyzed leukemia deaths in Utah among children fifteen years of age or younger and uncovered 152 deaths in northern Utah where 119 were expected, or a 1.3-fold increase; and in the high-fallout counties in southern Utah, where far fewer people lived, they catalogued 32 deaths where 13 were expected, or 2.4 times higher than anticipated.[10] "Children born in southern Utah between 1951–1958 experienced 2.4 as many deaths from leukemia as children born before and after above-ground bomb testing," Lyon explained.[11] Selecting the five southern Utah counties closest to the NTS, Lyon and colleagues reported a level 3.4 times normal.[12] "The statistical implications of the study were clear. The chance of a child dying from radiation-induced leukemia was greater the closer the youngster lived to the test site, and, it could be inferred, the chance for an adult was somewhat the same," Philip Fradkin observes.[13] Joseph Lyon confided to congressional investigators in 1997 that the study garnered attention among medical researchers but particularly among government officials: "The publication of this paper immediately made us the focus of an intense effort by the Federal Government to disprove our findings. . . . For example, our study was reanalyzed four times at substantially more cost than was spent on the original study."[14]

What Joseph Lyon did not know is that he and his colleagues inadvertently replicated research from fourteen years earlier. Indeed, at the time, no one knew of prior research examining leukemia in Utah because it had been shelved under pressure from AEC officials. In 1960, the US Public Health Service began examining leukemia and thyroid abnormalities downwind of Nevada, and in 1965, Edward S. Weiss circulated a draft report illustrating that between 1950 and 1964 the two southwestern-most Utah counties, Iron and Washington, exhibited a statistically significant threefold excess of leukemia among people under nineteen years of age and a 1.5 times increase in the risk of leukemia among residents of all ages.[15] Overall, Weiss found twenty-eight leukemia deaths where nineteen were expected.[16] Commenting on Weiss's discovery, Lyon told congressional investigators, "It was found exactly where they thought it would be and exactly the population."[17]

Weiss's research was debated in a September 1965 meeting at the White House with AEC and PHS officials and staff from the office of President Johnson's scientific adviser, and there was a memorandum from AEC officials to the surgeon general stressing the methodological weaknesses of Weiss's study but also the public-relations implications.[18] From this point forward, Weiss's leukemia study lay dormant. "The decision was made not to publish the paper because the study was based on a small number of deaths, and the officials of the Atomic Energy Commission did not wish to alarm the public unduly with inconclusive findings," Lyon explains.[19]

Irrespective of the strengths and weaknesses of Weiss's research, a statistically significant excess number of childhood leukemia deaths in high-fallout counties was deserving of additional study—which is exactly what did not happen within either the PHS or the AEC. It would be difficult to find a more glaring example of willful ignorance. There simply were some things AEC officials did not want to know, and PHS officials acquiesced to ensure that they did not learn of those things. And neither the public nor the scientific community knew of Weiss's study until 1979, when the *Washington Post* obtained a copy through a Freedom of Information Act request.[20]

In 1984, Joseph Lyon and colleagues documented a statistically significant excess number of leukemia deaths in northern Utah, based on data from the DOE that had previously been published in the journal *Science*. After learning of Lyon's unpublished study, however, the DOE revised the data and then determined they merited classification status. The revision lowered the estimate of fallout in northern Utah attributable to testing in Nevada and upped the proportion attributable to Soviet testing, and as a result, the association between leukemia in northern Utah and fallout from Nevada dropped from statistical significance. When Lyon and colleagues requested access to the new data to continue their analysis, the request was denied—because the revised data were classified.[21] Walter Stevens and Joseph Lyon and colleagues published a study in 1990 documenting a 7.8-fold excess of leukemia deaths in southern Utah among people less than twenty years of age when they had been exposed during the years of open-air testing.[22] And in 1993, they observed a threefold excess number of thyroid neoplasms in areas of Utah characterized by the heaviest fallout exposure.[23]

Other researchers have studied the link between testing in Nevada and leukemia immediately downwind, but the results are contradictory for adults, whereas for children, the statistical evidence is more assured: fallout claimed a number of young lives.[24] As Sandra Steingraber notes, "Documenting whether a cancer cluster exists in a small community of only a few thousand or a few hundred inhabitants is statistically difficult work, and it is at this level where the fiercest arguments fly."[25] It is a signature effect of the risk society. There are bits and pieces of verifiable knowledge, at times, amid conflicting results that, at times, are shaped by political considerations. There are materials and substances that prior generations did not confront and unintended risks arising from innovations in science and technology. There is uncertainty and plausible deniability, and there are things that powerful actors in society prefer not to know. And in the end, the toll imposed on the public due to testing in

Nevada will never be completely clear, even though it is clear that there has been a very real impact. Worse still, it is difficult, if not impossible, to definitively link the disease suffered by any given individual to fallout from Nevada. This is a source of frustration for the afflicted and their loved ones and a rhetorical opportunity for powerful actors intent on fostering doubt. Epidemiological research looks for above-background rates of disease over a period sufficient to iron out fluctuations in the data. More cases than expected, after adjusting for all other relevant considerations, may warrant the inference that they are due to radioactive fallout. It sounds simple but in practice is maddeningly complex.

Permeability of the Body Downwind

Evidence of ill effects is inferred from the numbers but experienced by the individual, the family, and the community. Abstract and methodical procedures clash with the experience of living with a sick child. Parents put on a brave, optimistic demeanor, travel to and from distant hospitals, dole out medicine in measured increments, and brace for the future.[26] They engage in retrospective cause-and-effect reasoning and parse official statements and word-of-mouth for clues. And they confront the unequivocal statements of people in positions of formal, authoritative status, statements that often hide the tenuous assumptions and shaky inferences lying underneath. In time, they come to question the credibility of institutional structures in society.[27]

Detonations in Nevada did not entail the diffusion and dilution of substances within a natural world separate from humanity, as things came back around, again and again. In crucial respects, nature continually challenged the prevailing discursive constructions of AEC and military officials long after the flash and shockwave subsided. "Things assemble," Ian Hodder stresses.[28] And things push back. They diffract, filter, channel, transport, concentrate, evade, and surprise. Boundaries between the human and nonhuman are more imaginative than real. Hybridity is the rule rather than the anomaly.

Andrew Pickering's conception of a dialectic of resistance and accommodation, in which humanity and nature are embroiled in a performative and open-ended dance, is a more prudent lens from which to assess the contours of rational planning.[29] How we conceptualize our relationship to nature shapes how we act in the world, and the pretense of command and control is guaranteed to overlook the web of connections forged through goal-directed activity but obscured by the hubris of unrealistic assumptions. Environmental risk is derived from failure to maintain the stability of human-nonhuman configurations such that the force and efficacy of entities, processes, and things exceed what was originally intended. Radioiodine, cesium-137, and strontium-90 did not diffuse and dilute harmlessly into the vastness of nature but drifted across pastures, where they were ingested by grazing cows that concentrated them in their milk. Fresh milk on the table completed the circuit between a technocracy that was resistant to public oversight and those who lived downwind. And it was a *contingent materiality*. Children and lambs were most at risk.

Stacy Alaimo refers to "trans-corporeality" in describing the movement of organic and nonorganic substances within and across the body. "Trans-" denotes movement, and "corporeal" refers to the physicality of the body. Alaimo recounts sending a sample of hair for laboratory analysis, the flinch of yanking one's hair out not as jarring as learning of the traces of matter embedded within. Alaimo's hair harbored traces of mercury. At this point, the guessing game begins, as one sketches "maps of transit." How, where, and to what end do vestiges of substances become entangled with the body? "When I received my results, I imagined various routes that mercury may have taken to my body (tuna sandwiches in childhood? Dallas air pollution?)."[30]

Substances, synthetic compounds, radioactivity pass through and concentrate in the body. Whether derived from plastics, pesticides, herbicides, fugitive industrial emissions, cosmetics and other consumer items, or low-level radioactivity, matter exemplifies material efficacy and force. Consistent with Ulrich Beck's risk society thesis, trans-corporeality reveals the dialectical forces of science and technology, in which the more robustly the nonhuman world is viewed as inert, the more trenchant the unintended rebound effects. Institutional decisions and nondecisions are expressed at a personal level, as progress comes with trade-offs defined in terms of "permissible dose" and "established thresholds." Cartesian dualism appears quaint in an age when detectable traces of matter are inscribed in blood, tissue, and hair, present in one's progeny before they are born. Detachment and distance hardly describe the human-nonhuman relationship, as "the human is always the very stuff of the messy, contingent, emergent mix of the material world," Alaimo clarifies.[31]

"The dominion of the senses, which had sufficed in traditional times, no longer protects us," Steve Matthewman observes.[32] And caught in the maelstrom between epistemology and ontology, public relations and the etiology of disease, contamination and strategic ignorance, lies a genre that Alaimo calls "material memoir," in which individuals seek meaning amid environmentally induced illness.[33] Material memoirs make evident slow environmental violence. They add flesh and bone to crescive controversies that might otherwise escape attention, and they underscore a shift in subjectivity as people find themselves situated in between those who do not understand the lived experience of illness and the organized irresponsibility of institutional structures in society—organized irresponsibility demonstrating that as science and technology have become pervasive in society, institutional mechanisms of accountability and redress lag far behind.[34] This in-betweenness forces a redefinition of self, nature, and society. While material memoirs are motivated by the desire to make sense of one's illness or that of family and friends, easy answers are rarely forthcoming, and the sense of betrayal is amorphous. "True awareness eludes most, if not all, members of risk society," Alaimo observes.[35]

Environmental illness is never simply the experience of any given individual but is rooted in cultural attitudes, beliefs, and prevailing social organization in society. The very act of sketching out the linkages between lived experience and society gives material memoirs weight, as powerful interest groups prefer that environmental illness remain a private burden. Material memoirs question the prevailing orthodoxy of the undisciplined modification of nature under the banner of progress and reticence to confront the unintended harmful effects.[36]

As Steingraber recounts, even after decades of regulation and legislation, few people do not live "downstream" of industrial production in some manner.[37] In *Full Body Burden*, Kristen Iversen recalls growing up in the shadow of the Rocky Flats Nuclear Weapons Plant near Arvada, Colorado. At times sudden and dramatic, as with a fire in 1969, contamination was more often slow and incremental. "The problem with Rocky Flats is not just a smoking chimney or a hole in the dike. The weapons plant is like a bag filled with ultrafine sand—a bag filled with millions of glittering, radioactive specks too tiny to see—and the bag has been pricked with pins."[38] In *Refuge: An Unnatural History of Family and Place*, Terry Tempest Williams contemplates faith, land, and loss. Growing up in Utah during the downwind years, she asks questions that many Utahans ponder and looks to her genealogy for answers: "I belong to a Clan of One-Breasted Women. My mother, my grandmothers, and six aunts have all had mastectomies. Seven are dead. The two who survive have just completed rounds of chemotherapy and radiation. . . . This is my family history."[39]

It is impossible to definitively attribute Martin Bordoli's leukemia to fallout, but that does not indicate that fallout was definitively not the cause. Difficulty establishing cause and effect is not tantamount to evidence of its absence, but it does provide the opportunity for powerful interest groups to engage in the organizational production of denial and of ignorance.

Amid manufactured uncertainty, however, Beck does not retreat into nihilism but suggests greater democratic deliberation and "subpolitical" movements are key to crafting a greater balance between innovation and accountability.[40] "Subpolitics" refers to activity that seeks to insert itself into the routines and practices of formalized political policy and practice. Social movements that are concerned with risk and public accountability are an expression of subpolitical activity and evidence of mistrust and skepticism in an age of organized irresponsibility. Subpolitics endeavors to pry open dialogue and debate to include a broader range of perspectives, values, and possibilistic thought. Beck argues, "We are living in an age of technological fatalism, an 'industrial middle ages,' that must be overcome by more democracy—the production of accountability, redistribution of the burdens of proof, division of powers between the producers and the evaluators of hazards, public disputes on technological alternatives. This in turn requires different organizational forms for science and business, science and the public sphere, science and politics, technology and law, and so forth."[41]

Beck stresses that subpolitical activity is driven by "reflexive modernization," in which the public, law, and government are increasingly concerned with the contradictions of science and technology. It is an evolution, a "new modernity," engaging in self-confrontation and attuned to the paradoxes of national security and a higher standard of living.[42] Reflexive modernization is not conceived, Beck notes, as embracing irrationality but as a more expansive, holistic, and participatory rationality that is suitable to the novel risks plaguing modernity.

Joint Committee on Atomic Energy Hearings, 1957

The Bravo detonation in March 1954 demonstrated the hazards of radioactive fallout, and apprehension reverberated over time. "Like the radioactive cloud that had swept over the Pacific, the fallout debate could not be contained: it spread beyond government circles," Richard Hewlett and Jack Holl explain.[43]

And over the course of 1950s, the JCAE adopted a more assertive stance.[44] In May 1957, the JCAE, through a subcommittee chaired by Representative Chet Holifield of California, conducted hearings to gather data on fallout and attendant hazards to public health.[45] It was a "scientific seminar" involving fifty experts from around the country.[46] The JCAE wrestled with the question of whether the stratosphere served as a repository in which strontium-90 was trapped for an extended period, so that upon redeposition, it constituted less of a hazard. Further, there was debate regarding the uniformity thesis, suggesting that when strontium-90 did eventually descend from the stratosphere, it deposited around the world in an even, unvarying manner. Consistent with the *diffusion premise*, suggesting that nature acts as a buffer absorbing the assaults of humanity, AEC officials argued for long-term stratospheric holdup and the uniformity thesis.

Willard Libby was an AEC commissioner in 1957 and a key participant at the JCAE hearings. He was an esteemed scientist who earned a PhD in chemistry, participated in the Manhattan Project, taught at Princeton University and the University of Chicago before joining the AEC, and was awarded the Nobel Prize in 1960 for research on radiocarbon dating. "Our policy is to discover the truth about fallout and to make it public," he assured the members of the JCAE.[47] However, Libby consistently pushed the long-term stratospheric holdup argument—at one point insisting that strontium-90 was trapped for up to ten years before descending back to earth. It was an overly optimistic assessment contributing to the commission's penchant for discursive momentum.

By the late 1950s, research indicated that strontium-90 can descend from the stratosphere in less than eighteen months.[48]

At the 1957 JCAE hearings, Libby went to great lengths to sustain the organizational production of willful and strategic ignorance regarding the hazards of strontium-90. Indeed, his testimony epitomized the effort to sustain belief in an idea that lacked empirical support. Under pointed questioning, he was derailed by his own obfuscation and refusal to admit mistake, all while introducing irrelevant side issues so that the debate became dizzyingly convoluted. Indeed, his testimony was contradictory and, at times, nonsensical.

Confronted with the testimony of Lester Machta, a meteorologist at the US Weather Bureau, Libby maintained his commitment to the uniformity thesis, backed down, redoubled his commitment, backed down slightly—all while dodging the substance of the committee's central concerns. And this is after briefly disavowing that he ever advanced the uniformity thesis amid questioning from Senator Clinton P. Anderson of New Mexico:

> SENATOR ANDERSON: But do you believe that strontium 90 is deposited uniformly over the world?
> DR. LIBBY: No, sir; and I never said that, sir. What I said was different—that the stratospheric components of the fallout might very well be uniform, and for the time being until we know better, we would assume so.[49]

That strontium-90 disperses uniformly in the stratosphere—as opposed to eventual uniform distribution over the earth's surface—is not the argument that Libby had been articulating in public speeches and academic journal articles or the assertion that he eventually settled on as his testimony progressed. Indeed, after denying that he said what he clearly said, he turned around and said it: "The stuff that goes upstairs into the stratosphere stays so long that it has time to mix pretty much all over the world."[50] In August 1957—two months after the hearings concluded—Libby continued to advance the uniformity thesis in tandem with the ten-year stratospheric holdup assumption.[51] He repeated these assertions in January 1958.[52] Libby contended in August 1957, "The *stratospheric fallout*, in contrast, takes years. We are not completely certain, but it appears that an average time of something like ten years, or perhaps somewhat less, is a reasonable figure, and during this time the distribution becomes nearly world-wide. . . . For air-fired megaton weapons our present indication is that the fallout is almost world-wide, and for reasons of simplicity and in the absence of better information at the present time, we work on the model that this is a uniform distribution, over the entire world, of material that falls from the stratosphere."[53]

Better information was available, and it had been discussed in detail at the JCAE hearings by Lester Machta, who urged those who were assembled to think of the stratosphere as a bathtub filled with water and radioactive debris as a blob of ink. Vigorous stirring ensures that the ink diffuses throughout the tub, but tepid mixing means that it remains concentrated as it sinks toward the bottom (or troposphere). Machta presented data illustrating that the temperate zone of the Northern Hemisphere received two to three times more strontium-90 deposition than the Southern Hemisphere did.[54] It came down more heavily in a latitudinal band encompassing the United States, Europe, and Russia. The ink did not vigorously mix in the bathtub. Moreover, Machta insisted that there was probably significant local variability of strontium-90 on the ground after drifting down to the troposphere, where it was then scavenged by uneven rainfall patterns.[55] In contrast, Libby suggested that Machta and colleagues were inadvertently measuring "young fallout," or debris that had not reached the stratosphere and had not had time to "mix," an assumption for which he offered no supporting data.[56]

Senator Anderson confronted Libby with a copy of a speech Libby had given one year earlier. Anderson inquired,

SENATOR ANDERSON: I am only trying to say, why do you always say "uniformity," when the experience shows nonuniformity?

DR. LIBBY: The experience does not show nonuniformity on the stratosphere, Senator, definitely.

SENATOR ANDERSON: I did not say that. You twisted it to mean something else. I said: "Why do we always have to assume uniformity in the fallout pattern of these materials, when all of the experience shows nonuniformity of fallout?" I am not talking about what is upstairs.

DR. LIBBY: We certainly should not do that, and we have never done
  that Senator; and I think it is important to make that clear.
SENATOR ANDERSON: You mean this speech does not assume that?
DR. LIBBY: That speech is correct as far as I know.
SENATOR LIBBY (READING): "We work on a model that this is a
  uniform distribution over the entire world."
DR. LIBBY: Stratospheric, sir.[57]

Outside the halls of Congress, Libby frequently advanced the uniformity thesis before there were data on the issue and continued to do so after the data indicated that it was erroneous.[58] And over the course of the 1950s, he was forced to continually revise downward the stratospheric sequestration of radioactive debris: from ten years to seven to five to less than five to less than a year, depending on how high in the stratosphere debris ascended. In 1960, the same year he received the Nobel Prize, he coauthored an academic journal article documenting that stratospheric holdup could be as little as eight months for polar debris and less than five years at the equator.[59] In 1968, Libby coauthored a study illustrating a stratospheric residence time of around 1.6 years.[60]

Reality "is what resists," Latour argues.[61] First the sheep and then strontium-90: when confronted with biophysical-material resistance, AEC officials defaulted to the organizational production of willful and strategic ignorance.

In an editorial in the *Saturday Review* in August 1957, Representative Chet Holifield asserted that when faced with the hazards of atomic testing, AEC officials consistently sought to "play it down" or withhold information that was vital to public dialogue and debate. "It has been my experience that a Congressional investigation is often the only way to make the Atomic Energy Commission come out into the open. We literally squeeze the information out of the agency." Moreover, Holifield accused AEC officials of the selective release of data to promote continued testing rather than balancing the competing imperatives of weapons development and safeguarding public health.[62] In a memoir published in 1970, Senator Anderson also declared that he had come to distrust AEC officials over the course of the 1950s and voiced concern that the commission treated fallout as if it were a political issue rather than a public health threat.[63]

The unstated implication of Libby's testimony before the JCAE was that strontium-90 was responsive to technocratic management and control, even as other experts introduced data illustrating the opposite. Nature is not uniform. The world, not to mention radioactive fallout, is a messy, splotchy, and emergent Jackson Pollack painting. In daring to advance a nonrepresentational style, the twentieth-century abstract expressionist painters articulated a more ontologically authentic depiction of the world.[64] And it is one that is resistant to technocratic manipulation and control.

"Discourse" refers to the meanings, metaphors, statements, stories, documents, and images that, in tandem, construct a particular definition of the situation, but representations of the world accord with existing empirical patterns to varying degrees. The biophysical and material effects of nature were overtaking the AEC's preferred definition of the situation by 1957, and Libby's testimony before the JCAE was an effort to sustain the commission's narrative of strategic ignorance. He did not advance a sustainable argument, but he probably helped delay a voluntary moratorium on testing.

"Radiation Is Not New to Our Lives"

In preparing for Operation Plumbbob, the AEC released another pamphlet. If the downwinder pamphlets are sleight-of-hand, the ruse becomes more recognizable by examining all three versions.[65] Together, the downwinder pamphlets total over twenty-nine thousand words, and table 9.1 delineates the top ten most frequently occurring ones. Frequency generally signals the significance of a given word in the production of meaning within a text.[66] "Radiation/radioactive/radioactivity" is the most frequently occurring word cluster. A central aim of the downwinder pamphlets is to construct and sustain the meaning attributed to radioactive fallout. "Test/tests/testing" is the next most recurrent word cluster, along with "fallout/fall-out" and "exposure/expose/exposes/exposures." From this, one might conclude that AEC officials were not reticent to discuss testing, radioactive fallout, and public exposure levels. Indeed, given the top ten frequently occurring words, one might be inclined to applaud AEC officials for their forthright effort to discursively engage with residents downwind. It is

TABLE 9.1. Top Ten Frequency Words

	Word	Frequency
1	radiation /radioactive/radioactivity	465
2	test/tests/testing	332
3	fallout/fall-out	229
4	exposure/expose/exposes/exposures	146
5	site	141
6	level/s	120
7	Nevada	114
8	effects/effect	98
9	air	93
10	not	90

Note: All function words—the, and, a, an, and so forth—are omitted in this analysis.

instructive, however, to examine those words clustering around "radiation" that impart meaning. The downwinder pamphlets reduce the potentially multiple meaning(s) of radioactive fallout, but in doing so, what is then discursively crafted?

Table 9.2 lists the top ten co-occurring words to "radiation," ranging from one to five words to the left and right.[67] "Exposure/expose/exposes" and "level/s" frequently co-occur with "radiation." Typically, they do so in a manner insisting that radiation can be measured and is being monitored and that these measurements indicate that the levels detected are well within the established guidelines. "Nuclear" is the third most frequently co-occurring word with "radiation," generally just preceding as an adjective describing radiation.

"Background" and "natural" frequently co-occur with "radiation" as the AEC strived to naturalize its occurrence. The 1953 pamphlet observes, "Man always has lived in a sea of nuclear radiation."[68] To further naturalize radiation and neutralize public concern, the commission equates radioactivity with electricity, noting, "Very few of us can explain electricity, although we have learned to live with it and to use it. Even fewer can explain nuclear radiation."[69] Moreover, "Light, heat and radio waves are familiar kinds of radiation."[70]

Consistent with the paternalistic and instrumentally rational posture adopted, the commission stresses the imperative to harness radioactivity

TABLE 9.2. Top Ten Words That Co-Occur with "Radiation"

	Word	Frequency
1	exposure/expose/exposes	35
2	level/s	34
3	nuclear	27
4	background	24
5	effects/effect	22
6	fallout	17
7	natural	14
8	amount	12
9	external	12
10	people	10

Note: All function words—the, and, a, an, and so forth—are omitted in this analysis.

in the service of humanity. The employment of radiation in dental X-rays and the diagnosis of disease is discussed, as are the industrial applications. The commission stressed, "So, radiation is not new to our lives. In this atomic age we are living on a more familiar basis with it. It is important that we try to understand it, accept it, and use it. It is also important that we respect its powers, so that we will be guided by knowledge and not be blinded by fear of the unknown."[71]

Though "radioactive fallout" is a key term in the downwinder pamphlets, it is a circumscribed conception. This is exemplified by those words that do not appear or are used infrequently. The relative occurrence of potentially "antagonistic" words reveals the manner in which the downwinder pamphlets embody a particular configuration of meaning posing as a universal representation of the issue at hand.

As depicted in table 9.3, of the 29,254 words in the downwinder pamphlets, "cancer" appears only four times. "Leukemia" and "thyroid" cancer, recognized effects of radiation exposure by the 1950s, are each mentioned just once. Cow's "milk," the overarching link between testing and ingestion of radioiodine as well as strontium-90, only appears twice. Signifiers indicating the temporal lag between exposure to radioactivity and the expression of disease, such as "chronic," "long-term," and "cumulative," are sparsely invoked. This is indicative of a focus on the acute

TABLE 9.3. Frequency of Antagonistic Words

	Word	Frequency
1	cancer	4
2	leukemia	1
3	thyroid	1
4	chronic	0
5	longterm/long-term	2
6	cumulative	1
7	milk	2
8	contamination/contaminated	3
9	danger	2
10	decontamination	1

Note: All function words—the, and, a, an, and so forth—are omitted in this analysis.

occurrence of symptoms and avoidance of the implications of crescive exposure. Consideration of the variant impact of fallout on women, infants, or children is absent in the downwinder pamphlets. Potentially antagonistic words such as "contamination/contaminated," "danger," or "decontamination" are also scarce. Such declarations would have undercut the meaning that AEC officials sought to construct.

The commission asserted that in curbing Soviet aggression, time is short, but so too is the presence of radioactivity in the environment. Long-lived radionuclides such as strontium-90 and cesium-137 are acknowledged, but residual radioactivity is portrayed as ephemeral and transient. The AEC declared, "As the cloud moves on, it becomes dispersed and usually within a few hours is no longer visible, having spread into an air mass. With each minute, its radioactivity loses strength. . . . The path of fall-out is narrow at the test site and in the nearby region, widens to hundreds of miles as it moves on, and tends eventually to be distributed uniformly over the earth's surface. It does not constitute a serious hazard to any living thing outside the test site."[72]

The pamphlets insist that radioactivity is endemic to continental testing but is natural, ubiquitous, and amenable to understanding and control. Continental testing is an efficient use of time, labor, and materials, as well as essential to national security amid a looming Soviet threat. Reason and rational thought have unlocked the secrets of atomic structure and guide testing at the NTS to ensure that it is done with due regard for the health and safety of downwind residents. National security, in turn, is predicated on the imposition of radioactive fallout, but the AEC is uniquely prepared to minimize the risks through methodical preparation, planning, and measurement. "Radiation is not new to our lives," and the best option for downwind residents is acquiescence to AEC authority and expertise.

"Safety Testing" in Nevada

A troubling concern was whether an atomic detonation could unintentionally be initiated by an asymmetrical conventional explosion, as might occur when a bomber crash lands, and there were questions regarding the biological effects of plutonium dispersal. Safety testing was conducted between 1955 and 1958, including five shots during the Plumbbob series, and in 1963 at the Tonopah Test Range in central Nevada to answer such questions.[73] It contributed to more robust failsafe systems in the United States' atomic weapons stockpile but spread plutonium far and wide. One truism of the atomic age is that things, energetic things, migrate farther than generally anticipated.

Project 57 in April 1957, just prior to Operation Plumbbob, was one such effort. Designation as a "safety test," Annie Jacobsen argues, was intended to hide from the public the fact that it entailed plutonium dispersal and was potentially quite the opposite of "safe."[74] Plutonium-239 has a half-life of twenty-four thousand years. Released into nature, it is not going anywhere anytime soon. It is also one of the "most radiotoxic materials in the world" if inhaled or ingested.[75] "Plutonium is the darling and the demon of the nuclear age," Kristen Iversen explains.[76]

A sixteen-square-mile area was cordoned off for Project 57 beyond the northeast boundary of the NTS and just five miles from the secret airbase at Groom Lake, the infamous Area 51. Over seventy beagles were placed in the contaminated area to study the effects of chronic exposure to plutonium.[77] The menagerie also included burros, sheep, and rats in cages. Vehicles were dispersed about the desert, and concrete sidewalks were built to study decontamination procedures. The vehicles, equipment, protective suits, and other materials were buried onsite, a barbed-wire fence was constructed, and the area was largely forgotten for two decades. Project 57 was conducted beyond the boundaries of the NTS, Jacobsen contends, so it could be designated a military operation and sidestep information disclosure rules, to the extent they existed, pertaining to the AEC as a civilian governmental organization.[78]

In the spring of 1957, however, a problem arose eliciting considerable discussion within the AEC. A reporter from Los Angeles was visiting towns downwind of the NTS, asking questions, and seeking information regarding radioactive contamination. Paul Jacobs published a story titled "Clouds from Nevada" in the Reporter on May 16, just prior to the start of Plumbbob. Jacobs wrote a perceptive critique of AEC operations in Nevada. He argued that the AEC failed to collect data on external gamma radiation in many places, while insisting that the data revealed no cause for alarm, even as the data collected were classified, despite no reason for being classified other than to avoid public oversight. From the beginning, Jacobs asserted, AEC officials assured the public that there was no threat to public health, and yet by 1957, they were scrambling to reduce offsite fallout. Moreover, he highlighted the death of Martin Bordoli, the tribulations of mining and ranching families living just beyond the test site, the sheep deaths of 1953, and the shelter-in-place directive at St. George due to shot Harry.[79]

In April 1958, Jacobs published an article in the Reporter titled "The Little Cloud That Got Away," describing a recent safety test at the NTS in which radioactive debris was detected in Los Angeles.[80] Project 58 involved four safety experiments, and it was Coulomb-C on December 9 that produced the cloud that Jacobs described. Coulomb-C had a nuclear yield of half a kiloton, and the cloud rose to thirteen thousand feet while unexpectedly heading west-southwest. The problem, as always, Jacobs stressed, was information, or the lack thereof, and he accused the AEC of downplaying the issue as it did fallout from weapons testing.[81]

In 1971, the AEC sampled soil throughout Utah to document strontium-90, and the results uncovered levels of plutonium that could not be accounted for by Nevada and global atmospheric weapons testing alone—suggesting that safety testing dispersed plutonium far downwind.[82] Of note, plutonium was higher in samples taken from major population centers including Salt Lake City and Provo. In 1974, soil was sampled from areas in Utah, Nevada, Wyoming, and Idaho and again documented excess plutonium in Utah where the bulk of the state's population resides. Lower levels were observed in southwestern Utah, where external gamma radiation was consistently higher from weapons testing. This suggested that the excess plutonium was not unfissioned material from detonations but from safety testing. Twelve of the twenty-six safety tests conducted in Nevada occurred when wind trajectories exhibited a northeasterly pattern, depositing material over north-central Utah.[83]

The desert may be better suited to weapons testing given the lack of precipitation, but it is less ideal for dispersal experiments. Rain means greater vegetation, more stable soil, and less chance that a contaminant will be resuspended by wind over time. The Great Basin stretches from the NTS to Salt Lake City and is infamous for its winds and lack of rain. Research in 1998 sampled soil and attic dust in areas of southern Nevada and Utah and detected excess plutonium, including to the south and west of the NTS, where it was least expected. "Thus safety tests, contrary to popular opinion, contributed significant amounts of plutonium to Nevada and Utah," James Cizdziel and colleagues observe. They further conclude that some of the plutonium detected was due to "wind driven resuspension."[84] The most radiotoxic substance in the world is migrating. This is consistent with research published in 2003 documenting that a high proportion of excess plutonium sampled is "retained for decades in the upper few centimeters of soil in Nevada's desert environment."[85] Even when atomic yields were low or nonexistent, there were experiments in Nevada constituting a risk to public health.

Operation Plumbbob, May–October 1957

Operation Plumbbob was a flurry of detonations amid a looming moratorium on weapons testing. It consisted of twenty-nine shots over five months, outlined in table 9.4. It was the longest series conducted in Nevada, with the highest overall explosive yield injected into the atmosphere. More cesium-137 was deposited around the country during Plumbbob than during any other series in Nevada.[86] Simon and Harry in 1953 deposited more cesium-137 than any other detonations, but although Plumbbob tower and balloon shots deposited less on average, they occurred one after another: Boltzmann, Wilson, Priscilla, Hood, Diablo, Shasta, Doppler, and Galileo occurred in a rapid-fire

TABLE 9.4. Operation Plumbbob Detonations

Name	Type	Yield
Boltzmann	tower	12 kilotons
Franklin	tower	140 tons
Lassen	balloon	.5 tons
Wilson	balloon	10 kilotons
Priscilla	balloon	37 kilotons
Coulomb-A	surface (safety test)	0 kilotons
Hood	balloon	74 kilotons
Diablo	tower	17 kilotons
John	rocket	2 kilotons
Kepler	tower	10 kilotons
Owens	balloon	9.7 kilotons
Pascal-A	shaft (safety test)	56 tons
Stokes	balloon	19 kilotons
Saturn	tunnel (safety test)	0 kilotons
Shasta	tower	17 kilotons
Doppler	balloon	11 kilotons
Pascal-B	shaft (safety test)	1 gram
Franklin Prime	balloon	4.7 kilotons
Smoky	tower	44 kilotons
Galileo	tower	11 kilotons
Wheeler	balloon	197 tons
Coulomb-B	surface (safety test)	300 tons
Laplace	balloon	1 kiloton
Fizeau	tower	11 kilotons
Newton	balloon	12 kilotons
Rainier	tunnel	1.7 kilotons
Whitney	tower	19 kilotons
Charleston	balloon	12 kilotons
Morgan	balloon	8 kilotons
Total		343 kilotons

Source: DOE, *United States Nuclear Tests: July 1945 through September 1992* (Las Vegas: Nevada Operations Office, 2000).

manner.[87] Indeed, the levels of cesium-137 deposited on the public in 1957 was unlike anything ever undertaken in Nevada. New strategies were employed to reduce offsite fallout, but the scale of the Plumbbob series ensured that it would leave its mark on the downwind towns and much of the country.

"A problem with the Plumbbob series was not long in coming. In fact, it occurred on the first shot," Fradkin notes.[88] Boltzmann was relatively small, at twelve kilotons, and detonated atop a five-hundred-foot tower. Fallout was predicted to go eastward.[89] As outlined in figure 9.1, it went every direction but east in the hours after detonation. Lower-level debris moved northwest over Washington state. Fallout at twenty thousand feet made a circle over Oregon and Idaho. Debris around thirty thousand feet bent westward and passed over San Francisco, California, before reversing course and passing near Los Angeles, making a loop over Utah, dipping down around El Paso, Texas, and then crossing over Pennsylvania before reaching the Atlantic. A hot spot—with radiation measured seven times higher than the surrounding countryside—was detected

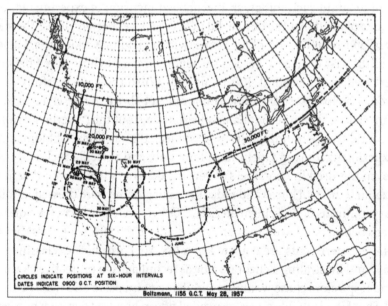

Figure 9.1. Shot Boltzmann fallout trajectory over the United States. (Gordon M. Dunning, "Fallout from Nuclear Tests at the Nevada Test Site," AEC, May 1959)

sixty miles north-northwest of the test site.[90] There probably had been many hot spots downwind of Nevada, but this one was captured by off-site monitors. Six years after testing began in Nevada and twelve years after Trinity, AEC and military officials were still struggling with the uneven and capricious behavior of fallout. Radioactive contamination flowed over the landscape like invisible snowdrifts.

Several shots in the Plumbbob series reaffirmed the hot spot phenomenon, which was then addressed at the JCAE hearings in May 1959. AEC officials downplayed the issue. A summary report of the 1959 hearings recounts, "Some testimony suggested that higher radiation levels in limited areas do not, any more than other nonuniformities, increase the overall hazard of fallout to the world's population."[91] This was hardly the point. Hot spots increase the risks imposed on someone, even if not everyone. As with the averaging of data over a large area or an extended period or the liabilities of a single threshold of gamma radiation exposure, that hot spots did not increase the *worldwide* risk of radiological contamination was sleight-of-hand.

Franklin was next, and estimated at two kilotons, it was a "fizzle," at only 140 tons.[92] The weapons laboratories and AEC officials were reticent to admit such things, an inevitable dimension of experimentation, and explained that the device did not "meet its expected yield." The shot was later refired as Franklin Prime, with a yield of 4.7 kilotons. Lassen was the first device in Nevada suspended from a helium-filled balloon tethered to the ground. It, too, fizzled. Wilson was successfully fired from a balloon suspended five hundred feet above the desert, and then came a large balloon shot named Priscilla, a stout thirty-seven kilotons. Around twelve hundred pigs were subjected to a variety of biomedical experiments during Operation Plumbbob, and Priscilla included over seven hundred in containers at varying distances from ground zero to examine thermal burns and radiation injury. Some were dressed in military uniforms of differing materials.[93] Others were placed in fox holes or pens behind large window panels to examine the effects of glass shards. These experiments are depicted in a US Air Force film declassified in 1997 and illustrating a chaotic scene as technicians in protective clothing and respirators chase wounded and dying animals, to and fro.[94]

Shot Hood, on July 5, was suspended fifteen hundred feet high by a captive balloon, and it was the largest-yield atmospheric shot ever detonated

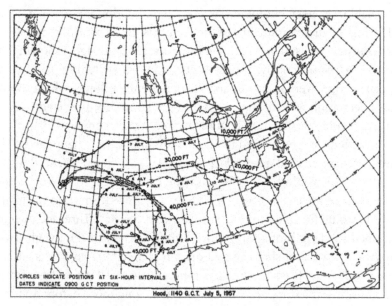

Figure 9.2. Shot Hood fallout trajectory over the United States. (Gordon M. Dunning, "Fallout from Nuclear Tests at the Nevada Test Site," AEC, May 1959)

on the continent—at seventy-four kilotons. It was also the first thermo-nuclear device fired in Nevada.[95] Figure 9.2 depicts fallout from Hood over time at variant heights. Few detonations better embody the fluidity of fallout. Upper-level debris moved eastward over Colorado and Nebraska, before dipping down over Oklahoma and Texas and then returning to Colorado, before moving across the Midwest. Although the fireball did not touch the ground, Hood was one of the top contributors of cesium-137 deposition around the country.[96] As demarcated in figure 9.2, it did so in a manner reminiscent of a child's playful scribbling on paper.

Heat amassing at ground level pushes air upward, wind shear results from the abrupt shifting of air flow, and clear air turbulence stems from currents of air converging and displaying chaotic state conditions; if the latter were visible, it would resemble swirling eddies in a rocky stream. "The smooth flow that early meteorologists envisioned for the upper atmosphere does not exist," Jan DeBlieu explains. "Instead the air above us is filled with waves and crashing breakers." "Watch the clouds; they are evidence of our atmosphere's complex restlessness."[97]

Diablo was next, but there was a problem. Loudon Wainwright notes, "During the final minute one of the scientists in the control room stared at his instruments in disbelief. On the console board, lights showing the orderly activation of the experiments had been switching from red to green, as they were supposed to. Now, suddenly, voltages which should have been high were dropping off. 'It's not gonna go,' the scientist said, his calm voice clearly audible in the room. 'It's not gonna go.'"[98]

Electricity surged toward the detonator, but nothing happened. "The desert remained cold, pitch black, and very, very quiet," Richard Miller explains. "It would have to be disarmed manually. To make matters worse, since the elevator winch had been removed, whoever disarmed Diablo would first have to climb the 500-foot tower."[99] Three technicians climbed the length of a forty-story building to reach the twenty-two-by-twenty-two-foot shot cab and disarm Diablo.[100] Two weeks later, it was successfully fired. Despite a comparatively small yield of seventeen kilotons, it was a dirty shot. Diablo ranks among those detonations accounting for significant population exposure to gamma radiation, and it deposited a notable proportion of cesium-137 throughout the country.[101]

Shot John on July 19 was a two-kiloton air-to-air missile designed to be deployed against Russian bombers, but it is best known for the five Air Force officers standing at ground zero as it detonated overhead, along with a videographer.[102] Film footage depicts their nervous energy as they peer into the sky, and a sign reading, "Ground Zero Population: 5," is staked into the sand. A jet enters the frame, a missile streaks across the sky, and a diffuse explosion ripples 18,500 feet overhead. The men squint in the flash of light and then crouch as the shock wave comes to earth, followed by a rumbling from above.[103] Assembling Air Force officers below an exploding atomic missile was a public-relations ploy to illustrate that high-altitude atomic warfare would not pose a radiological risk to civilians peering up from below, and no radioactivity was detected at ground-level.[104]

A ten-kiloton tower shot named Kepler followed John, less than a week later, and then Owens, a ten-kiloton balloon detonation the day after that. It was a brisk pace. August was particularly busy: Stokes, a nineteen-kiloton balloon shot; Saturn, a safety experiment; Shasta, a seventeen-kiloton tower detonation; Doppler, an eleven-kiloton balloon shot; a safety experiment named Pascal-B; the five-kiloton Franklin

Prime shot; and then Smoky on August 31, at forty-four-kilotons, it was the largest tower shot ever fired in Nevada.

Perhaps the strangest event of the Plumbbob series involved a disappearing four-inch-thick steel cap, welded atop a four-foot-diameter pipe extending five hundred feet into the ground. The cap weighed approximately one ton. Pascal-B was a safety experiment with an unanticipated partial nuclear yield. A high-speed camera captured a single image as the cap propelled skyward, allowing for calculation of its velocity: around 150,000 miles per hour and beyond the escape velocity of Earth's gravity.[105] If correct, it was the fastest human-made object ever detected. Some people speculate that it ascended into orbit and constitutes the first human-made object to do so, beating Sputnik by less than three months.[106] Others surmise that it was vaporized as it ascended skyward.[107] Either way, it has never been found.

Smoky sat atop a seven-hundred-foot tower, shown in figure 9.3, as several thousand troops waited eight miles distant. They then advanced on ground zero. Miller observes,

> After the detonation, the troops on the hillside were loaded into trucks and taken through the smoke and dust to within several hundred yards of ground zero to inspect the damage. . . . At the hypocenter, surrounded by a glazed area of greenish glass, lay the melted and twisted remnants of the tower. About 2,000 feet away and to the west, another combat group had begun a steady march to the hypocenter, the green glass crunching beneath their boots. At 300 feet from the hypocenter, the men stopped and the man leading the group saluted the twisted, smoking tower.[108]

Smoky inundated the shot area with radioactivity but particularly to the southeast, while the troops, in tandem with radiological survey teams, approached from north and west of ground zero.[109] An elongated area recording 300 roentgens per hour, an extraordinarily hot reading, stretched more than two thousand yards southeast of ground zero, and a large sweeping arc at 50 roentgens per hour stretched for nearly three miles.

At ground level, conditions were calm, but velocity increased with height and winds pushed the mushroom cloud westward at over forty miles per hour.[110] Lower-level debris moved to the southeast but

Figure 9.3. Smoky, seven-hundred-foot tall tower shot. (Courtesy of the National Nuclear Security Administration)

then circled back across southern Nevada and northern Arizona, before returning to the test site and moving northeast over Wyoming, South Dakota, and Missouri.[111] Another segment moved east toward St. George, Utah. Smoky deposited radioiodine over the area, and the dose to infants and children drinking fresh milk was second only to that deposited by Harry. "These two events, together, constituted roughly 80–90% of the total dose summed across all test events that produced a measurable thyroid dose to St. George residents," research uncovered in 1996.[112] AEC officials did not advise people to avoid milk consumption after either shot. More broadly, aerial surveys revealed that debris from Smoky remained "readily detectable" seven hundred miles from ground zero five days after detonation.[113]

In the wake of Smoky, Kermit Larson and colleagues at the University of California at Los Angeles examined jackrabbits from the NTS outward for nearly five hundred miles to assess levels of strontium-90 uptake, and near Rock Springs, Wyoming, they stumbled on a "biological hot spot." Jackrabbits had levels of strontium-90 in their bone marrow comparable to those proximate to ground zero. Rainout, presumably, brought down debris from Smoky. Follow-up research was not conducted to assess the radioactive burden absorbed by children, or lambs, in the Rock Springs area. And Larson and colleagues' report on the biological uptake of radionuclides from Operation Plumbbob was not issued until 1966. With the report running over six hundred pages in length, the section constituting a potential public-relations problem is brief and couched in a highly technocratic language, but the lesson is clear: the "biological availability of fallout" from shot Smoky extended at least five hundred miles from ground zero. The report also documents biological hot spots in the aftermath of Boltzmann (78 miles from ground zero), Diablo (60 miles from ground zero), and Shasta (172 miles from ground zero).[114]

In 1980, at a workshop conducted by the DOE in preparation for *Allen et al. v. United States*, Larson admitted in regard to his team's analysis of jackrabbits at Rock Springs, "It sort of shook us up."[115] At the time, it was assumed that larger particles would fall out well before Wyoming and that the smaller particles that persisted constituted less of a hazard. Radioactive jackrabbits at Rock Springs undercut such assumptions. Another truism of the atomic age is that the biological uptake of things, energetic things, is more substantial at greater distances than generally anticipated.

The Centers for Disease Control (CDC), in 1980, documented an excess number of leukemia cases among the 3,200 Army troops participating in maneuvers during shot Smoky, and subsequent research again recorded leukemia cases above the expected background rate.[116] Controversy surrounding the health effects among troops at shot Smoky elicited a broader examination of military participants in Nevada and the South Pacific during the era of atmospheric detonations. In 1995, the President's Advisory Committee on Human Radiation Experiments (ACHRE) counted around 220,000 US military personnel participating in at least one open-air detonation. ACHRE concluded that there were no consistent patterns illustrating increased cancer among atomic veterans overall but did identify "a number of suggestive findings" in reference to particular detonations, including shot Smoky. However, the committee noted that information on troop participation and individual-level gamma radiation exposure was sparse and that research was thus beset by many uncertainties. Of the 17,062 military personnel participating in the 1953 Upshot-Knothole series, for example, only 2,282 were issued film badges to record their personal exposure.[117]

If August was hectic, then September was a mad dash: Galileo, an 11-kiloton tower shot; Wheeler, a 197-ton balloon shot; Coulomb-B, a safety experiment; Laplace, a 1-kiloton balloon detonation; Fizeau, an 11-kiloton tower shot; Newton, a 12-kiloton balloon detonation; Rainier, a 1.7-kiloton tunnel shot; Whitney, a 19-kiloton tower detonation; and Charleston, a 12-kiloton balloon detonation.

And amid all of it were the troops providing support and logistics, observing from trenches, and parachuting and running toward the twisted wreckage of a tower as if at war. They were often as much a focus of study as the device detonated. Indeed, psychologists, sociologists, and physiologists "swarmed" Camp Desert Rock for shot Galileo.[118] In its immediate aftermath, troops disassembled and reassembled their weapons, navigated an obstacle course, and lofted grenades over a wall, under the watchful eye of Army psychologists.

The 1.7-kiloton Rainier shot, nestled among the bustle of September, was one of the most important experiments of the Plumbbob series and a harbinger of things to come. Drilling 790 feet horizontally into Rainier Mesa at the northwestern edge of the NTS, it was the first detonation in which fallout was designed to be contained underground. The prevailing theory was that the intense heat would sequester radioactivity within the melted rock. "There was a low rumbling as the mesa seemed to jump slightly, then settle back in a barely detectable ripple that quickly spread over the face of the mesa. A few rocks fell down the side of the mountain," Miller explains.[119] No radioactivity was detected, nor did Rainer jump-start a major earthquake. Rainier demonstrated the viability of a self-contained underground shot.

Morgan, an eight-kiloton balloon shot on October 7, marked the end of Operation Plumbbob. It did not generate notable offsite radioactivity, and it was a tepid end to a long and hazardous five months.

AEC and military officials typically did not possess a holistic conception of the complexity and connectivity of nature, but atmospheric testing forced a broader reckoning that the atomic technocracy could hardly stave off forever. This reckoning began at Trinity. And the unintended consequences of accelerated testing during Operation Plumbbob generated their own controversies, which were a focus of congressional hearings two years later, including evidence of radioactive wheat in the Midwest. In October 1957, Governor Orville Freeman of Minnesota established a committee composed of two dozen state experts to examine evidence of fallout affecting the state's wheat and other agricultural crops.[120] They catalogued surprisingly high levels of strontium-90, and this broadened the chorus of critics accusing AEC officials of withholding information and downplaying the hazards of fallout.

Baby Teeth, Project Sunshine, and a Moratorium on Testing, 1958–1962

In August 1958, in the journal *Nature*, Dr. Herman M. Kalckar proposed an "International Milk Teeth Radiation Census" to assess strontium-90 uptake among children worldwide. Given its chemical similarities to calcium, strontium-90 is absorbed in deciduous teeth. Herein lies a record of the geopolitical posturing of the US and the Soviet Union in the mouths of babes. "If a continued general trend toward a rise in radioactivity in children's teeth were ascertained, it might well have important bearings on national and international policy," Kalckar argued.[1] It did have bearing on national and international policy and, in time, dislodged the prevailing AEC narrative regarding open-air testing.

The worldwide census that Kalckar envisioned never materialized, but the Greater St. Louis Citizens' Committee for Nuclear Information (CNI) took up the challenge domestically. The name is particularly apt. W. K. Wyant Jr. explains, "The view prevailed that what really was needed was information. It was felt that too many people—the politicians, the military and the oracles speaking 'ex cathedra' from the Atomic Energy Commission—were taking decisive attitudes on the basis of indecisive information, or none."[2]

CNI was a consortium of faculty from Saint Louis University, Washington University, and community members. The biologist Barry Commoner was a founding participant, pushing forward the organization's objectives before the public while volunteers catalogued and prepared baby teeth for analysis. Many were mothers concerned for the health of their children. CNI produced brochures and a newsletter and reached out to St. Louis–area schools, libraries, Cub Scout troops, dentist offices, and civic organizations large and small. Children who mailed in a tooth received a button declaring, "I Gave My Tooth to Science," and by 1959, CNI was receiving hundreds of teeth each week.[3] Over 320,000 teeth were collected before the project ended in 1970.[4]

Dr. Louise Reiss served as the project director from 1959 to 1961 and published the first results in the journal *Science* in 1961. Children born in 1951 and 1952 had stable strontium-90, but levels increased in 1953 and again in 1954, consistent with increasingly high-yield detonations by the United States and the Soviet Union.[5] Subsequent research found a thirty-fold increase in strontium-90 levels between 1951 and 1963, in tandem with the increasing pace of the global arms race.[6]

Few parents forget the drama of coaxing loose a tooth from their child's mouth, but by the late 1950s this right of passage was of scientific and political consequence. It was a means of obtaining information that the atomic-industrial complex could not contain, control, or classify. The levels of strontium-90 catalogued were not known to be hazardous, but neither was it clear that they were benign as children aged. What was clear is that the increasing presence of strontium-90 suggested to many people that the debate over atmospheric testing could no longer be dominated by a select few whose authority was derived from their ties to the technocracy from which the problem arose. Even in a place as far removed from Nevada and the Pacific Proving Grounds as St. Louis, Missouri, strontium-90 found its way into the water supply, pasture, and soil. The survey made manifest a contaminant composed of energetic matter derived from the prerogatives of the atomic state. And as a distinctive marker of the atomic age, that contaminant was drifting far and wide.

The baby tooth survey revealed trans-corporeality. It made the world feel smaller. Indeed, it underscored the reduction of *ontological distance*. Particularly after the invention of thermonuclear devices, technology compressed the distance between the uncontainable material and biophysical implications of militarism while forcing the public to wrestle with an invisible contaminant whose effects are difficult to substantiate empirically. The reduction of ontological distance amid the extended *temporal distance* between exposure to strontium-90 and the expression of disease presented significant scientific as well as political challenges. Geography and even social class no longer offered sanctuary from US militarism. "The Atomic Energy Commission turned me into an environmentalist," Barry Commoner remarked.[7]

"Matter is always ongoing historicity," Karen Barad notes.[8] Material and biophysical movement and change are at the forefront of human activity. They open up new courses of action but also defy expectations and preferred courses of action. That strontium-90 arose as a persistent signature of the Cold War arms race was an unforeseen contradiction intruding on the AEC's best laid plans. The struggle to assess its vibrancy and fluidity was of historical import and, in time, a material and biophysical force disrupting a technocratic order that the US public, alone, could not dislodge.

Project Sunshine

Few Cold War research projects so vividly highlight the permeability of the body as Project Sunshine, a clandestine effort to obtain bodies and body parts for analysis. It was founded on Project Gabriel, established by the AEC in 1949, which examined the potential health effects of fallout in the wake of nuclear war. Project Gabriel's principal conclusion was that strontium-90 constituted the most hazardous by-product of the fission process.[9] This conclusion gained urgency in 1952 with the detonation of Ivy Mike in the Marshall Islands, the world's first thermonuclear device, with a yield of ten megatons (ten thousand kilotons). The Soviets detonated a thermonuclear device less than a year later. Project Sunshine began at a meeting of AEC officials and affiliated scientists at the RAND Corporation in Santa Monica, California, in 1953. Project Sunshine began just prior to Castle Bravo, and the capacity to harness the materiality of nature clearly outpaced understanding of the ecological and biophysical connections that were inadvertently set in motion. Eileen Welsome clarifies, "The group decided the only way they could properly ascertain worldwide hazards from fallout was by collecting and analyzing plants, animals, and human tissue from the four corners of the Earth. Thus was born Operation Sunshine, one of the most bizarre and ghoulish projects of the Cold War. The source of its name is a matter of debate, but some say it was derived from the fact fallout, like sunshine, covered the globe."[10]

Project Sunshine was conducted in strict secrecy because of the public-relations implications. AEC officials did not want the public to know that they were collecting bodies for analysis, and a central problem discussed at the RAND conference was the difficulty of obtaining human "samples," particularly from outside the United States, without attracting attention. Participants confronted another conundrum: What happened to the radioactive debris from Ivy Mike? Little fallout was detected by monitoring stations across the Pacific. Attendees speculated that the remainder either came down where there were no monitoring stations or remained suspended above the troposphere—something that had never occurred before. Willard Libby suggested that debris was mixing in the stratosphere and that when it came back to earth—which he speculated could be as long as ten years—it was unlikely that there would be significant terrestrial hot spots.[11] Nature provided an advantageous buffer. "The problem, however, was that Libby's model of stratospheric fallout was pure speculation," Emory Jessee notes.[12] Indeed, the classified Project Sunshine report of 1953 clearly stated the tenuousness of assuming even, smooth redeposition of strontium-90: "First, we fear that the concept of uniform world-wide contamination has little meaning and that the necessary assumptions for such a calculation are unrealistically simple. The contamination undoubtedly will occur unevenly—in 'blobs' over large areas—mainly because of large differences in localized fallout concentrations."[13]

Nonuniformity would demonstrate the difficulty of assessing the ebb and flow of a hazard that was global in scope but localized with regard to the risk it presented. And wide variance suggested variance that was uncatalogued through scattered gummed filters or rainwater collection stations. In contrast, uniformity implied that data collected at far-flung places were good enough, as there was little disparity to capture. Uniformity of redeposition was a simplifying assumption that was consistent with the preferences of the prevailing technocracy, rather than established through empirical observation.

Several key operating assumptions were articulated at the RAND conference that remained the subject of debate in the years to come. These included (1) that strontium-90 displays uniform redeposition over the Earth's surface; and (2) that strontium-90 is primarily absorbed by the growing bones of children.[14] Both assumptions were erroneous.

In January 1955, the AEC held a "Biophysics Conference" in Washington, DC, to review Project Sunshine data gathered over the preceding thirteen months. John Bugher, director of the AEC's Division of

Biology and Medicine, began the meeting by noting that in the wake of the Castle Bravo series, "events have overtaken us."[15] AEC commissioner Willard Libby stressed the challenges in assessing radiostrontium deposition encompassing much of the globe. This included analysis of descent from the stratosphere and incorporation into milk and other food products, retention in the oceans and human consumption of fish, and runoff from soil into creeks and rivers. Of particular concern was the rate of absorption in human bone as weapons yields multiplied. Libby remarked, "By far the most important is human samples. We have been reduced to essentially zero level on the human samples. I don't know how to get those, but I do say that it is a matter of prime importance to get these and particularly in the young age group."[16]

Attendees reviewed a July 1954 Project Gabriel report indicating that analysis had so far focused on milk and cheese samples, calves, lambs, and fifty-five stillborn children's remains from Chicago, one from Utah, three from India, and three adult legs from Massachusetts. Those who were assembled agreed that more samples should be acquired through personal contacts with medical professionals, including hospital directors and pathologists. To sustain the secrecy of the program, the cover story was that they were needed to study exposure to naturally occurring radium. In turn, Libby endorsed "body snatching." Dr. J. Laurence Kulp, of Columbia University, declared that personal contacts were in place from which to obtain human cadavers from Vancouver, Houston, New York, and Puerto Rico: "Down in Houston they don't have all these rules. They claim that they can get virtually and they intend to get virtually every death in the age range we are interested in that occurs in the City of Houston. They have a lot of poverty cases and so on."[17]

The sleight-of-hand was twofold: (1) AEC officials did not want the public to know they were collecting human remains for analysis; and (2) people outside the AEC supplying human remains were told that it was needed to assess the effects of something other than strontium-90.

In a meeting of the General Advisory Committee to the AEC in August 1953, Libby confided that obtaining human samples covertly was difficult but that stillborn babies "are often turned over to the physician for disposal" and constituted a "practical source."[18] Their remains were cremated, and the ashes were subjected to radiochemical analysis. The dead contributed to assessment of matter that could not be measured in the living.

Attendees at the January 1955 Biophysics Conference wrestled with the legality of their efforts, potential public-relations consequences, and the unique problems of obtaining human samples from abroad. Libby explained that a law firm had been hired to investigate the "law of body snatching" but the results were not "encouraging." "It shows you how very difficult it is going to be to do legally." John Bugher noted that when procuring "human material" from outside the United States, it was necessary to have a contact at the local level "who got paid by the sample."[19]

Participants at the 1955 Biophysics Conference also wrestled with the question of whether adults absorbed strontium-90. Indeed, the data acquired over the preceding thirteen months suggested that adult bone did not register "essentially zero assay," as previously assumed.[20] Libby was surprised. The strontium-90 hazard encompassed more than simply the bones of children:

COMMISSIONER LIBBY: The only ones that are of interest are the young ones. Is this true?

DR. BUGHER: No.

COMMISSIONER LIBBY: Why are adults interesting? There is no strontium in them.

DR. KULP: There is some strontium in the 30 year olds. It cuts off at about 40.

COMMISSIONER LIBBY: Really? How many Sunshine units?

DR. KULP: .1, .2.

COMMISSIONER LIBBY: That is very frightening.[21]

Further, during the 1955 Biophysics Conference, there was intermittent discussion in reference to the "iodine story." John Bugher contended that the Castle series of 1954 revealed that radioiodine could be a "very significant problem." Later in the meeting, Robert A. Dudley, of the AEC's Division of Biology and Medicine, described how the commission was first alerted to the issue by British scientists who detected iodine-131 in children's urine due to fallout from Nevada. Subsequently, Dr. Kermit Larson of the University of California at Los Angeles detected radioiodine in animals around the NTS, and research found iodine-131 in human samples in the wake of the Castle series as well as Russian detonations in the fall of 1954. Dudley remarked, "All the time we have been more or less ignoring it. It is certainly something we want to consider more than we have done in the past." He articulated for those who were assembled that scaling the radioiodine data based on concurrent strontium-90 data led to the estimation of "huge doses to the thyroid." "Presumably the thyroid is not nearly as radiosensitive as the skeleton, and a different time scale is involved. It is a problem which we want to consider," Dudley admitted. And as the meeting progressed, J. Lawrence Kulp, of the Lamont Observatory, assured the group that human thyroid samples would be collected under the auspices of Project Sunshine to better evaluate iodine-131 uptake over time.[22] This discussion is significant, as it suggests that AEC officials were aware of the potential threat posed by iodine-131 as early as 1955. Nonetheless, little attention was devoted to the "iodine story" until 1962.

The 1955 conference was characterized by a specialized, technocratic language of exposure levels and laboratory procedures but also a familiarity born of thirty participants who were well-educated and well-placed men. This familiarity born of common social position is best articulated by Forest Western of the AEC. At the end of a technical overview regarding strontium-90 uptake in the general population, Western pondered,

> I want to say just for my own entertainment more or less that I personally do not think that it would be possible for the human race to wipe itself out of existence by methods of this sort. I think you will find a few Eskimos or a few Patagonians or a few people in some isolated part of the earth who will keep the race going. *They might not populate the earth with just the descendants we would like to see. They might not be highly civilized like we are.* They might not know anything about atomic warfare, for example. But I think the concept of wiping the race out with nuclear weapons is a little bit far-fetched.[23]

Subsequent research indicated that many assumptions informing Project Sunshine were far-fetched. This included the notion that adults do not absorb strontium-90, the length of time it remains sequestered in the stratosphere, and the uniformity with which it descends to earth.

Indeed, nature held a number of surprises that were at odds with the detached and static conceptions of the world articulated at the Biophysics Conference. These dynamics challenged the expertise of even the "highly civilized" individuals within the atomic-industrial complex.

The 1955 Project Sunshine conference endeavored to assess radio-strontium uptake around the world. The intent was noble. The methodology was unethical, even by the standards prevailing at the time, and AEC officials feared that it was illegal.[24] Nonetheless, Project Sunshine procured nearly nine thousand human samples, including intact skeletons and six hundred fetuses from near and far.[25] This would not have been feasible if scientists at prestigious universities around the country had not offered their expertise and technical resources to such a covert undertaking, not to mention the medical professionals delivering "human material."[26]

Running parallel to Project Sunshine was the effort to define strontium-90 uptake in terms of a "sunshine unit." The July 1954 Project Gabriel report documented that the fifty-five stillborn infants from Chicago had an average strontium-90 content of 0.14 sunshine units per gram of calcium, the infant from Utah 0.19 sunshine units, and stillborn remains from India an average of 0.043 sunshine units.[27] The term is a notable example of what Ulrich Beck labels "symbolic detoxification."[28] This entails rhetorical erasure of toxicity as opposed to biophysical and material erasure. Symbolic detoxification may ease the public scrutiny that organizational actors confront but does little to reduce the disease and disability that are eventually expressed. If sunshine is ubiquitous and yet harmful only if one receives too much, then strontium-90 was analogous. Americans needed to become accustomed to it and manage their exposure through dietary adjustments, as one applies sunscreen at the beach. In no small part due to the ridicule provoked by "sunshine unit," however, it was later renamed "strontium unit."

In 1956, Willard Libby disclosed the existence of Project Sunshine to assure the public that steps were being taken to measure strontium-90, and data gathered under the auspices of Project Sunshine appeared in studies published in academic journals. The particulars as to how the human samples were collected were rarely identified. In 1957, Merril Eisenbud reasoned, "From its formation in a nuclear detonation until it is absorbed by man, the path of a Sr^{90} atom is long and tortuous. . . . The phenomenology of Sr^{90} in fallout can be described only in the combined languages of all of the principal combined sciences: geophysics, physical chemistry, biophysics, and biological chemistry."[29]

By "long and torturous," Eisenbud was referring to the pathways whereby strontium-90 is absorbed in soil, water, vegetation, milk, and the human body, but the same can be said of public disclosure of Project Sunshine. Many details were not uncovered until 1995. These details encompass deception, cover stories, and body snatching. Science, politics, and the macabre intertwined as the arms race accelerated over the course of the 1950s. In 1994, President Bill Clinton established the Advisory Committee on Human Radiation Experiments (ACHRE) to untangle the history of government-sponsored human radiation studies during the Cold War. The Advisory Committee had unprecedented access to the archives of federal agencies and in October 1995 handed the president a nine-hundred-page report outlining not just the activities conducted under the auspices of Project Sunshine but a litany of other research projects in which informed consent of participants was not obtained, the potential risks were not fully discussed, and the family of deceased "samples" were not told that their loved ones were obtained for biomedical research.[30] The US General Accounting Office (GAO) released its own report in 1995 and listed fifty-nine tissue-analysis studies, including Project Sunshine, sponsored by the AEC and, later, the Department of Energy. More than fifteen thousand people were administered radioactive substances in some manner, and many of these studies employed deception and misinformation to solicit participation.[31]

In the Advisory Committee's report to President Clinton in 1995, it cautioned against judging activities of the past from the moral and ethical frameworks prevailing in the present.[32] However, Project Sunshine and other studies violated the tenets of informed consent established by the AEC as well as the 1947 Nuremburg Code, which was created in response to Nazi biomedical experimentation during World War II.[33]

Atmospheric atomic detonations enrolled the entire world in a biological experiment without informed consent or voluntary participation, although some participated more than others. Project Sunshine found that children had higher levels of strontium-90 in their bones as they were growing at a time when the two major superpowers were dispersing it with abandon. Human samples covertly harvested from near and far evoke images of people, in whole or in part, treated as objects for dissection, as opposed to subjects with value in themselves. Human radiation experiments conducted on US citizens have a "Buchenwald touch" irrespective of the ends to which the research was applied.[34]

An overarching lesson of Project Sunshine was an evolving awareness that it is was possible to contaminate much of the globe from a limited number of origination points. Stontium-90 was an unintended tracer descending back to earth relatively quickly when detonations were measured in kilotons and debris ascended to the troposphere, but detonations measured in megatons ascended into the stratosphere, where debris circled the globe before drifting into the troposphere, where it then returned to earth with rainfall. As expansive as the planet may be, atomic military technology heralded a reduction of ontological distance that was jarring in its implications. Joseph Masco observes, "*Bodysnatching, baby bones, genetic mutations, sunshine units*—these are the terms of a new American modernity based not only on technoscience but on managing the appearance of the bomb. Project Sunshine can be read as an official articulation of nuclear fear, a tacit recognition that a new tactile experience of the world was being created by the distribution of nuclear materials into the environment."[35]

This new tactile experience made a mockery of dualisms such as nature/culture, organic/inorganic, human/nonhuman, and near/far, all while illustrating the dependence of the public on the fiduciary conduct of AEC and military officials. Adults, children, infants—all become objectified specimens for scientific analysis. Subjects become objects to gauge the materiality of energetic matter that cannot be seen, tasted, heard, or smelled. A survivor of the Chernobyl disaster articulated this experience: "What is there to say about exposure to radiation that cannot be seen nor smelled nor heard nor touched nor tasted? Those of us who have been in its eerie neighborhood have resembled objects, onto which certain effects have been inflicted, as opposed to subjects in control and aware of what is going on."[36]

Hardtack II and the Beginning of the End

The Teapot series of 1955 was more restrained than Upshot-Knothole two years earlier but was followed by Operation Plumbbob, which constituted a dramatic uptick in fallout. The Hardtack II series in September–October 1958 witnessed the pendulum swinging back to a more restrained schedule. By 1958, the era of open-air detonations was slowly coming to an end. Hardtack II comprised thirty-seven safety and weapons development tests in less than two months, but many were conducted in tunnels or vertical shafts underground. Of those atop a tower or suspended from a balloon, the largest included Socorro at six kilotons, Sanford at around five kilotons, and De Baca at just over two kilotons.[37] In crucial respects, the tunnel and deep vertical detonations that constituted operations in Nevada for the next three decades had begun. There would be intermittent venting of radioactive fallout detected offsite, most notably the Baneberry shot in December 1970, but not the brusque and weighty imposition of fallout downwind, with one exception: the Sedan cratering shot of 1962.

On October 31, 1958, the United States and the United Kingdom embarked on a voluntary moratorium on open-air and underground testing, as did the Soviet Union several days later. Halloween 1958 signaled a hopeful new turn in the tense and dangerous global arms race. It was a welcome respite for downwind residents who were weary of fallout and those concerned with rising strontium-90 levels in milk. AEC officials and key actors from the weapons labs were staunchly opposed to the moratorium and urged President Eisenhower to soften the agreement to allow for underground detonations or to abandon the proposal altogether.[38] The issue, however, was now much broader than what the cloistered technocracy orbiting the AEC could contain.

The Soviets broke the moratorium in August 1961, and the AEC resumed detonations one month later. It ignited a feverish new round of testing. A greater volume of explosive yield was released into the atmosphere over the next two years than at any time during the atmospheric era, although US open-air detonations were conducted exclusively in the South Pacific.[39] The Soviets conducted the 50-megaton (50,000-kiloton) Tsar Bomba in October 1961. It was detonated above the uninhabited island of Novaya Zemlya in the arctic, and it was the largest atomic weapons test ever conducted—three thousand times larger than the bomb dropped over Hiroshima. It was less a scientific experiment than a political statement. Indeed, it was generally viewed worldwide as an unnecessary and reckless provocation. "The bomb's yield (which might actually have been 57 megatons, but why quibble?) was ten times the combined total of all the explosives used during the Second World War," Gerard DeGroot explains.[40] Many detonations conducted by the AEC between 1961 and 1962 were in the Johnston and Christmas Islands in the South Pacific. The unofficial end of US atmospheric testing occurred in October 1962 with an 8.3-megaton (8,300 kiloton) airdrop in the Johnston Islands. From July 1962 until 1992, all detonations in Nevada occurred underground.[41] But even as detonations were going underground, long-simmering issues were arising within the AEC.

Radioiodine: An Eleven-Year Oversight

The AEC did not conduct systematic sampling of milk for radioiodine in the downwind towns or anywhere else in the 1950s.[42] The combination of fission, wind, elevation, precipitation, topography, grass, milk, and human bodies was not envisioned as meaningful, and this was compounded by failure to appreciate that many downwind residents drank fresh milk from their own cows or a local herd.[43] And although fallout exhibited unpredictability, once radioiodine returned to earth, there were opportunities to intervene. When warranted, AEC officials could have instructed the public to consume powdered milk and farmers to use feed that had been sheltered from the elements. To minimize economic losses, contaminated milk could have been diverted to the production of cheese and other dairy products, so that by the time it reached consumers, the radioiodine had long decayed. Unlike strontium-90 or cesium-137, radioiodine was characterized by *disconnectability* due to its half-life of eight days.

Gordon Dunning was the most high-profile AEC official working to assess the hazards to public health, and he was the most visible defender of the commission's operations in Nevada. In 1955, Dunning acknowledged the hazards of radioiodine in the *Scientific Monthly* but insisted that the levels of I-131 dispersed across the country were too low to produce detectable effects in humans. He did not address the concentrating effects of radioiodine in the smaller thyroids of children (or lambs).[44] In 1957, Dunning prepared a technical report for the JCAE depicting the milk pathway and recent research, based on the nuclear-reactor accident at the Windscale Works facility in Sellafield, England, that same year.[45] He described the surprisingly high intensity of radioiodine with an external gamma measurement of 40 roentgens—a dose as high as "tens of thousands of reps" and presenting a particular hazard to a fetus or baby. "Additional evaluation will be given this problem."[46] In 1959, he again prepared a technical report for the JCAE, and in this one, he calculated that infants drinking milk laced with radioiodine could be exposed to very high doses in the aftermath of a nuclear strike. "Milk as a food item should be avoided until the iodine activity levels dropped to acceptable levels, or canned or powdered milk (prepared before the fallout occurred) should be substituted," Dunning advised.[47] But such measures were never undertaken downwind of Nevada.

The nuclear-reactor accident at Sellafield should have been a *tipping point* that disrupted the prevailing definition of the situation for AEC and military officials. It did so for many people outside the commission. At the 1959 congressional hearings examining the hazards of nuclear war, at which Dunning presented his technical analysis, Dr. E. B. Lewis, a professor of biology from the California Institute of Technology, remarked, "It has become increasingly clear since 1957 that the radioiodine in fallout are a special hazard to infants and children." Lewis outlined research documenting radioiodine in fresh cow's milk and the relevance for past and future weapons testing: "Whether radioiodine levels as low as these are capable of producing thyroid cancer is not known and may not be known for many years."[48]

In 1962, Dr. Harold Knapp recognized that Dunning's 1959 report to Congress included equations for estimating radioiodine in milk, given data on external gamma radiation. To assess their validity, Knapp learned of gamma radiation and iodine-131 measured in tandem at two points in 1953, Alamo and Caliente, Nevada, as described in Dunning's 1959 report. Knapp approached Dunning seeking access to the data but was told that there were no such data. He later approached Dunning a second time and was given the same answer: no such data exist. "Then it developed that Dr. Dunning's office was the only one which had the reports from the Offsite Radiological Safety Organizations, from which all the data on the Alamo-Caliente milk-gamma relations derive," Knapp explained.[49]

Measurement of radioiodine and gamma radiation from the same pasture then made it possible to estimate levels of radioiodine in milk when one only had data on gamma radiation by way of proportional scaling, and Knapp's calculations showed that doses to children and infants could be alarmingly high even when levels of gamma radiation had not exceeded the established threshold. Over successive series, children in the downwind towns may have accumulated thyroid doses in the hundreds of rads.[50] Knapp stumbled on something that AEC officials found disturbing, with regard less to public health than to public relations. Discursive momentum and the inadequacies of operational conduct were catching up to the commission. "It was not exactly enthusiastically received, let me say it that way," Knapp confided to congressional investigators in 1979.[51]

Knapp's discovery should not have come as a surprise. Dunning's analysis and the work of British researchers in the wake of the Windscale accident laid the groundwork for Knapp's calculations.

Knapp submitted a draft report in September 1962 with a memorandum addressed to Charles Dunham, director of the Division of Biology and Medicine. He warned that children downwind of Nevada could have received hundreds of rads concentrated in the thyroid. "Under these circumstances we managed to miss the obvious in a haystack."[52] Scarcely a week earlier, Dr. Ralph Lapp estimated the thyroid dose to infants and children in Troy, New York, due to shot Simon in 1953, in the journal *Science*.[53] Lapp built on the lessons learned from the Windscale accident, but he did not turn his attention to children immediately downwind of Nevada. Had he done so, the numbers would have been even more alarming, and Knapp suggested to Dunham that it was only a matter of time before Lapp or someone else took this additional step. It would be better to get out front of the issue before someone outside the commission did so, Knapp stressed.[54]

Dunham did not share Knapp's proactive approach. In October 1962, he forwarded a short, barbed review of Knapp's report: "The draft is amateurish to a degree. Its inadequacies, i.e., almost total absence of thoughtful basis for many sweeping assumptions, uncritical use of analogy and models, are such as to make me reluctant to waste any more people's time on this draft. In fact what has been thrown together here can be looked on as an invitation to others to write the document for the author. I will not assist in such an enterprise beyond what has already been done."[55]

Dunham's position was clear: Knapp's research was a dead end. "When I got the first official memorandum strongly criticizing my paper, I felt the only honorable out it left me other than to resign was to do the work more completely," Knapp explained. "I would say that the people who were most immediately effected [sic] and responsible proceeded immediately to try to find as many reasons as possible why my conclusions might not be true."[56]

Science advances through dialogue and debate, but after Knapp distributed a revised draft, it was determined that the report contained restricted data.[57] This would keep Knapp's research from making the rounds, particularly beyond the AEC. Restricting a report that threatened public relations rather than national security was a stopgap measure, and by the summer of 1963, AEC officials established a committee to review Knapp's research and write an addendum to accompany the report's release to the public. To expedite publication, Knapp agreed to delete a crucial "punch line" that AEC officials feared would alarm residents in one particular town.[58]

In June 1963, Knapp forwarded a memorandum to Charles Dunham containing a response to one of the report's most vocal critics: Gordon Dunning, then deputy director of operational safety. Dunham's terse critique of the first draft paled in comparison to Dunning's review of a subsequent draft. The jab-counterjab is notable, but so too is the subtext: Knapp refused to acquiesce to Dunning's desire to protect the AEC's standing with the public above and beyond all other considerations. Dunning was frustrated. Knapp was frustrated. Dunning was worried that the report called into question eleven years of assurances that testing was not a public health hazard. Knapp was worried that such assurances were ill founded. And Dunning was in a difficult position. Knapp employed calculations that he devised but had not applied to fallout downwind of Nevada. And now Dunning was questioning the validity of research following the path he laid out but had not trod.

In the memo critiquing a revised draft, Dunning asserted, "Firstly, one of the principal bases for the author's conclusions is attempting to relate one set of measurements (gamma dose rate measurements) to an entirely different parameter (iodine-131 in milk)." Knapp countered, "All of science involves relating one set of measurements to an entirely different set of measurements." Dunning proclaimed, "Secondly, *before* the paper is published (if this is decided) sufficient time should be given to prepare a fully documented report as to what measures were and are being taken to conduct a safe operation in Nevada." Knapp responded, "If the Commission hasn't by this time a fully documented report as to what measures were and are being taken to conduct a safe operation in Nevada, one wonders if it will ever get one." Dunning suggested, "Whether or not the conclusions of the paper are true may be debatable—this paper does not provide the scientific basis for a decision one way or another." Knapp remarked, "I would be pleased to let the scientific community be the judge of this assertion." Dunning shifted his footing: "Since writing the above, there has come to mind one possible 'out' to the situation. Let the Commission tell Dr. Knapp in a matter-of-fact and bland manner that the Commission interposes no objection if he, as an individual scientist, wishes to publish his paper." Knapp countered, "It wouldn't work, Dr. Knapp might find out about this suggestion and go up in smoke. He is said to be mighty sensitive by now to this type of scheming." Dunning then advised, "The above approach will make it clear that the Commission is not trying to suppress the paper." Knapp curtly replied, "No comment." Knapp then pushed Dunning to worry less about how to "manage the news" and more about the real-world implications. Knapp insisted, "It looks mightily silly to suddenly become so doubtful just because when the theory is applied, we suddenly realize we have all been missing the obvious in a haystack for over 10 years, and that the public relations impact of this oversight are painful to ponder."[59]

To Dunning's dismay, an ad hoc committee recommended publication of Knapp's research, but a summary of the committee's concerns would accompany the report—not at the end but at the very beginning, before Knapp's text: "The reactions of the individual committee members to the general nature and utility of the report were not in accord, ranging all the way from favorable to completely negative." Several of the ad hoc committee's concerns were rooted in the AEC's haphazard data collection. The commission rarely collected gamma radiation measurements on pastures in tandem with data on levels of radioiodine in milk originating in the same area, and Knapp's report was criticized for not being based on milk samples and gamma dose readings collected in the same area or, indeed, pasture. Nonetheless, the technical review also conceded, "The committee agree that the author's mathematical treatment is expertly and meticulously done." Moreover, they reasoned that Knapp's technique for estimating the radioiodine burden was "probably qualitatively correct" and that the report would probably inform future research efforts.[60]

Nonetheless, Knapp's report did not motivate AEC officials to rewrite the standards for radioiodine exposure of infants, children, or the public in general. Despite the controversy, it was concluded that the existing standards were sufficient. There was trepidation within the AEC that changing the standard would invite questions, such as, What was wrong with the old standard?[61] AEC officials were beset by the momentum of omission. Public relations and public health were again entangled, to the detriment of the latter.

Although Knapp's 1963 report did not spur change in the radioiodine exposure standard, it generated greater concern within the AEC. Indeed, Knapp's research had an impact within the commission, at the same time that officials were struggling with the question of whether it should be published. The AEC established the Ad Hoc Working Group on Radioiodine in the Environment (Working Group), which met from March 5 to 7, 1963, and released its report on April 25. Knapp's revised draft was dated March 1, and the final version was published in the summer of 1963. The Working Group was unambiguous in its conclusion: "The release of radioiodine from key activities of the U.S. Atomic Energy Commission has become one of the principal, if not the principal, controlling factor in terms of environmental contamination. This conclusion is based on data and theoretical calculations obtained during the past year." People within the AEC recognized the validity of Knapp's research, even as others sought to suppress it. The Working Group's report recommended a litany of new procedures, enhanced sampling for radioiodine, and continued research into the issue.[62] Consternation within the AEC had less to do with the technical merits of Knapp's report than with the public-relations implications of an eleven-year oversight. It is not clear that Knapp understood the impact his report had within the AEC. Soon after it was released, he resigned his position.

After Knapp left the commission, his research was published in 1964 in the journal *Nature*, with the punch line that had been omitted from his 1963 report included in the text: *As with the lambs, so too with the children: damage is greater the smaller the thyroid gland, and infants residing in and around St. George in the wake of shot Harry in May 1953 may have received a dose ranging from 110 to 440 rads.* This was 150–750 times higher than the annual permissible dose at the time.[63]

Science can be corrupted or employed to root out corruption, and in 1979, Harold Knapp again found himself the nemesis of the AEC.[64] He did not welcome the role. In a January 1979 letter to the DOE, Knapp was conflicted. His career at the AEC was turbulent, and he bristled reviewing old reports and memoranda. "One would think that their problem was me and not the radioiodine," Knapp confided. "The resurrection of a 15 year old controversy has worried me in that it isn't at all clear what I should or should not do. . . . If all this makes somebody very unhappy, please tell them I'm sorry. It's just that I couldn't really figure out what else to do."[65]

In 1979, Knapp explained to congressional investigators that operational conduct in Nevada was founded on a simplifying assumption: unless the external gamma radiation dose was high, approximately 20 rads or more, there was no need to worry about internal emitters of radioactivity.[66] His research while employed at the AEC undercut this assumption. What Knapp unearthed was evidence of high doses of radioactivity in particular internal organs among select segments of the population—the kind of complexities that AEC officials preferred to know less rather than more about.

Radioiodine contamination entailed a significant ecological failure of foresight or an inability to recognize the extent to which the sociotechnical system was intertwined with ecological processes—despite dispersing fallout over much of the country. It constituted a large, complex organization developing a risky technology that was entangled with broader ecological systems and thus expanded complexity, contingency, and risk. Charles Perrow argues that such accidents entail "an interaction of systems that were thought to be independent but are not because of the larger ecology."[67] More perplexing still is that AEC officials did not appreciate the risks of I-131 even as medical researchers documented its carcinogenic consequences over the course of the 1950s. While the AEC was dispersing radioiodine around the country, the medical sciences literature was discussing the implications of overexposure within clinical settings.

Could the AEC Have Known about the Hazards of Radioiodine?

It is difficult to envision a scenario in which AEC officials could not have known that radioiodine was a hazard before 1962. Rather, they were beset by a *material-discursive bind* of their own making and reticent to confront a contaminant whose effects only unfold over a significant period.[68]

Soon after iodine-131 was discovered in the 1930s, it was employed as a tracer introduced into the body to assess uptake in the thyroid gland, but by the late 1940s, researchers were warning that its administration could unwittingly contribute to cancer years later.[69] In 1956, Dr. J. E. Rall published a warning in the *American Journal of Medicine* directed at practicing physicians: "Radioactive iodine should be administered with discretion." It was a warning that AEC officials should have heeded. Rall remarked, "It should be borne in mind, therefore, that the administration of any tracer dose of radioiodine represents a calculated risk. . . . Radioiodine tracers to children especially should be viewed with considerable caution. The ease of administration of radioiodine and the lack of any acute effects of even large doses cannot help but *encourage carelessness*."[70]

The message was clear: young thyroid glands are radiosensitive. Moreover, women are more likely to develop radiation-induced thyroid cancer than men are.[71] By the early to mid-1950s, physicists understood that iodine-131 was present in abundance in the wake of an atomic detonation, and medical researchers understood that administration of iodine-131 could produce delayed carcinogenic effects. In 1955, research observed that radioiodine was passed from lactating mothers to their babies, and by 1957 research documented that radioiodine detected in human and bovine thyroids was correlated in time with detonations in Nevada and in the Pacific.[72] There were numerous experimental studies on animals utilizing iodine-131 in the 1950s—many funded by the AEC. Moreover, the survivors at Hiroshima and Nagasaki constituted a unique body of data on the delayed effects of radioactivity, and through the 1950s, they were illustrating excess thyroid cancer, particularly pronounced among those younger than twenty years of age when exposed. Leukemia was the first cancer identified beyond the expected background rate, but as Arthur Schneider and Elaine Ron note, "Thyroid cancer was the first solid tumor to be linked to radiation exposure from the atomic bombings in Hiroshima and Nagasaki."[73]

The nuclear-reactor accident at the Windscale Works facility in 1957 released between 16,200 and 27,000 curies of radioiodine, and in response, the United Kingdom curtailed milk consumption to reduce the risks to public health. Approximately 150 million curies scattered over the United States during atmospheric testing in Nevada, and not once did the AEC advise the public not to consume milk.[74] Failure to do so constitutes one of the most significant examples of organizational mistake and misconduct in US history.

AEC officials overlooked the hazards of radioiodine due to a number of factors. Despite the dramatic ecological situatedness of open-air testing, they were scarcely cognizant of the ecological consequences of injecting radionuclides into the atmosphere. Testing was ecologically situated, but the interpretive framework of AEC and military officials was not. Further, it was not in the interests of people at key positions within the sociotechnical system to unearth empirical evidence threatening the momentum and sunk costs invested in the NTS. Knowledge regarding radioiodine lay dormant. It was yet another instance of the organizational production of willful ignorance.

Moreover, the *average man logic* of a single standard for external gamma radiation allowed for a degree of environmental violence imposed on select segments of the population. It was not attuned to biophysical differences by age and gender. An average man logic, in literal terms, combined with the variable effects of internal emitters meant that certain segments of the population confronted a greater risk. "I do not recall any standard ever having been made for internal emitters which would have applied in the same way the external standards did around the test site," Knapp explained to congressional investigators in 1979.[75] Indeed, there was no such standard.

As Beck notes, too often exposure standards are tainted by political interest groups in society. The permissible dose is not always based on the best available science but is an unstated compromise between public health and the objectives of industry or the military-industrial complex. The permissible dose or "acceptable levels," biased by the parochial objectives of particular interest groups, Beck argues, normalize the imposition of hazardous substances and materials on others: "The really rather obvious demand for non-poisoning is rejected as *utopian*. At the same time, the bit of poisoning being set down becomes *normality*. It disappears behind the acceptable values. Acceptable values make possible a *permanent ration of collective standardized poisoning*. They also cause the poisoning they allow not to have occurred, by declaring the poisoning that did occur harm*less*."[76]

Knapp's research in 1962 was a long-overdue disruption of interpretive understanding. When this happens, things that were previously unseen or unappreciated become progressively apparent, and existing frames of reference become less stable and taken for granted. A tipping point amid a succession of empirical effects suggesting an accumulating disaster is a shift in perspective such that a retreat to the prior point of view is problematic, if not impossible. Once a decisive disruption occurs, there is shock and bewilderment as what was there all along is suddenly apparent.[77] Knapp initiated, in Barry Turner's words, a conception of "the discrepancy between the way the world is thought to operate and the way it really does."[78]

Radioiodine and Public Health

What is defined in the popular imagination as a disaster is predicated on the characteristics of the event and its impact on society, not least of which is the disruption of routine thought and behavior. As Lee Clarke explains, other characteristics include the "body count" as well as evocative visual images disseminated in the mass media.[79] "Crescive" and "disaster" are not typically assumed to relate to each other. Due to the crescive and mundane characteristics of radioiodine contamination from open-air testing, it remains generally unacknowledged. It occurred in biophysical and material terms but barely in social and cultural terms.

In 1983, the US Congress mandated study of radioiodine from continental open-air testing, and fourteen years later, the National Cancer Institute published a dose-reconstruction study illustrating imposition of I-131 in every county in the contiguous US. The downwind communities encountered the most intense deposition of I-131; but it was widely dispersed throughout those states just east and north of the NTS, and hot spots occurred throughout the Midwest.[80]

TABLE 10.1. US Health Effects from Open-Air Atomic Testing

	Additional cases of cancer	Estimated deaths
1. Thyroid cancer due to radioiodine	50,000	2,500
2. External gamma radiation; comprised of various cancers	22,000	11,000
3. Leukemia due to internal irradiation, primarily strontium-90	1,000	550
4. Internal irradiation due to various other radionuclides	6,000	3,000
Total	79,000	17,050

Source: Data summarized from Institute for Energy and Environmental Research (IEER), *Fact Sheet on Fallout Report and Related Maps* (Takoma Park, MD: IEER, 2002), http://ieer.org. The IEER data are based on analysis of the 1997 NCI data; and CDC and NCI, *A Feasibility Study of the Health Consequences to the American Population of Nuclear Weapons Tests Conducted by the United States and Other Nations* (Atlanta: CDC, 2002), www.cdc.gov.

Table 10.1 summarizes the estimated health effects of atmospheric testing between 1951 and 1962. The NCI estimated between 11,300 and 212,000 additional cases (geometric mean of 49,000) of thyroid cancer, primarily due to Nevada detonations. Researchers at the Institute for Energy and Environmental Research (IEER) then estimated 2,500 deaths, extrapolated from the 1997 NCI study.[81]

In 1998, Congress authorized study of additional radionuclides from NTS as well as global atmospheric testing, and in 2002, the CDC and NCI released a study estimating 22,000 additional cases of cancer in the United States due to external gamma radiation from the NTS and global open-air testing between 1951 and 1962.[82] IEER researchers then estimated that approximately 11,000 of these cases would be fatal.[83] The downwind communities often encountered gamma radiation from testing in Nevada, but the burden is more pronounced in those parts of the US subject to frequent rainfall and thus terrestrial deposition from noncontinental atmospheric testing. In turn, NTS testing accounts for the bulk of radioiodine, but global atmospheric testing is responsible for greater overall levels of long-lived radionuclides such as cesium-137 and strontium-90 dispersed throughout the US.[84]

Further, IEER research indicates an additional 550 fatal cases of leukemia, primarily due to strontium-90 exposure, and approximately 3,000 deaths due to internal irradiation from various other radionuclides, such as cesium-137, carbon-14, and tritium. As outlined in table 10.1, overall excess cancers in the US due to NTS and all other open-air testing is approximately 79,000, of which about 17,050 were predicted to be fatal.[85]

The data presented in table 10.1 are probably a conservative estimate. The overall impact of atmospheric testing on public health remains the subject of controversy, and it may be impossible to fully assess. Whatever the toll, it is not insubstantial.

11

Sedan, *Silent Spring,* and the Reenchantment of Nature

In July 1962, a thermonuclear device buried deep beneath a windswept playa propelled rock and soil three hundred feet skyward. Three seconds after detonation, hot gases burst through the ascending alluvial dome as the debris cloud climbed to twelve thousand feet.[1] When the dust settled, the experiment was deemed a success, given the immense crater left behind at the Nevada Test Site.[2] Sedan was a step forward in Project Plowshare. The goal was to demonstrate the viability of atomic devices for the construction of canals, ocean ports, mountain passes, open-pit mining, rerouting of rivers, creation of lakes, even underground to enhance the recoverability of oil and natural gas deposits.

The physicist Edward Teller insisted Project Plowshare would prove that there were no limits to the modification of nature in the service of humanity.[3] His unabashed optimism was reminiscent of the English philosopher Francis Bacon, more than three centuries earlier. Bacon invoked the Latin phrase *natura vexata*, or nature vexed and forced out of its natural state, in depicting how best to uncover the secrets hidden within it.[4] Prying, twisting, and wrenching, he stressed, were necessary to construct a body of instrumental knowledge in a cumulative manner. It was a new intellectual attitude insisting that nature must be acted on strategically, and advances in technology, "the mechanical arts" in seventeenth-century parlance, were the linchpin of progress.[5] "While Socrates had equated knowledge with virtue, Bacon equated knowledge with power. Its practical usefulness was the very measure of its validity," Richard Tarnas explains.[6]

At 104 kilotons, the Sedan device was powerful—seven times larger than the bomb dropped over Hiroshima.[7] The crater it created is evidence of the ability to move millions of tons of earth in the blink of an eye. The debris cloud was 50 percent larger than predicted, however, and fallout soon traversed the boundaries of the test site.[8] It was anticipated that virtually all of the radioactivity would remain confined within the onsite displaced rock and soil, but Sedan necessitated limited evacuations north of the NTS.[9] "The dust was so thick at Ely, Nevada, 200 miles away, that the streetlights had to be turned on at four in the afternoon," Dan O'Neill notes.[10] Sedan also elicited the concern of state health officials in Utah in regard to radioiodine, as well as depositing significant radioactivity in the Midwest.[11] Eleven years after testing began in Nevada and numerous lessons learned the hard way, Sedan was one of the dirtiest detonations with regard to population exposure to radioactive fallout.[12]

In September 1962, *Silent Spring* arrived on bookshelves. Examining the proliferation of pesticides, and DDT in particular, Rachel Carson highlighted fallout, both radioactive and chemical, to describe the indiscriminate imposition of harmful substances on the public, substances with deleterious effects unbounded across space and persisting tenaciously over time. Her argument for more prudent engagement with the nonhuman world was entwined with public concern over the excesses of the atomic state. It is an anxiety that is often overlooked when writers recount the rise of the contemporary US environmental movement.[13] In a letter penned in 1958, Carson confided, "I suppose my thinking began to be affected soon after atomic science was firmly established. Some of the thoughts that came were so unattractive to me that I rejected them completely, for the old ideas die hard, especially when they are emotionally as well as intellectually dear to one. It was pleasant to believe, for example, that much of Nature was forever beyond the tempering reach of man—he might level the forests and dam the streams, but the clouds and the rain and the wind were God's."[14]

In the atomic age, the clouds, the rain, and the wind serve as transit for radioactive substances provoking injury in an ambiguous, crescive manner. In this sense, Sedan and the publication of *Silent Spring*—just months apart in 1962—are indicative of a distinctly modern juxtaposition. The former is a testament to the capacity of science and technology to remake the natural world. "We will change the earth's surface to suit us," Edward Teller declared.[15] The latter is an indictment of the myopia underlying the reckless modification of nature.

Plowshare exemplifies the hubris entrenched within the atomic state, but *Silent Spring* was a cultural turning point accentuating issues of increasing public concern. From a contemporary perspective, it can be difficult to appreciate the radical departure *Silent Spring* presented. As William Leiss argues, "Humanity's entitlement to mastery over nature is a subterranean theme that runs throughout the collective consciousness of the modern era."[16] Carson did not reject science and technology but clearly questioned this subterranean theme.[17] She helped spark greater recognition of the implications of science and technology untethered to democratic processes. In crucial respects, Carson anticipated key tenets of Ulrich Beck's risk society thesis by several decades.

Carson died in 1964, ravaged by cancer at just fifty-six years old, but *Silent Spring* has an enduring appeal. The Sedan crater also endures. It was added to the National Register of Historic Places in 1994, is notably depicted at the National Atomic Testing Museum in Las Vegas, Nevada, and is the high point of public tours of the NTS. *Silent Spring* argues that modernity is characterized by scientific and technological prowess but an antiquated cultural worldview, and the socio-ecological principles that Carson highlighted are important lessons for understanding the intertwining of human societies and the nonhuman world. These principles include human connectivity with, as opposed to detachment from and omnipotence over, nature and the vibrancy of nonhuman entities, processes, and things in effecting change. Moreover, Carson argued that reenchantment with the natural world is imperative to confronting unintended ecological degradation and change.

"We're Going to Work Miracles"

In laying a foundation for the experimental manipulation of nature, Francis Bacon first had to persuade his seventeenth-century contemporaries that such conduct was not an affront to Christian principles.[18] He went further by insisting that it was essential to return humanity to its rightful dominion over the Earth, which was torn asunder when Eve partook of the forbidden fruit in the Garden of Eden. Increasingly efficient and efficacious technological means for remolding nature not only lead to the elimination of human misery and toil, Bacon mused, but are favored by God. Science and technology are instrumental and redemptive, pragmatic and transcendent, utilitarian and sacred. David Noble observes, "Increasingly, in the inspired imagination of the time, man's contribution to creation loomed ever larger in the scheme of things.

Despite their caveat about the necessity of humility, and despite their devout acknowledgement of divine purpose in their work, the scientists subtly but steadily began to assume the mantle of creator in their own right, as gods themselves."[19]

A detached nonhuman world that must be wrenched from its natural state is a "controlling image" guiding scientific rationalism, Carolyn Merchant argues.[20] Imposition on nature and wresting *her* secrets by force in the service of power and control constituted a seductive stance embraced by Bacon's contemporaries and those who came after, irrespective of their methodological differences.[21] René Descartes advanced a deductive, mathematical stance very different from Bacon's inductive experimental approach. Nonetheless, he similarly regarded nature as set apart from the human.

Demonstrating the viability of atomic devices for peaceful objectives was "redemption," and its highest profile adherent was the man who, along with Stanislaw Ulam, pioneered the hydrogen or thermonuclear bomb.[22] Edward Teller traversed the "is" position of a scientist working to understand the natural world and the "ought" position of a tireless policy advocate. It was not enough to disentangle the atomic structure of matter when there was a world to remake, and Teller represented a way of conceiving of the relationship between humanity and nature with a long pedigree. Teller proclaimed, "We would be unfaithful to the tradition of Western Civilization if we shied away from exploring what man can accomplish, if we failed to increase man's control over nature. . . . Every citizen, whether he is a politician or a farmer, a businessman or a scientist, has to carry his share of the greater responsibility that comes with greater power over nature. But a scientist has done his job as a scientist when that power has been demonstrated."[23]

Project Plowshare was established in 1957 and received funding for nearly two decades. The name derived from the Old Testament book of Isaiah, which speaks of turning weapons of war into instruments of peace. Project Plowshare was Manifest Destiny by atomic means. It would deliver economic prosperity through the unprecedented command and control of nature.[24] It was conceived and managed by officials at the Lawrence Radiation Laboratory (LRL) in California, under the auspices of the AEC. The LRL was established in 1952 and, in tandem with Los Alamos, remains a crucial node in the atomic state. Teller himself was instrumental to the creation of LRL and served as director from 1958 to 1960. He was not alone in lobbying on behalf of Project Plowshare, but he was its most enthusiastic and high-profile proponent.

Atomic geoengineering encompassed excavation, as exemplified in digging a harbor, open-pit mining, carving out a mountain tunnel or railroad cut, and digging underground to generate steam to produce electricity, capture rare isotopes, and promote the recoverability of oil and natural gas reserves—akin to hydraulic fracturing, or "fracking." The intent was not simply to recontour the surface of the Earth but to enhance access to natural resources that were not otherwise exploitable using conventional technology.

Beginning in 1961, the AEC conducted twenty-seven detonations under the auspices of Plowshare, the majority at the NTS.[25] Edward Teller worked to sell Congress and the US public on the advantages of atomic geoengineering while downplaying the liabilities.[26] The central shortcoming was radioactivity, and the promise of "clean" devices was a key rhetorical strategy. Thermonuclear devices employ a fission trigger to initiate the fusion of hydrogen atoms, and the fission process generates more radioactivity than the fusion does. Cleaner devices are, in part, predicated on minimizing the fission/fusion ratio.[27] "Thus the possibility of clean nuclear explosions lies before us," Teller and Albert Latter insisted in 1958.[28] Despite advances in this regard, clean nuclear devices remained a "propaganda tool."[29] The rhetoric consistently outpaced the material, technological advances.

In addition to the promise of material prosperity, Teller warned of the dire geopolitical consequences if a Soviet program dedicated to the peaceful uses of nuclear devices became more attractive to the leaders of developing societies.[30] Further, in describing Plowshare, Teller and Latter invoked a distinctly Baconian spirit of adventure:

> The spectacular developments of the last centuries, in science, in technology and in our everyday life, have been produced by a spirit of adventure, by a fearless exploration of the unknown. When we talk about nuclear tests, we have in mind not only military preparedness but also the execution of experiments which will give us more insight into the forces of nature. Such insight has led and will lead to new possibilities of controlling nature. There are many specific political and military reasons why such

experiments should not be abandoned. There also exists this very general reason—the tradition of exploring the unknown.[31]

To further knowledge of atomic geoengineering and demonstrate its promise, the AEC announced an audacious plan in 1958 to dig a harbor on the northwestern Alaskan coastline utilizing six thermonuclear devices. Project Chariot was introduced with considerable fanfare as well as reassurances of safety, and Teller made frequent trips to Alaska to confer with business leaders and state officials. Given that Chariot was not explicitly tied to national security and was not a time-sensitive project, researchers at the University of Alaska proposed a series of studies to assess the impact of fallout on the local ecology and subsistence patterns of Inuit villages. And research indicated that despite living in one of the most remote locales in the world, villagers in and around Point Hope, Alaska, had already ingested surprisingly high levels of strontium-90 and cesium-137. This pointed to the widespread impact of Soviet and US atmospheric testing. Further research suggested that the pathway of contamination involved lichen, which absorb atmospheric particles, the ingestion of lichen by caribou, and the consumption of caribou meat by people living far from the bustling centers of modernity.[32]

Local Inuit villagers, with the assistance of biologists and landscape ecologists at the University of Alaska, mounted a protracted campaign in opposition to Project Chariot. "This little-known episode represents the first successful opposition to the U.S. nuclear establishment," O'Neill notes.[33] It provoked a retreat to the sequestered confines of the NTS.[34] In lieu of a harbor on the Alaskan coast, Plowshare proponents settled for a cratering shot at the AEC's outdoor laboratory. It would be the largest detonation on the US mainland, with the potential for localized fallout due to the intertwining of radioactivity with alluvial sediment.

Indeed, fallout moved north before arching northeast over Utah and then over South Dakota, Iowa, and Indiana.[35] Approximately 7 percent of all fallout imposed on the US public from continental testing is attributable solely to Sedan. Further, it was one of the dirtiest detonations with regard to the deposition of cesium-137 as well as radioiodine.[36] In response to the nuclear accident in Sellafield—releasing between 16,200 and 27,000 curies of radioiodine—the United Kingdom restricted milk consumption to reduce the risks to public health.[37] The Sedan

Figure 11.1. The Sedan crater at the Nevada Test Site.
(Courtesy of the National Nuclear Security
Administration)

detonation released around 880,000 curies of radioiodine, but the AEC
never advised the public to avoid milk.[38]

Deep within the NTS lies Yucca Flat. It is scarred by hundreds of
underground detonations that elicited not mushroom clouds but
subsidence craters due to the obliteration of rock far below the surface.
Few places so embody nature vexed and tortured as this two-hundred-
square-mile dusty desert playa. One imprint left behind on Yucca
Flat, however, is unlike the others. The Sedan crater is pictured in figure 11.1. The dirt road at the lower right provides a point of reference,
and the depressions in the distance are subsidence craters. The Sedan

crater measures 1,280 feet in diameter and is 320 feet deep.[39] Standing on a platform built to enhance the view while keeping the careless from careening over the edge, one can inspect the bottom. "Tumbleweeds fill the base of the crater like dust bunnies in the corners of an ill-kept apartment," Phil Patton explains.[40]

Silent Spring and the Reenchantment of Nature

While Francis Bacon was optimistic at the forefront of the Scientific Revolution, three centuries later, the German social scientist Max Weber was ambivalent. In 1917, amid the carnage of World War I, Weber spoke at Munich University. "Science as a Vocation" pondered the forward march of formal-instrumental rationality, or the conscious, methodical calculation of the most efficacious and efficient means in the accomplishment of a particular end. It is goal-oriented behavior, and it generally neglects whether the goal is reasonable in relation to prevailing ethics and values. Science epitomizes formal-instrumental rationality, but it cannot, Weber declared, articulate a coherent overall meaning of the world. Quoting Leo Tolstoy, Weber confronted his audience: "Science is meaningless because it gives no answer to our question, the only question important for us: 'What shall we do and how shall we live?'"[41] Rachel Carson again raised this question in 1962.

In advocating for the advancement of the mechanical arts and power over nature, Baconian scientific rationalism privileged the *how* over the *why*. That something worked with regard to its practical usefulness was more important than whether it was compatible with prevailing cultural values. In displacing animism, mythology, contemplative philosophy, and, in time, religion from the center of Western culture, meaning was set adrift. How to live, particularly in relation to nature, appears a trite and irrelevant consideration next to the promise of acting in a strategic and productive manner. "'How' has become our 'why,'" Morris Berman stresses. From the sixteenth century onward, European thought stressed, "to do is the issue, not to be."[42]

What the Scientific Revolution bequeathed to humanity was a view of nature as pliable and amenable to a breakdown of its constituent parts to reassemble in pursuit of practical aims. It is a mechanistic conception of the nonhuman world, and it is a vision as otherworldly as any animistic notion of spirits residing in trees, lakes, or mountains that Enlightenment scholars so virulently scorned.[43]

The persistent entanglement of human activity with a vibrant natural world is a central lesson of *Silent Spring*. It is the principle of connectivity and therefore contingency, complexity, and change. Connectivity may appear simplistic, but it is profound in its implications. Feedback, equilibrium, disruption, succession, reciprocity, nonlinearity, mutation, and emergence stem from connectivity and run counter to a worldview predicated on separation, linearity, reductionism, and mechanistic determination. Indeed, some people have questioned whether, as a scientific discipline, ecology is "subversive," given it focuses on processes that the prevailing industrial and militaristic order prefers to ignore.[44] Ecology thus presents an alternative approach that is at variance with centuries of Western thought. "The applications of the other sciences are particulate, specialized, based on the solution of individual problems with little if any attention to side effects and practically uncontrolled by any thought of the larger whole," Paul Sears observes.[45]

Rachel Carson highlighted connectivity amid the heedless and unrestrained manufacture and marketing of synthetic herbicides and pesticides. She depicted connectivity not simply as a principle of the natural world but as impinging on the community, the family, and the individual. And she tied her argument to the controversy over strontium-90: "Strontium 90, released through nuclear explosions into the air, comes to earth in rain or drifts down as fallout, lodges in soil, enters into the grass or corn or wheat grown there, and in time takes up its abode in the bones of a human being, there to remain until his death."[46]

Recognition of connectivity is key to making sense of the hidden costs of unrestrained science and technology, but *Silent Spring* also notes the often slow and cumulative pace of contamination. Strontium-90 and other long-lived radionuclides, as well as DDT and other synthetic chemicals, can cause acute injury but more often injury only over an extended period. "Slow violence" is the phrase Rob Nixon employs in describing adverse consequences imposed in an incremental manner, characterized by ambiguous biophysical effects, a scarcity of spectacular visual images sufficient to garner sustained media attention, and uncertainty in regard to the geographical boundaries of impact. He argues that Carson clarified a new typology of injury. "By slow violence I mean a violence that occurs gradually and out of sight, a violence of delayed destruction that is dispersed across time and space, an attritional violence that is typically not viewed as violence at all."[47]

The degree to which cause and effect are visible scientifically and socially depends, in part, on the temporal distance between exposure and the recognition of disease, and nonvisibility is a crucial factor underlying the capacity of powerful actors to evade accountability for past and present activities.[48] Organizational actors have few incentives to consider the interests of the public when the passage of time insulates them from decisions made or those abandoned.

Silent Spring is a lyrical defense of the natural world, but it is also rooted in the laboratory, epidemiological, and biological field research cataloguing the unintended consequences of synthetic herbicides and pesticides. Carson demarcated the contrast between what the chemical industry asserted as matters of fact and the empirical observations of numerous researchers focused on a variety of issues. Her penchant for synthesis revealed, despite the long interval between exposure and the onset of disease, that there was compelling evidence to conclude that nature and the human body were at risk. In crucial respects, then, *Silent Spring* articulated the unsustainable definition of the situation that the chemical industry perpetuated. As Raymond Murphy argues, we engage in the social construction of nature such that it is imbued with symbolic meaning, but when such meaning is at odds with real dynamics of nature, it provokes an "echoing back," illustrating the limits of knowledge and control.[49] Carson underscored the reassuring rhetoric of industry versus the empirical record, and she bristled at the "little tranquilizing pills of half truth" fed to the public.[50]

Humanity's "war against nature" is thus self-defeating.[51] Connectivity ensures that things come around again and again. Increasingly potent synthetic chemicals kill more than simply that for which they are marketed, readily move through the food chain, provoke injury in a crescive manner, and promote evolutionary resistance. Genetic adaptation forces farmers to apply more of the same pesticide or search for an even more potent alternative. War against nature, in turn, is never-ending. Carson thus illuminated what Bacon could not envision: undisciplined, heedless *natura vexata* ultimately eventuates in humanity harassed and overrun (*hominum vexata*).

Human Agency and Nonhuman Materiality

What Francis Bacon, René Descartes, and others left behind was a dynamic and vibrant inorganic and organic world with which humanity was immutably intertwined. Prior to the Scientific Revolution, nature was viewed in contradictory ways, but it was not viewed as dead or inert.[52] Nature provided sustenance as well as capricious destruction by wind, water, fire, drought, and other chaotic forces: "Nature as female, virgin, mother, and witch," Merchant explains.[53] This cultural stance permeated daily life and shaped how humanity interacted with nature—enabling some material practices while constraining others. The Scientific Revolution, in contrast, was founded on a conception of how to profitably exploit the characteristics of nature, an approach distinct from that informing the vast expanse of human history.[54]

Francis Bacon anticipated that by wresting from nature her veiled secrets, humanity could attain an "empire over things."[55] But what if the "things," both organic and inorganic, are not inert, pliable, or necessarily docile? What if they are vibrant and unruly? In making sense of the echoing back of nature, it is crucial to recognize the materiality of the nonhuman or its propensity to provoke effects and to initiate change. Indeed, *Silent Spring* is brimming with the materiality of nature amid the myopia of humanity.

The Sedan crater is evidence of human agency in tandem with the materiality of nature. Human agency was key, but so too was the force of matter; and fallout from Sedan forged connections over space and time between the animate and the inanimate, the nonhuman and the human, the institutional and the individual. Fallout contained energy in a variety of chemical concoctions and the materiality to provoke ecological and biophysical change in both an acute and crescive manner.

While Descartes viewed the nonhuman world as "a lifeless machine," Carson argued that such an ontological perspective was short-sighted, as it obscured the contingency, complexity, and recalcitrance of nature.[56] She pointed to the reckless optimism of Baconian scientific rationalism and the myopic pursuit of absolute command and control over nature, which not only proves illusionary but results in harmful long-term effects imposed on the individual body as well as the integrity of ecological systems. These deleterious effects are unintended, discursively denied even in the context of opposing scientific evidence, and existing beyond the boundaries of prevailing legal norms of accountability. Carson declared, "As crude a weapon as the cave man's club, the chemical barrage has been hurled against the fabric of life—a fabric on the one hand delicate and destructible, on the other miraculously tough and resilient, and capable of striking back in unexpected ways. These extraordinary capacities of life have been ignored by the practitioners of chemical control who have brought to their task no 'high-minded orientation,' no humility before the vast forces with which they tamper."[57]

Toward Reenchantment

Silent Spring is a repudiation of the disenchantment of nature. Following Weber, disenchantment is predicated on the conviction that "one can, in principle, master all things by means of calculation."[58] Carson illustrated not only the majesty and wonder of the natural world but also that it is not as Bacon, Descartes, and Teller envisioned it to be. Together, the empirical demonstration of the irrationality of formal-instrumental rationality—on its own terms—as well as the valuation of nature beyond simply instrumental objectives presents a compelling argument for reenchantment. "But it seems reasonable to believe—and I do believe—that the more clearly we can focus our attention on the wonders and realities of the universe about us the less taste we shall have for destruction of our race," Carson contended.[59]

Formal-instrumental rationality—based on the premise that nature can be purged of uncertainty—engenders unintended, ambiguous consequences calling into question mastery itself. Such consequences go beyond mistakes with regard to technically correct reasoning. Rather, because of nature's very complexity, as embodied in connectivity and materiality, there are enduring mysteries resistant to calculation, let alone control, and demanding a greater degree of human humility and restraint. This does not then imply that the effort to understand nonhuman entities and processes is in vain, but control and exploitation as an end in itself is folly, as humanity is embroiled in a fluid engagement with nature. And failure to envision reciprocity and feedback contributes to the "secondary consequences," or side effects, of human activity. In important respects, Carson sought to instill an *ecological rationality,* motivated not by the wholesale abandonment of formal-instrumental rationality but by its reformulation to account for ecological dynamics.[60] The objective was not to stifle purposive human activity but to moderate the destructive unintended consequences.

Rather than dead and meaningless, moreover, *Silent Spring* argues that nature is transcendent and deserving of valuation in itself, an end worthy of consideration as opposed to simply a means to human objectives. Carson thus asserts a substantive or value-rational argument as a counter to a purely instrumental and technical worldview. While instrumental rationality is concerned with how best to achieve a given end, substantive rationality involves inquiry into the prudence of its attainment.[61] It is concern with the "why" as opposed to simply the "how." Substantive rationality thus exists in reference to values shaping thought and social activity, and *Silent Spring* stands as a counterweight to the forward march of formal-instrumental rationality at the expense of a more coherent overall meaning of the world. Carson laments, "Have we fallen into a mesmerized state that makes us accept as inevitable that which is inferior or detrimental, as though having lost the will or the vision to demand that which is good?"[62]

From a contemporary perspective, atomic geoengineering appears "absurd," if not "insane," but it was conceived at a time when advances in science and technology appeared limitless, even as the contradictions were increasingly difficult to ignore.[63] As with radioactive fallout, so too with DDT—introduction into society provided evidence of socioecological complexities that were not fully anticipated beforehand. The concurrent activities of the atomic state and the chemical industry promised control over nature, and yet rather than dominion, there was the production of novel risks and reverberating ecological changes, as Beck evocatively argues.[64] As enacted through atomic technology, the unforeseen is expressed on a correspondingly grander scale.

Exhibiting a Baconian optimism, Edward Teller advanced a technocratic agenda based on a merging of the militaristic state, corporate contractors, and university scientists—all beholden to advancement of technological systems with the potential for widespread and hidden costs imposed on the public. And yet Teller downplayed the public health implications of exposure to radioactive fallout and was a vocal opponent of the Limited Test Ban Treaty of 1963, which drove US and

Soviet detonations underground. He exhibited a particulate, specialized perspective quite at odds with Carson's ecological worldview.

Modernity is characterized by scientific and technological prowess but an antiquated cultural worldview. Sedan and publication of *Silent Spring*, just months apart in 1962, are indicative of this tension. Sedan embodied nature forced out of her liberty: *natura constricta et vexata*, or nature contorted and interrogated.[65] Carson articulated not simply an ethic of concern for the nonhuman world but an ontological perspective at odds with the prevailing industrial and militaristic order.[66]

In retrospect, Sedan was the high point of Project Plowshare.[67] It would continue to receive funding and conduct detonations, but the public was losing faith in the AEC by 1962. This was less a fear of all things nuclear than an evolving awareness of the commission's reckless imposition of fallout on society. In tandem, Carson argued that oversight of science and technology cannot be left solely to corporate and governmental actors. The public has a stake in and a right to contribute to such debate. "If the Bill of Rights contains no guarantee that a citizen shall be secure against lethal poisons distributed either by private individuals or by public officials, it is surely only because our forefathers, despite their considerable wisdom and foresight, could conceive of no such problem," Carson argued.[68]

The drive to dominate is shaped by a cultural worldview and associated material and technological practices, and *Silent Spring* sketches an alternative that is less oppositional and detached. The challenge is to learn from and evolve with the fluid and, at times, recalcitrant materiality of nature. It requires a different type of mastery. "Now, I truly believe that we in this generation, must come to terms with nature, and I think we're challenged as mankind has never been challenged before to prove our maturity and our mastery, not of nature but of ourselves," Carson insisted.[69] In this regard, she questioned a singular focus on "to do" at the expense of "to be," an obsession with "how" to the detriment of "why?" Carson did not deny the capacity of scientific inquiry to improve the human condition but did scrutinize the unstated presuppositions, the reductionist image of domination, guiding Western science and technological innovation (*scientia vexata*).

Today, the Sedan crater is the highlight of public tours of the NTS, and *Silent Spring* remains an evocative repudiation of the prevailing

Western conception of the relationship between humanity and the non-human world. It is a corrective to brash and ill-considered instrumentally rational action that neglects the connectivity and vibrancy of the nonhuman, and it calls for emotional and aesthetic reenchantment to foster a more prudent and insightful cultural worldview. Remnants of brash Baconian optimism remain, of course. Indeed, they are on display at the National Atomic Testing Museum in Las Vegas, Nevada. Collective remembrance of the past is often sanitized, but when expressed at museums, this is easily misconstrued as an unassailable depiction of an otherwise complicated past. The museum serves to erase the mistakes and misconduct accompanying open-air testing and the experiences of those who have lived downwind. It is a powerful reassertion of a preferred institutional narrative. This is not uncommon. At memorials, museums, and tourist sites around the United States, tension over the interpretation of Cold War militarism is prevalent, and the impulse to sanitize and simplify is pervasive.

Conclusion

The Legacy of Atomic Testing in the Great Basin

Just below the timberline on Wheeler Peak, rising thirteen thousand feet in eastern Nevada, reside some of the oldest living creatures on Earth. Gnarled, twisted bristlecone pines up to five thousand years old. It is a staggering persistence anywhere, let alone in such a harsh environment. But the secret to longevity is austere isolation. It is cold much of the year. The air is wearing thin. The soil holds scarce nutrients. Few insects, few interlopers of any kind, are common here. And bristlecone pines grow slowly. Life lies deep within and is barely discernable except in tuffs of needles sparingly displayed. In an age valuing the swift and flamboyant, the fleeting and garish, they embody the beauty of tenacity. Bristlecone pines know how to survive.

In 1964, Donald Currey, a graduate student at the University of North Carolina at Chapel Hill, attempted to obtain a core sample from a stately, weather-beaten bristlecone pine that local conservationists named Prometheus. Careful boring allows one to measure a tree's rings without causing lasting harm, but the first auger got stuck, as did the second. With the blessing of US Forest Service officials, Currey cut down the tree and hauled it into the lab. Examining a cross-section of Prometheus's trunk, the count reached 1,000, 3,000, 4,000, and then 4,844 rings.[1] Currey then realized that he had taken a chainsaw to one of the oldest living things on the planet. Prometheus had been growing from around 2880 BC as the ground shifted around it and wind-blown sand polished its wood to a hard, smooth sheen. For five millennia, it stood high on Wheeler Peak, only to be reduced to a lifeless stump in just hours.

The name is perhaps the strangest part of the story. "Prometheus" refers to the Greek god who stole fire from Zeus and gave it to humanity and, thus, symbolizes the pursuit of knowledge through science. For his theft, Prometheus was condemned to an endless cycle of punishment, as each day, bound to a rock, an eagle tears his liver from his body. The tree was so named because of its sweeping branches reaching out at extreme angles, like Prometheus writhing in pain.[2] And it was felled in the name of science.

The incident did not go unnoticed, but Currey and Forest Service officials anticipated that someone would soon document an even older bristlecone and that the controversy would abate. This has only happened once, and it took nearly half a century. Eric Rutkow explains, "Thirty-year-old Donald Currey had unintentionally felled the most ancient tree ever discovered—an organism already wizened when Columbus reached Hispaniola, middle-aged when Caesar ruled Rome, and starting life when the Sumerians created mankind's first written language."[3]

It was a tragic lapse in judgment, but it led to greater awareness of those things lost amid the heedless degradation of nature. In 1986, Great Basin National Park was founded, and it encompasses the stand from which Prometheus was harvested. In turn, Prometheus's sacrifice contributed to a shift of interpretive understanding and productive social and political change. Past mistake and misconduct, should we address them in a forthright manner, can lead us to do better. Figure c.1 depicts a Great Basin bristlecone pine.

Gazing over the expanse of the Great Basin, high on a rugged mountaintop, even bristlecone pines are not immune to the unanticipated rebound effects of sociotechnical prowess, as climate change threatens their slow, resolute tenacity. Timberlines are shifting upward in a warming world, and it is not clear that bristlecone pines will adapt accordingly. There is no longer an outside, as opposed to an inside, of human influence. It is difficult to fully appreciate what this means and the ethical quandaries it presents. For millennia, bristlecone pines stood beyond the twists and turns of human history, but things have changed, in a multitude of ways, since Trinity.

"Prometheus" also appears in the subtitle of Mary Shelley's 1818 novel *Frankenstein; or, The Modern Prometheus*. Dr. Victor Frankenstein cobbles together body parts pilfered from the grave to create reanimated life: "So much has been done, exclaimed the soul of Frankenstein—more, far more, will I achieve; treading in the steps already marked, I will pioneer a new way, explore unknown powers, and unfold to the world the deepest mysteries of creation."[4]

Figure c.1. Great Basin bristlecone pine. (Courtesy of the US National Park Service)

In the Frankenstein of Hollywood movies, the result is a monstrous entity bringing death and destruction—a rambling, relentless transgression. This misses a key point of Shelley's novel, Langdon Winner argues: "Suggested in these words is, it seems to me, the issue truly at stake in the whole of Frankenstein: the plight of things that have been created but not in a context of sufficient care."[5] Dr. Frankenstein's sin lay not in innovation but in abandonment. He left the creature to roam among the villagers. Dr. Frankenstein's brilliance eclipsed his sense of responsibility.

The AEC embarked on the unprecedented production of environmental risk over eleven years of open-air detonations in Nevada. The commission's failure lies less in the risks undertaken than in neglecting to adequately confront the hazards imposed on the public. Too often, AEC officials did not take actions within their purview to reduce the hazards of radioactive contamination. The AEC excelled at surmounting difficult scientific and technological problems but fell short in creating a "context of sufficient care." The legacy of open-air detonations, however, has contributed to productive change. The public is now less willing to acquiesce to the cloistered objectives of organizational actors, both corporate and governmental, and demands greater consideration of those things that are lost absent public deliberation and debate. Indeed, amid the production of novel risks, the public has little choice but to take account of organizational decision-making, lest the unintended consequences sweep through town and deposit on homegrown gardens, school yards, or lush pasture and then milk on the table.

The Lessons of Atmospheric Testing in Nevada

Harry, Simon, Nancy, Smoky, and Sedan are some of the monikers underscoring the turbulent history of open-air detonations in Nevada and the efficacy and recalcitrance of nature. AEC and military officials adopted a number of operational assumptions that proved erroneous and contributed over time to mistakes that were the incentive to misconduct and cover-up. The most significant blind spot was failure to recognize the contingencies of injecting radioactive debris into the atmosphere. Shot after shot attested to the performative engagement of AEC operations and a vibrant nonhuman realm. Fallout flowed over the landscape like invisible snowdrifts. Hot spots appeared with abandon. Moreover, testing continually violated the diffusion premise: the idea that nature is a vast repository in which substances harmlessly diffuse and dilute. Things, energetic things, continually came back around, showed up where they were least expected, and reached farther geographically than anticipated and in a more intense state.

Pressure from the armed services to utilize the NTS for a variety of purposes that were not originally stipulated when it was created made the situation worse. Soon after inauguration, Los Alamos's outdoor laboratory was home to elaborate troop maneuvers and effects testing and, in tandem, a greater reliance on tower detonations that pulled up vast amounts of alluvial soil. The escalation of activities between Operation Ranger in 1951 and Plumbbob just six years later could not be more dramatic.

Materiality is the capacity of organic and inorganic entities, processes, and things to engage in action-reaction and therefore to effect or resist transformation and change. Harnessing the efficacy of uranium is the linchpin of the atomic age. It allows for unprecedented productive and destructive possibilities. But it is a challenge to sustain the boundaries of sociotechnical systems that are predicated on a rock whose atoms are drops teetering on an edge. Environmental risk results from failure to maintain the stability of human-nonhuman configurations. We harness uranium but are similarly captured, as resources must be devoted to ensuring the durability of sociotechnical arrangements in society. Atomic technology entails a dance of human and nonhuman agency that must be tightly choreographed, lest things, energetic things, are left to roam among the villagers.

And environmental violence is indirect. Plausible deniability is thus endemic amid trans-corporeality, or the movement of matter within and through the body. Nature is a medium of interaction. Powerful organizational actors, both corporate and governmental, alter the landscape as they harness and exploit the materiality of nature. And when the durability of sociotechnical arrangements breaks down, the marginalized in society generally bear the brunt, although few truly find refuge anymore.

Open-air testing demonstrated that it is possible to contaminate large swaths of the planet, to varying degrees, from just a handful of origination points. The diffusion premise means little in an age when science and technology compress the ontological distance between industry or the military-industrial complex and the public—even as the temporal distance between cause and effect is long indeed. The materiality set in motion by state and economic actors now reaches farther than ever while encroaching on the sovereignty of the individual body. The planet does not simply seem smaller; it is smaller. Slow environmental violence proliferates amid brilliant advances in science and technology and necessitates a "context of care" that institutional structures in society are scarcely prepared to address. The compression of ontological distance demonstrates that the public, now more than ever, has a stake in the oversight of novel sociotechnical innovation.

The unanticipated challenges of containing fallout led AEC and military officials to consistently promise a level of control that they could not achieve in practice. In turn, AEC and military officials often downplayed provisional or situational ignorance, a temporary deficit of knowledge, to project the reassuring image that testing could be conducted at the highest standards. They engaged in willful and strategic ignorance when consequences arose suggesting that control was tenuous. Willful ignorance is the deliberate refusal to acknowledge the validity of new information that may be disadvantageous: dismissing the scabbing and lesions on sheep grazing in the shadow of the test site, the metallic taste in the air as shot Harry moved through town, radioactive rain drenching upstate New York. Actively producing uncertainty and doubt where it is not reasonably, and empirically, justified constitutes a more strategic or manufactured ignorance: suppressing information on the dire effects of radioiodine on newborn lambs, coercing experts hired by the government to change their testimony, "fraud upon the court."

Despite, or perhaps precisely because of, far-reaching innovations in science and technology, we live in the "thick of things." Socionatural hybridity, the intertwining of the human and the nonhuman, the social and the natural, has not abated but expanded with innovation. And yet we continue to conflate manipulation with mastery and decoding with dominion.

Nature does not act with intention or malice, but nature acts. And downwind of Nevada, the inability to contain fallout resulted in a tightening material-discursive bind as public relations outpaced pragmatic management of radioactive contamination. The hazards inherent to many materials and substances cannot be assessed in the laboratory and only become apparent as they are utilized in practice, but as sociotechnical innovations prove instrumentally useful, and profitable, interest groups increasingly have incentives to contest evidence of injury that arises. Too often, a material-discursive bind promotes the organizational production of doubt, diversion, and denial.

Erroneous operational assumptions rooted in a reticence to acknowledge that the nonhuman world is teeming with complexity, movement, and pushback contributed to failures of foresight, and the capacity of AEC officials to restrict and classify memoranda, reports, and data made it nearly impossible for people outside the technocracy to scrutinize such conduct. It required congressional inquiry and civil litigation, twenty-five years later, to pry open the documents detailing the inner workings of the sheep investigation and, indeed, many other aspects of AEC operations between 1951 and 1962. Contorted science and the control of information, in tandem, were key to the commission's ability to sustain an advantageous definition of the situation. Atmospheric testing was motivated by national security concerns, but it was never exclusively dictated by national security concerns. The atomic-industrial complex had a momentum of its own and, at times, pursued its own parochial aims and its own perpetuation.

Atomic technological rationality is expressed in the employment of uranium and its derivatives in pursuit of key objectives contributing to social-organizational complexity. The goal is to increasingly interweave atomic technology within the relatively enduring social structural patterns of society. This hinges on constructing islands of sociotechnical stability approximating prevailing expectations of Cartesian dualism. But the fundamental separation and distance between the human and the nonhuman at the core of Cartesian duality is not how the world is but how it is made to appear.[6] It is a superficial gloss. When things go wrong, we blame human error, when it is more often a reminder that islands of sociotechnical stability demand a mutual disciplining of non-human materiality and organizational conduct that is difficult to sustain over time—particularly when organizational actors adopt a fast-and-loose approach to environmental hazards.

Modernity is characterized by the production of novel risks, and the individual, family, and community are "not in a position to recognize, evaluate, or thwart a potentially negative course of action."[7] Trust is expected of the public. But competency and fiduciary conduct are expected of organizational actors, both corporate and governmental.

There is an overarching vulnerability on the path to the assumed transcendence of the natural world. We now live in a risk society. That powerful actors will abdicate their responsibility to ensure the collective welfare and instead focus on narrow, parochial considerations, without accountability or redress, is what the public fears.

"Radiophobia" denotes public concern over ionizing radiation impeding rational decision-making and contributing to counterproductive behavior and policies. There is reason to be mindful of radiophobia and the social and political problems it creates, but those who proclaim its liabilities rarely consider the social and cultural dynamics that lay underneath. It is declared a "serious but curable mental disorder," and it is postulated as due to a lack of education or adoption of the linear no threshold theory or images of the destruction at Hiroshima and Nagasaki or opportunistic nongovernmental organizations inflaming public emotion.[8] This does not get at the essence of the problem, and to suggest that radiophobia is a mental illness absent consideration of the legacy of the atomic-industrial complex is naïve. It makes little sense to belittle the public for excessive fear of radiation—as if that itself is the problem—when it is an expression of prior organizational and institutional failures of foresight and reticence to meaningfully embrace public concerns in the present.

Even as the roots of radiophobia are often neglected, so too is consideration of how it functions to curtail public critique of nuclear technology. Radiophobia may identify injurious consequences worthy of consideration, but it is also employed in a condescending and patronizing manner. It is rooted in a deficit of trust in powerful actors in society as well as public resistance to technocratic decision-making that ignores broader values and concerns. Whether describing atmospheric atomic testing or the civilian nuclear-power industry, corporate actors and government officials have, at times, neglected to demonstrate that public health and safety is a predominant focus and have pushed technocratic solutions on the public at the expense of alternative modes of conduct.

Revisionist History at the National Atomic Testing Museum

The final atmospheric detonation in Nevada occurred in July 1962.[9] Little Feller I was a small atomic charge fired from a rocket launcher. It was a tepid end to eleven long years of open-air shots in the Great Basin. Social debate and contestation over the interpretation and meaning of open-air testing in Nevada, however, was only beginning and, in important respects, continues to this day.

The organizational production of willful and strategic ignorance also continues to this day. The National Atomic Testing Museum (NATM) in Las Vegas, Nevada, opened in 2005 and recounts open-air and underground detonations at the NTS. Some things are prominently displayed, while others are not. Central to the museum's creation is the nonprofit Nevada Test Site Historical Foundation (NTSHF), established in 1998. The NTSHF secured funding for construction of the museum and serves as its parent organization. This includes responsibility for the collections in the eight-thousand-square-foot permanent exhibits and the two-thousand-square-foot alternating exhibit space. The museum received $2.5 million in federal funds and $3 million in private donations, a substantial proportion coming from defense contractors with ties to the NTS. The themes and exhibit texts were devised by the NTSHF, the DOE, and affiliated DOE contractors, and, despite additional review by independent consultants, Matt Wray notes, "At this point, it is fair to say that the museum's narrative has essentially been scripted by the DOE."[10] The museum demonstrates the laudable role played by tens of thousands of civilian and military personnel in the technological experimentation enacted at the NTS, but the overall tone is "justification, necessity, even celebration," Patrick McCray observes.[11]

Museums entertain, educate, and commemorate people, places, and events credited with unique social meaning. In doing so, they may sanitize the interpretation put forward for public consumption. There is a tension endemic to preserving and interpreting the past, with the propensity to privilege celebration over discomforting reflection.[12] Reticence to confront the contested implications of past social activity, moreover, serves to deny the lived experiences of less empowered segments of the population and mask the lessons to be learned. Absent is forthright discussion of radioactive fallout persisting tenaciously through time in the bodies and minds of rural downwind residents. The NATM engages in "revisionist" historical interpretation of atmospheric testing that is unjustified given the available historical evidence, and it exhibits an expert-driven and instrumentally rational narrative allowing for little consideration of the mistakes and misconduct accompanying AEC operational conduct.

The NATM raises difficult questions as to the boundary between entertainment and history and whether museums have a responsibility to present multiple viewpoints, however uncomfortable, or simply the collective memory of politically connected groups in society that then passes for "history." Should the public be content with museums offering "a distorted and self-serving awareness of the past?"[13]

The goal of the NTS Historical Foundation is to "encourage the development and public exchange of views regarding the Nevada Test Site and its impact on the nation."[14] In this regard, it is time for the museum to evolve. Amid thousands of square feet, there is no mention of forty-four hundred dead sheep during the Upshot-Knothole series of 1953, associated lawsuits, and the controversy regarding the AEC's efforts to cover up this incident. There is scarce discussion in reference to the epidemio-logical research examining the health burden imposed on downwind residents and indeed the public more broadly, in particular, childhood leukemia deaths and thyroid cancer. The museum should embark on a substantive overview of shot Simon raining out over upstate New York and the commission's belated response to radioactive fallout in southern Utah from shot Harry. Further, the NATM does not mention the 1984 judicial opinion by Judge Bruce S. Jenkins in *Allen et al. v. United States*, the results of congressional inquiry convening in 1979, or the Radiation Exposure Compensation Act (RECA), passed by the US Congress in 1990.

One step forward is to permit downwind residents to assemble ma-terials and interpretive panels of their choosing. Native Americans liv-ing downwind of the NTS and military veterans who participated in open-air testing, moreover, are deserving of greater recognition at the NATM. Opening up space for noninstitutional actors may be unnerving for those who are committed to a reassuring and self-congratulatory in-terpretation, but fostering respectful dissent would demonstrate that the museum values inclusivity as well as the pursuit of historical truthful-ness. Privileging the collective memory of politically connected groups at the expense of multiple viewpoints and uncomfortable historical evi-dence diminishes democratic dialogue and debate. It obscures the les-sons of the past that may inform contemporary issues and concerns. Past mistake and misconduct, should we address them in a forthright manner, can lead us to do better.

A Spare and Stubborn Beauty

The Great Basin evokes a "spare and stubborn beauty."[15] It is burnt, abrasive, and tough but easily scarred. One can effortlessly create lasting change here even if dominion is elusive. It provides space for experimentation and renewal, but hubris frequently prefigures disaster here. Nothing east of the hundredth meridian prepares one for such contradiction.

"In the old days, in blizzardy weather, we used to tie a string of lariats from house to barn so as to make it from shelter to responsibility and back again. With personal, family, and cultural chores to do, I think we had better rig up such a line between past and present," Wallace Stegner counsels.[16] In linking back to the past, how would we know it if we saw it? In connecting present to past, Stegner is resolute: it is time to move beyond self-aggrandizing Western mythology, as it makes it difficult to see clearly. For Stegner, this mythology encompasses the trope of rugged individualism when much of Western life is rooted in cooperation and community, or insistence that "rain follows the plough," when a more prudent course of action is respecting the limits of aridity. Mythology encompasses many things, but confusing militarism and weapons development with progress and patriotism must, surely, make the list.

Few people understood humanity's resentment of the contradictory nature of the desert better than John C. Van Dyke. As the nineteenth century gave way to the twentieth, he published *The Desert*, an alternative cultural interpretation in which the arid and the desolate are valued for more than simply what can be mined and stripped from the earth or pitied for what cannot be grown there, a place where aridity is its enduring charm. The desert is also, at times, unforgiving. It is indelibly untamable and deserving not simply of aesthetic and ecological appreciation but of respect. Van Dyke insisted,

> The feeling of fierceness grows upon you as you come to know the desert better. The sunshafts are falling in a burning shower upon rock and dune, the winds blowing with the breath of far-off fires are withering the bushes and the grasses, the sands drifting higher and higher are burying the trees and reaching up as though they would overwhelm the mountains, the cloud-bursts are rushing down the mountain's side and through the torn arroyos as though they would wash the earth into the sea. . . . One can almost fancy that behind each dome and rampart there are cloud-like Genii—spirits of the desert—keeping guard over this kingdom of the sun.[17]

NOTES

INTRODUCTION

1. Richard G. Hewlett and Jack M. Holl, *Atoms for Peace and War, 1953–1961: Eisenhower and the Atomic Energy Commission* (Berkeley: University of California Press, 1989).
2. The site was originally the Nevada Proving Grounds but was renamed in 1955. It was subsequently renamed the Nevada National Security Site in 2010.
3. DOE, "Discussions with Frank Butrico Monitor at St. George, Utah, May 1953," Deposition at Nevada Operations Office, Las Vegas, Nevada, August 14, 1980, NTA Accession No. NV0011593; *Allen et al. v. United States*, 588 F. Supp. 247 (D. Utah 1984).
4. DOE, "Discussions with Frank Butrico."
5. Richard L. Miller, *Under the Cloud: The Decades of Nuclear Testing* (Woodlands, TX: Two Sixty, 1991).
6. Philip L. Fradkin, *Fallout: An American Nuclear Tragedy* (Tucson: University of Arizona Press, 1989); Carole Gallagher, *American Ground Zero: The Secret Nuclear War* (New York: Random House, 1993).
7. DOE, "Discussions with Frank Butrico."
8. DOE, 3; "Mercury" refers to AEC headquarters at the NTS.
9. *Allen et al.*, 588 F. Supp. 247.
10. AEC, "AEC Meeting Minutes No. 865, Subject: Fallout Resulting from May 19 Shot," May 21, 1953, NTA Accession No. NV0072212; "Discussion of Criteria for Determining Whether a Test Detonation is Fired," May 21, 1953, NTA Accession No. NV0072210; Hewlett and Holl, *Atoms for Peace*; AEC, "Fourteenth Semiannual Report of the Atomic Energy Commission," July 31, 1953, NTA Accession No. NV0727877.
11. John G. Fuller, *The Day We Bombed Utah: America's Most Lethal Secret* (New York: New American Library, 1984), 16.
12. Monroe A. Holmes, "Data on Sheep Losses—Cedar City, Utah, Area," June 17, 1953, NTA Accession No. NV0020399; see also Monroe A. Holmes, "Compiled Report on Co-operative Field Survey of Sheep Deaths in S.W. Utah (Cedar City)," December 31, 1953, NTA Accession No. NV0020585.
13. F. H. Melvin, Bureau of Animal Industry, to Dr. Simms, June 8, 1953, NTA Accession No. NV0020403; Holmes, "Data on Sheep Losses."
14. Testimony of Kern Bulloch, in US Congress, Subcommittee on Oversight and Investigations of the House Committee on Interstate and Foreign Commerce, Health and Scientific Research Subcommittee of the Labor and Human Resources

Committee, and the Senate Committee on the Judiciary, *Health Effects of Low-Level Radiation*, joint hearing, 96th Congress, 1st session, April 19, 1979, vol. 1, 228–229.

15. R. Jeffrey Smith, "Scientists Implicated in Atom Test Deception," *Science* 218, no. 4572 (1982): 545.

16. AEC, "Report on Sheep Losses Adjacent to the Nevada Proving Grounds," January 6, 1954, NTA Accession No. NV0020422.

17. AEC, "Meeting of Livestockmen and AEC Officials, January 13, 9:30 a.m., Fire House, Conference Room, Cedar City, Utah (sheep)," January 13, 1954, NTA Accession No. NV0025557; Barton C. Hacker, *Elements of Controversy: The Atomic Energy Commission and Radiation Safety in Nuclear Weapons Testing, 1947–1974* (Berkeley: University of California Press, 1994).

18. Terrence R. Fehner and F. G. Gosling, *Battlefield of the Cold War: The Nevada Test Site*, vol. 1 (Washington, DC: DOE, 2006), 176.

19. AEC, "Report of the Committee to Study Nevada Proving Grounds" (Tyler Committee), February 1, 1954, NTA Accession No. NV0061646.

20. Donald K. Grayson, *The Great Basin: A Natural Prehistory* (Berkeley: University of California Press, 2011).

21. Chip Ward, *Canaries on the Rim: Living in the Downwind West*, (New York: Verso, 1999), 40.

22. Stephen Trimble, *The Sagebrush Ocean: A Natural History of the Great Basin*, 10th anniversary ed. (Reno: University of Nevada Press, 1999).

23. Terry Tempest Williams, *An Unspoken Hunger: Stories from the Field* (New York: Vintage Books, 1994), 25.

24. John McPhee, *Basin and Range* (New York: Farrar, Straus and Giroux, 1980), 46.

25. Trimble, *Sagebrush Ocean*, 20.

26. Ellen Meloy, *The Last Cheater's Waltz: Beauty and Violence in the Desert Southwest* (New York: Holt, 1999); Rebecca Solnit, *Savage Dreams: A Journey into the Hidden Wars of the American West* (Berkeley: University of California Press, 1994).

27. Mary Austin, *The Land of Little Rain* (Boston: Houghton Mifflin, 1903), 3.

28. Wallace Stegner, *Beyond the Hundredth Meridian: John Wesley Powell and the Second Opening of the West* (New York: Penguin Books, 1992).

29. Patricia Nelson Limerick, *Desert Passages: Encounters with the American Deserts* (Albuquerque: University of New Mexico Press, 1985).

30. Fehner and Gosling, *Battlefield of the Cold War*

31. Solnit, *Savage Dreams*, 4.

32. John Beck, *Dirty Wars: Landscape, Power, and Waste in Western American Literature* (Lincoln: University of Nebraska Press, 2009).

33. Charles Bowden, *Blood Orchid: An Unnatural History of America* (New York: North Point, 2002).

34. Howard G. Wilshire, Jane E. Nielson, and Richard W. Hazlett, *The American West at Risk: Science, Myths, and Politics of Land Abuse and Recovery* (New York: Oxford University Press, 2008), 158.

35. David Gessner, *All the Wild That Remains: Edward Abbey, Wallace Stegner, and the American West* (New York: Norton, 2015).

36. Wallace Stegner, *The Sound of Mountain Water* (Lincoln: University of Nebraska Press, 1980).

37. A. Costandina Titus, *Bombs in the Backyard: Atomic Testing and American Politics*, 2nd ed. (Reno: University of Nevada Press, 2001), xiii.

38. Diane Vaughan, "The Dark Side of Organizations: Mistake, Misconduct, and Disaster," *Annual Review of Sociology* 25 (1999): 273.

39. Lee Clarke, "The Disqualification Heuristic: When Do Organizations Misperceive Risk?," *Research in Social Problems and Public Policy* 5, no. 1 (1993): 289–312.

40. William R. Freudenburg, "Risk and Recreancy: Weber, the Division of Labor, and the Rationality of Risk Perceptions," *Social Forces* 71, no. 4 (1993): 909, 919.

41. Robert Jacobs, "Nuclear Conquistadors: Military Colonialism in Nuclear Test Site Selection during the Cold War," *Asian Journal of Peacebuilding* 1, no. 2 (2013): 173.

42. Miller, *Under the Cloud*.

43. Andrew Feenberg, *Questioning Technology* (New York: Routledge, 1999).

44. Andrew Pickering, *The Cybernetic Brain: Sketches of Another Future* (Chicago: University of Chicago Press, 2010), 17.

45. Tom Zoellner, *Uranium: War, Energy, and the Rock That Shaped the World* (New York: Penguin, 2010).

46. Ulrich Beck, *Risk Society: Towards a New Modernity* (Thousand Oaks, CA: Sage, 1992); Pickering, *Cybernetic Brain*.

47. AEC, *Report of Committee on Operational Future of Nevada Proving Grounds* (Santa Fe Operations Office, May 11, 1953), NTA Accession No. NV0128723.

48. Linsey McGoey, "The Logic of Strategic Ignorance," *British Journal of Sociology* 63, no. 3 (2012): 553–576.

49. AEC, *Report of the Committee to Study Nevada Proving Grounds*.

50. *Bulloch et al. v. United States*, 95 F.R.D. 123 (D. Utah 1982).

51. DOE, *United States Nuclear Tests: July 1945 through September 1992*, DOE/NV-209 (Rev. 15) (Las Vegas: Nevada Operations Office, 2000).

CHAPTER 1. URANIUM AND THE AGENCY OF NATURE IN THE RISK SOCIETY

1. Bruno Latour, *We Have Never Been Modern* (Cambridge, MA: Harvard University Press, 1993).

2. Noel Castree, "False Antitheses? Marxism, Nature and Actor-Networks," *Antipode* 34, no. 1 (2002): 118.

3. Timothy J. LeCain, *The Matter of History: How Things Create the Past* (New York: Cambridge University Press, 2017); Andrew Pickering, *The Cybernetic Brain: Sketches of Another Future* (Chicago: University of Chicago Press, 2010).

4. Tom Zoellner, *Uranium: War, Energy, and the Rock That Shaped the World* (New York: Penguin Books, 2009), ix.

5. Richard L. Garwin and Georges Charpak, *Megawatts and Megatons* (New York: Knopf, 2001); Zoellner, *Uranium*.

6. Zoellner, *Uranium*, ix–x.

7. Stephen I. Schwartz, *Atomic Audit: The Costs and Consequences of U.S. Nuclear Weapons since 1940* (Washington, DC: Brookings Institution Press, 1998).

8. Garwin and Charpak, *Megawatts and Megatons*.

9. Eiichiro Ochia, *Hiroshima to Fukushima: Biohazards of Radiation* (New York: Springer, 2014).

10. Zoellner, *Uranium*, quote at x.

11. Jane Bennett, "The Force of Things: Steps toward an Ecology of Matter," *Political Theory* 32, no. 3 (2004): 347–372.

12. Linda Nash, "The Agency of Nature or the Nature of Agency?," *Environmental History* 10, no. 1 (2005): 69.

13. Karen Armstrong, *The Battle for God: A History of Fundamentalism* (New York: Random House, 2000), 71.

14. Leona Marshall Libby, *The Uranium People* (New York: Crane Russak, 1979), 171.

15. Richard Rhodes, *The Making of the Atomic Bomb* (New York: Simon and Schuster, 1986), 611.

16. Alfred Romer, *The Restless Atom* (New York: Doubleday, 1960), 15.

17. Zoellner, *Uranium*.

18. H. G. Wells, *The World Set Free* (London: Macmillan, 1914).

19. Rhodes, *Making of the Atomic Bomb*.

20. Zoellner, *Uranium*, 30.

21. Andrew Pickering, *The Mangle of Practice: Time, Agency, and Science* (Chicago: University of Chicago Press, 1995).

22. Romer, *Restless Atom*, 7.

23. Rhodes, *Making of the Atomic Bomb*.

24. Zoellner, *Uranium*. Lead is not radioactive, but it is efficacious and indeed the "mother of all industrial pollutants," as it has damaging effects on the human central nervous system.

25. Garwin and Charpak, *Megawatts and Megatons*.

26. Robert Jacobs, "The Bravo Test and the Death and Life of the Global Ecosystem in the Early Anthropocene," *Asia-Pacific Journal* 13, no. 1 (2015): 1–17.

27. The phrase is taken from Moody E. Prior, "Bacon's Man of Science," *Journal of the History of Ideas* 15, no. 3 (1954): 348–370.

28. Prior.

29. Catherine Caufield, *Multiple Exposures: Chronicles of the Radiation Age* (Chicago: University of Chicago Press, 1990); Garwin and Charpak, *Megawatts and Megatons*.

30. Caufield, *Multiple Exposures*, 10.

31. Bruno Latour, "When Things Strike Back: A Possible Contribution of 'Science Studies' to the Social Sciences," *British Journal of Sociology* 51, no. 1 (2000): 107–123.

32. Latour, "When Things Strike Back," 115.

33. C. L. Comar, *Fallout from Nuclear Tests* (Oak Ridge, TN: AEC, 1966). "Half-life" refers to the length of time wherein radioactivity decays by half.

34. Andrew Pickering, "The Politics of Theory: Producing Another World, with Some Thoughts on Latour," *Journal of Cultural Economy* 2, nos. 1–2 (2009): 198.

35. Andrew Pickering, "Neo-Sigma: Art, Agency, and Revolution," *Contemporary Arts and Cultures*, November 9, 2017, https://contemporaryarts.mit.edu.

36. Raymond Murphy, *Leadership in Disaster: Learning for a Future with Global Climate Change* (Montreal: McGill-Queen's University Press, 2009), 42.

37. Pickering, *Mangle of Practice*.

38. Andrew Pickering, "The Ontological Turn: Taking Different Worlds Seriously," *Social Analysis* 61, no. 2 (2017): 139.

39. Pickering, *Mangle of Practice*, 199; Pickering, *Cybernetic Brain*.

40. Lee Clarke, *Worst Cases: Terror and Catastrophe in the Popular Imagination* (Chicago: University of Chicago Press, 2006); Charles Perrow, *Normal Accidents: Living with High-Risk Technologies* (Princeton, NJ: Princeton University Press, 1999); Scott D. Sagan, *The Limits of Safety: Organizations, Accidents, and Nuclear Weapons* (Princeton, NJ: Princeton University Press, 1993).

41. Pickering, *Mangle of Practice*, 17.

42. Pickering, "Ontological Turn."

43. Stephen Healy, "A 'Post-Foundational' Interpretation of Risk: Risk as 'Performance,'" *Journal of Risk Management* 7, no. 3 (2004): 277–296; Florian M. Neisser, "'Riskscapes' and Risk Management—Review and Synthesis of an Actor-Network Approach," *Risk Management* 16, no. 2 (2014): 88–120.

44. Perrow, *Normal Accidents*; Sagan, *Limits to Safety*.

45. Bruno Latour, "Is Re-modernization Occurring—And, If So, How to Prove It? A Commentary on Ulrich Beck," *Theory, Culture and Society* 20, no. 2 (2003): 36.

46. Eric Schlosser, *Command and Control: Nuclear Weapons, the Damascus Accident, and the Illusion of Safety* (New York: Penguin Books, 2013).

47. Ulrich Beck, *World at Risk* (Malden, MA: Polity, 2009), 45.

48. Ulrich Beck, "The Anthropological Shock: Chernobyl and the Contours of the Risk Society," *Berkeley Journal of Sociology* 32 (1987): 155.

49. Max Weber, *Economy and Society*, vol. 1, ed. Guenther Roth and Claus Wittich (Berkeley: University of California Press, 1978).

50. Beck, "Anthropological Shock," 154.

51. Beck, *World at Risk*, 42.

52. Ulrich Beck, *Risk Society: Towards a New Modernity* (Thousand Oaks, CA: Sage, 1992).

53. Ulrich Beck, "Foreword: Risk Society as Political Category," in *The Risk Society Revisited: Social Theory and Governance*, by Eugene A. Rosa, Ortwin Renn, and Aaron M. McCright (Philadelphia: Temple University Press, 2014), xx (emphasis in original).

54. Beck, *Risk Society*; Beck, *World at Risk*.

55. Beck, *Risk Society*; Beck, *World at Risk*.
56. Ulrich Beck, "Living in the World Risk Society," *Economy and Society* 35, no. 3 (2006): 332.
57. Beck, "Foreword," xv.
58. Ulrich Beck, "Risk Society Revisited: Theory, Politics and Research Programmes," in *The Risk Society and Beyond: Critical Issues for Social Theory*, ed. Barbara Adam, Ulrich Beck, and Joost Van Loon (Thousand Oaks, CA: Sage, 2000), 217.
59. Pickering, "Ontological Turn."
60. Pickering, *Cybernetic Brain*.
61. Timothy J. LeCain, *The Matter of History: How Things Create the Past* (New York: Cambridge University Press, 2017).
62. Raymond Murphy, *Rationality and Nature: A Sociological Inquiry into a Changing Relationship* (Boulder, CO: Westview, 1994).
63. Pickering, "Politics of Theory," 198.
64. Robert K. Yin, *Case Study Research: Design and Methods*, 3rd ed. (Thousand Oaks, CA: Sage, 2003).
65. Many documents utilized herein are accessible through the Nuclear Testing Archive in Las Vegas, Nevada, and OpenNet, a DOE database (www.osti.gov /opennet/). They can be located using the corresponding Nuclear Testing Archive (NTA) accession number.
66. Diane Vaughan, "The Dark Side of Organizations: Mistake, Misconduct, and Disaster," *Annual Review of Sociology* 25 (1999): 271–305.
67. Diane Vaughan, *The* Challenger *Launch Decision: Risky Technology, Culture, and Deviance at NASA* (Chicago: University of Chicago Press, 1996); Vaughan, "Organizational Rituals of Risk and Error," in *Organizational Encounters with Risk*, ed. Bridget Hutter and Michael Power (New York: Cambridge University Press, 2004), 33–66.
68. John Scott, *A Matter of Record* (Cambridge, MA: Blackwell, 1990), 34.
69. Diane Vaughan, "Theorizing Disaster: Analogy, Historical Ethnography, and the *Challenger* Accident," *Ethnography* 5, no. 3 (2004): 321.
70. Michael R. Hill, *Archival Strategies and Techniques* (Thousand Oaks, CA: Sage, 1993).
71. Robert Merton, *On Theoretical Sociology* (New York: Free Press, 1967), 39.
72. Stanley H. Hoffman, "International Relations: The Long Road to Theory," *World Politics* 11, no. 3 (1959): 346–377.
73. Merton, *On Theoretical Sociology*.
74. Murphy, *Rationality and Nature*, 170.

CHAPTER 2. TRINITY AND LESSONS NOT LEARNED

1. Amir D. Aczel, *Uranium Wars: The Scientific Rivalry That Created the Nuclear Age* (New York: Palgrave Macmillan, 2009); Stephen I. Schwartz, *Atomic Audit: The Costs and Consequences of U.S. Nuclear Weapons since 1940* (Washington, DC: Brookings Institution Press, 1998).

2. Jon Hunner, *Inventing Los Alamos: The Growth of an Atomic Community* (Norman: University of Oklahoma Press, 2004).

3. Rebecca Solnit, *Savage Dreams: A Journey into the Hidden Wars of the American West* (Berkeley: University of California Press, 1994), 135.

4. Richard Rhodes, *The Making of the Atomic Bomb* (New York: Simon and Schuster, 1986).

5. Hunner, *Inventing Los Alamos*, 65.

6. Rhodes, *Making of the Atomic Bomb*.

7. Alternative sites were considered, but Jornada del Muerto was close to Los Alamos and already under federal control, constituting a section of the Alamogordo Bombing and Gunnery Range. Today the Trinity site is secluded within White Sands Missile Range.

8. Gerard J. DeGroot, *The Bomb: A Life* (Cambridge, MA: Harvard University Press, 2005), 56.

9. Rhodes, *Making of the Atomic Bomb*, 653–654.

10. Robert Jungk, *Brighter than a Thousand Suns: A Personal History of the Atomic Scientists* (New York: Harcourt, 1958); Rhodes, *Making of the Atomic Bomb*.

11. Rhodes, *Making of the Atomic Bomb*.

12. Kenneth T. Bainbridge, *Trinity* (Los Alamos, NM: Los Alamos Scientific Laboratory, 1976).

13. Carl Maag and Steve Rohrer, *Project Trinity, 1945–1946* (Washington, DC: Defense Nuclear Agency, 1982).

14. Thomas E. Widner and Susan M. Flack, "Characterization of the World's First Nuclear Explosion, the Trinity Test, as a Source of Public Radiation Exposure," *Health Physics* 98, no. 3 (2010): 480–497.

15. Barton C. Hacker, *The Dragon's Tail: Radiation Safety in the Manhattan Project, 1942–1946* (Berkeley: University of California Press, 1987).

16. Joseph O. Hirschfelder and John L. Magee to Kenneth Bainbridge, "Danger from Active Material Falling from the Cloud," June 16, 1945, NTA Accession No. NV0311635. See also Louis Hempelmann and James F. Nolan to Kenneth Bainbridge, "Danger to Personnel in Nearby Towns Exposed to Active Material from the Cloud," June 22, 1945, NTA Accession No. NV0123776; Joseph O. Hirschfelder and John L. Magee to Kenneth Bainbridge, "Improbability of Danger from Active Material Falling from Cloud," July 6, 1945, NTA Accession No. NV0059903.

17. Hacker, *Dragon's Tail*, 84.

18. Bainbridge, *Trinity*; Maag and Rohrer, *Project Trinity*.

19. Hacker, *Dragon's Tail*, 84.

20. Hacker, *Dragon's Tail*; Rhodes, *Making of the Atomic Bomb*.

21. Rhodes, *Making of the Atomic Bomb*; Ferenc Morton Szasz, *The Day the Sun Rose Twice: The Story of the Trinity Nuclear Explosion, July 16, 1945* (Albuquerque: University of New Mexico Press, 1984).

22. DeGroot, *Bomb*, 58.

23. Bainbridge, *Trinity*.

24. Szasz, *Day the Sun Rose Twice*.

25. Rhodes, *Making of the Atomic Bomb*; Szasz, *Day the Sun Rose Twice*.

26. Rhodes, *Making of the Atomic Bomb*; Szasz, *Day the Sun Rose Twice*.

27. Rhodes, *Making of the Atomic Bomb*; Szasz, *Day the Sun Rose Twice*.

28. Leslie R. Groves, "Some Recollections of July 16, 1945," *Bulletin of the Atomic Scientists* 26, no. 6 (1970): 21–27.

29. J. M. Hubbard to J. R. Oppenheimer, "Long Range Weather Schedule," June 21, 1945, NTA Accession No. ALLA0000789; J. M. Hubbard to K. T. Bainbridge, memorandum, June 30, 1945, NTA Accession No. ALLA0003608; Szasz, *Day the Sun Rose Twice*, 71.

30. Szasz, *Day the Sun Rose Twice*, 74.

31. Szasz, 74.

32. Bainbridge, *Trinity*.

33. Rhodes, *Making of the Atomic Bomb*; Szasz, *Day the Sun Rose Twice*.

34. Morgan Rieder and Michael Lawson, *Trinity at Fifty* (Tularosa, NM: Human Systems Research, 1995).

35. Rhodes, *Making of the Atomic Bomb*, 659.

36. Rhodes; Szasz, *Day the Sun Rose Twice*.

37. Szasz, *Day the Sun Rose Twice*, 73.

38. Leslie R. Groves, "Memorandum for the Secretary of War," July 18, 1945, NTA Accession No. NV0039279.

39. Rhodes, *Making of the Atomic Bomb*; Szasz, *Day the Sun Rose Twice*. Even twenty-five years later, Groves did not view the situation in this manner and blamed Hubbard for an erroneous forecast. See Groves, "Some Recollections."

40. Rhodes, *Making of the Atomic Bomb*; Szasz, *Day the Sun Rose Twice*.

41. Marq de Villiers, *Windswept: The Story of Wind and Weather* (New York: Walker, 2006), 2–3.

42. Rhodes, *Making of the Atomic Bomb*; Szasz, *Day the Sun Rose Twice*.

43. Hacker, *Dragon's Tail*.

44. Rhodes, *Making of the Atomic Bomb*.

45. Szasz, *Day the Sun Rose Twice*.

46. Ferrell's description is contained in Groves, "Memorandum for the Secretary of War."

47. Rhodes, *Making of the Atomic Bomb*, 672.

48. Rhodes.

49. Jacob Darwin Hamblin, *Arming Mother Nature: The Birth of Catastrophic Environmentalism* (New York: Oxford University Press, 2013).

50. Stafford Warren to Major General Groves, "Report on Test II at Trinity, July 1945," July 21, 1945, NTA Accession No. NV0026157.

51. Hacker, *Dragon's Tail*.

52. Cyril S. Smith to Lieutenant Taylor, "Trinity Shot," July 25, 1945, folder 6, "Trinity," box 82, OSRD-S1 Committee, Record Group 227, US National Archives.

53. Lansing Lamont, *Day of Trinity* (New York: Scribner, 1965); Szasz, *Day the Sun Rose Twice.*

54. Widner and Flack, "Characterization of the World's First Nuclear Explosion."

55. Hacker, *Dragon's Tail*; Lamont, *Day of Trinity*; Widner and Flack, "Characterization of the World's First Nuclear Explosion."

56. Hacker, *Dragon's Tail*, 102–103.

57. Hacker, *Dragon's Tail*; Warren, "Report on Test II at Trinity"; Widner and Flack, "Characterization of the World's First Nuclear Explosion."

58. Warren, "Report on Test II at Trinity," 2.

59. Hacker, *Dragon's Tail*; Warren, "Report on Test II at Trinity"; Widner and Flack, "Characterization of the World's First Nuclear Explosion."

60. Maag and Rohrer, *Project Trinity*; Widner and Flack, "Characterization of the World's First Nuclear Explosion."

61. Widner and Flack, "Characterization of the World's First Nuclear Explosion," 493.

62. Lamont, *Day of Trinity*; Widner and Flack, "Characterization of the World's First Nuclear Explosion."

63. Hacker, *Dragon's Tail.*

64. Hacker; Widner and Flack, "Characterization of the World's First Nuclear Explosion."

65. Thomas Widner, "The World's First Atomic Bomb Blast and How It Interacted with the Jornada Del Muerto and Chupadera Mesa," *New Mexico Geological Society Guidebook, 60th Field Conference*, 2009, 425–428; Widner and Flack, "Characterization of the World's First Nuclear Explosion."

66. Widner and Flack, "Characterization of the World's First Nuclear Explosion."

67. Hacker, *Dragon's Tail*; Widner and Flack, "Characterization of the World's First Nuclear Explosion."

68. Widner and Flack, "Characterization of the World's First Nuclear Explosion."

69. Lamont, *Day of Trinity.*

70. Widner and Flack, "Characterization of the World's First Nuclear Explosion."

71. Wayne R. Hansen and John C. Rodgers, *Radiological Survey and Evaluation of the Fallout Area from the Trinity Test Site: Chupadera Mesa and White Sands Missile Range, New Mexico* (Los Alamos, NM: Los Alamos National Laboratory, 1985).

72. Jeremy Bernstein, *Plutonium: A History of the World's Most Dangerous Element* (Ithaca, NY: Cornell University Press, 2007).

73. Quoted in Bernstein, 105.

74. Warren, "Report on Test II at Trinity," 2–3.

75. The suite of papers was published in the October 2020 issue of *Health Physics.*

76. Susan Montoya Bryan, "Study: Cancer Cases Likely in Those Exposed to Atomic Test," Associated Press, September 1, 2020.

77. Elizabeth K. Cahoon, Rui Zhang, Steven L. Simon, Andre Bouville, and Ruth M. Pfeiffer, "Projected Cancer Risks to Residents of New Mexico from Exposure to Trinity Radioactive Fallout," *Health Physics* 119, no. 4 (2020): 478–493.

78. Cahoon et al., 488. Most excess cancers are likely to occur or have occurred in the New Mexico counties of Guadalupe, Lincoln, San Miquel, Socorro, and Torrance.

79. Harold L. Beck, Steven L. Simon, Andre Bouville, and Anna Romanyukha, "Accounting for Unfissioned Plutonium from the Trinity Atomic Bomb Test," *Health Physics* 119, no. 4 (2020): 504–516.

80. Widner and Flack, "Characterization of the World's First Nuclear Explosion."

81. Widner and Flack.

82. Zoellner, *Uranium*.

83. Ulrich Beck, *The Risk Society: Towards a New Modernity* (Thousand Oaks, CA: Sage, 1992).

84. Kai Erikson, *A New Species of Trouble: The Human Experience of Modern Disasters* (New York: Norton, 1995).

85. Toshihiro Higuchi, "'Clean' Bombs: Nuclear Technology and Nuclear Strategy in the 1950s," *Journal of Strategic Studies* 29, no. 1 (2006): 83–116.

86. Carolyn Merchant, *Autonomous Nature: Problems of Prediction and Control from Ancient Times to the Scientific Revolution* (New York: Routledge, 2016).

87. Author's site visit, October 7, 2017.

88. Iver Peterson, "40 Years Ago, the Bomb: The Questions Came Later," *New York Times*, July 16, 1985, C1.

89. Ellen Meloy, *Last Cheater's Waltz: Beauty and Violence in the Desert Southwest* (New York: Holt, 1999), 68.

90. Meloy, 32–33.

91. Meloy, 29.

CHAPTER 3. THE EMERGING ATOMIC SPECTACULAR, 1950–1951

1. Robert J. Duffy, *Nuclear Politics in America: A History and Theory of Government Regulation* (Lawrence: University of Kansas Press, 1997), 24.

2. Richard O. Niehoff, "Organization and Administration of the United States Atomic Energy Commission," *Public Administration Review* 8, no. 2 (1948): 91–102.

3. Byron S. Miller, "A Law Is Passed—The Atomic Energy Act of 1946," *University of Chicago Law Review* 15, no. 4 (1948): 799, 820.

4. Miller, 821; Richard G. Hewlett and Francis Duncan, *Atomic Shield, 1947–1952*, vol. 2 of *A History of the United States Atomic Energy Commission* (University Park: Pennsylvania State University Press, 1969).

5. Richard G. Hewlett and Oscar E. Anderson, *The New World, 1939–1946*, vol. 1 of *A History of the United States Atomic Energy Commission* (University Park: Pennsylvania State University Press, 1962).

6. Steven M. Neuse, *David E. Lilienthal: The Journal of an American Liberal* (Knoxville: University of Tennessee Press, 1997).

7. C. Wright Mills, *The Power Elite* (New York: Oxford University Press, 1956), 215.

8. David E. Lilienthal, *Change, Hope, and the Bomb* (Princeton, NJ: Princeton University Press, 1963), 115–116.

9. President Dwight D. Eisenhower, "Farewell Address to the Nation," January 17, 1961, National Archives and Records Administration.

10. Eugene Rabinowitch, "Five Years Later," *Bulletin of the Atomic Scientists* 7, no. 1 (1950): 3.

11. David E. Lilienthal, "Science and the Spirit of Man," *Bulletin of the Atomic Scientists* 5, no. 4 (1949): 98.

12. Paul Boyer, *By the Bomb's Early Light: American Thought and Culture at the Dawn of the Atomic Age* (Chapel Hill: University of North Carolina Press, 1994), 106.

13. Herbert Marcuse, *One-Dimensional Man* (Boston: Beacon, 1964).

14. William Leiss, "Technological Rationality: Marcuse and His Critics," *Philosophy of the Social Sciences* 2, no. 1 (1972): 34–35.

15. AEC, "Project 'Nutmeg," Armed Forces Special Weapons Project, December 31, 1948, NTA Accession No. NV0032697.

16. AEC, "Discussion of Radiological Hazards Associated with a Continental Test Site for Atomic Bombs," Los Alamos Scientific Laboratory, September 1, 1950, NTA Accession No. NV0051812; AEC, "Location of Proving Ground for Atomic Weapons," December 13, 1950, NTA Accession No. NV0750356.

17. A. Costandina Titus, *Bombs in the Backyard: Atomic Testing and American Politics*, 2nd ed. (Reno: University of Nevada Press, 2001).

18. AEC, "Discussion of Radiological Hazards."

19. Richard L. Miller, *Under the Cloud: The Decades of Nuclear Testing* (Woodlands, TX: Two Sixty, 1991).

20. AEC, "Memo for the President, Subject: AEC Has Made a Review of Possible Locations for the Required Additional Atomic Weapons Site," National Security Council, December 18, 1950, NTA Accession No. NV0304388.

21. AEC, "Documentation of Establishment of Continental Test Site," September 14, 1953, 12, NTA Accession No. NV0026308.

22. R. Miller, *Under the Cloud*, 80.

23. AEC, "Discussion of Radiological Hazards Associated with a Continental Test Site for Atomic Bombs," Los Alamos Scientific Laboratory, August 1, 1950, 10, 23.

24. Peter L. Bernstein, *Against the Gods: The Remarkable Story of Risk* (New York: Wiley, 1998), 8.

25. Eugene Rosa, "Metatheoretical Foundations for Post-normal Risk," *Journal of Risk Research* 1, no. 1 (1998): 28.

26. Eugene A. Rosa, Ortwin Renn, and Aaron M. McCright, *The Risk Society Revisited: Social Theory and Governance* (Philadelphia: Temple University Press, 2014), 15.

27. Bernstein, *Against the Gods*, 6.

28. Howard Ball, *Justice Downwind: America's Atomic Testing Program in the 1950s* (New York: Oxford University Press, 1986), 154; Scott Gibbs Monson, "*Allen v. United States*: Discretion Defined Downwind," *Utah Law Review* 2 (1985): 435–461.

29. *Allen et al. v. United States*, 588 F. Supp. 247 (D. Utah 1984).

30. Ball, *Justice Downwind*.

31. Gisele C. DuFort, "All the King's Forces, or, The Discretionary Function Doctrine in the Nuclear Age: *Allen v. United States*," *Ecology Law Quarterly* 15, no. 3 (1988): 477–502. Although *Allen et al.* was overturned, this should not be construed as evidence that the argument Jenkins sketched out lacks merit. Indeed, his memorandum opinion comprises a prescient overview of AEC mistakes and misconduct during the era of open-air testing in Nevada.

32. DuFort, "All the King's Forces"; Medora Marisseau, "Seeing through the Fallout: Radiation and the Discretionary Function Exemption." *Environmental Law* 22, no. 4 (1992): 1509–1538.

33. Marisseau, "Seeing through the Fallout," 1509.

34. A. Costandina Titus and Michael E. Bowers, "Konizeski and the Warner Amendment: Back to Ground Zero for Atomic Litigants," *Brigham Young University Law Review* 2 (1988): 387–408.

35. *Allen et al.*, 588 F. Supp. 247.

36. US Census Bureau, *U.S. Census of Population: 1960*, vol. 1, *Characteristics of the Population* (Washington, DC: US Government Printing Office, 1963).

37. Gerald J. DeGroot, *The Bomb: A Life* (Cambridge, MA: Harvard University Press, 2004), 26.

38. Ball, *Justice Downwind*; Philip L. Fradkin, *Fallout: An American Nuclear Tragedy* (Tucson: University of Arizona Press, 1989).

39. AEC, "Press Release: AEC Authorized to Use Las Vegas Bombing and Gunnery Range for Atomic Weapons Development Program," January 11, 1951, NTA Accession No. NV0030348.

40. Colonel Schlatter to Major Sturges, "Public Relations Conference Concerning Mercury," December 20, 1950, NTA Accession No. NV0029360; AEC, "Public Relations Plan Operations Buster and Jangle," July 24, 1951, NTA Accession No. NV0128200; AEC, "Tumbler-Snapper Information Plan," Santa Fe Operations Office, December 31, 1952, 1, NTA Accession No. NV0720393.

41. Terrence R. Fehner and F. G. Gosling, *Battlefield of the Cold War: The Nevada Test Site*, vol. 1 (Washington, DC: DOE, 2006). See also AEC, "Public Relations Plan," July 24, 1951.

42. Up through the Tumbler-Snapper series of 1952, test names were drawn from the military phonetic alphabet. Beginning in 1953, code names and nicknames were created and/or obtained from an official AEC list to designate a given test but not divulge any information about the device. See National Nuclear Security Administration, "How Nuclear Tests Got Their Names," 2015.

43. R. Miller, *Under the Cloud*.

44. DOE, *United States Nuclear Tests: July 1945 through September 1992* (Las Vegas: Nevada Operations Office, 2000).

45. Barton C. Hacker, *Elements of Controversy: The Atomic Energy Commission and Radiation Safety in Nuclear Weapons Testing, 1947–1974* (Berkeley: University of California Press, 1994).

46. Fehner and Gosling, *Battlefield of the Cold War*, 176.

47. Gladwin Hill, "3d Atom Test Lights Nevada Dawn: Peaks Stand Out in Weird Glare," *New York Times*, February 2, 1951, 1, 8.

48. Ball, *Justice Downwind*; Glen M. Feighery, "'A Light out of This World': Awe, Anxiety, and Routinization in Early Nuclear Test Coverage, 1951–1953," *American Journalism* 28, no. 3 (2011): 9–34.

49. William J. Kinsella, "Nuclear Boundaries: Material and Discursive Containment at the Hanford Nuclear Reservation," *Science as Culture* 10, no. 2 (2001): 163–194.

50. AEC, "Information Plan for Buster-Jangle," September 5, 1951, 1.

51. Hill, "3d Atom Test Lights Nevada Dawn."

52. AEC, "Public Relations Plan," July 24, 1951.

53. AEC, "Summary of Relations between the AEC and the Photographic Industry Regarding Radioactive Contamination from Atomic Weapon Tests, from January through December 1951," report by the Director of Military Applications, January 17, 1952, NTA Accession No. NV0072173.

54. AEC, "Public Relations Plan," July 24, 1951, 2, 13.

55. Hacker, *Elements of Controversy*.

56. Ball, *Justice Downwind*; Hewlett and Duncan, *Atomic Shield*.

57. Ball, *Justice Downwind*, 33.

58. AEC, "Notes on the Meeting of a Committee to Consider the Feasibility and Conditions for a Preliminary Radiological Safety Shot for Operation Windsquall," Los Alamos Scientific Laboratory, May 21–22, 1951, NTA Accession No. NV0750764. Note that Windsquall was renamed Operation Jangle.

59. Advisory Committee Staff to Members of the Advisory Committee on Human Radiation Experiments, "Risk to Downwinders from Government Consideration of Plans to Understand Fallout," March 8, 1995, 2, NTA Accession No. NV0750763.

60. Gaelen Felt to A. C. Graves, "Jangle Fallout Problems," June 28, 1951, NTA Accession No. NV0062743; T. L. Shipman to John C. Clark, "Special Rad Safe Problems—Operation Jangle," July 11, 1951, NTA Accession No. NV0750767; AEC, "Meeting of the Jangle Feasibility Committee," July 13, 1953, NTA Accession No. NV0720372.

61. Paul P. Kennedy, "Bombs Not Used in Nevada Blasts," *New York Times*, January 31, 1951, 14.

62. AEC, "Public Relations Plan," July 24, 1951, 6.

63. T. L. Shipman to A. C. Graves, "Summary Report to the Test Director: From the Rad Safe and Health Unit Operation Buster-Jangle," December 27, 1951, NTA Accession No. NV0122938.

64. Defense Nuclear Agency (DNA), "Operation Buster-Jangle, 1951," June 21, 1982, NTA Accession No. NV0760124.

65. Hacker, *Elements of Controversy*, 70.

66. R. Miller, *Under the Cloud*, 125.

67. DNA, "Operation Buster-Jangle."

68. Hacker, *Elements of Controversy*.

69. R. Miller, *Under the Cloud*.

70. H. F. Schulte to T. L. Shipman, "Preliminary Report on Buster-Jangle Fall-Out Program," Los Alamos Scientific Laboratory, December 15, 1951, NTA Accession No. NV0122093.
71. DNA, "Operation Buster-Jangle"; Hacker, *Elements of Controversy*.
72. Hacker, *Elements of Controversy*.
73. Schulte, "Preliminary Report," December 15, 1951, 6.
74. Shipman, "Summary Report," December 27, 1951, 1, 9.
75. Shipman, "Summary Report," 6.
76. Ball, *Justice Downwind*; Fradkin, *Fallout*.
77. Hill, "3d Atom Test Lights Nevada Dawn"; Gladwin Hill, "Atomic Boom Town in the Desert," *New York Times*, February 11, 1951, 158; Gladwin Hill, "Atomic-Age Sight-Seeing Way Out West," *New York Times*, April 15, 1951, 266.
78. Titus, *Bombs in the Backyard*.
79. Feighery, "Light out of This World."
80. Gladwin Hill, "Tactical Bomb Test 'Secrets' Open to Thousands but Not News Men," *New York Times*, October 18, 1951, 18; Hill, "Rumors on Atom Test Irk AEC, but Security Rules Yield Little Else," *New York Times*, October 26, 1951, 12.
81. William A. Gamson and Andre Modigliani, "Media Discourse and Public Opinion on Nuclear Power: A Constructionist Approach," *American Journal of Sociology* 95, no. 1 (1989): 1–37.

CHAPTER 4. THE ROAD TO UPSHOT-KNOTHOLE, 1952

1. AEC, "Troop Participation in Tumbler-Snapper," report by the Director of Military Application, March 31, 1952, NTA Accession No. NV0767725.
2. Shields Warren to Brigadier General K. E. Fields, "Staff Paper on Troop Participation in Operation Tumbler-Snapper," March 25, 1952, NTA Accession No. NV0767742. Note that Shields Warren is not related to Stafford Warren.
3. AEC Chairman Gordon Dean to General Loper, April 2, 1952, NTA Accession No. NV0767827.
4. DNA, "Operation Tumbler-Snapper, 1952," June 14, 1982, NTA Accession No. NV0018989.
5. Richard L. Miller, *Under the Cloud: The Decades of Nuclear Testing* (The Woodlands, TX: Two Sixty, 1991), 142 (emphasis in original).
6. Samuel Glasstone, *The Effects of Atomic Weapons* (Los Alamos, NM: Los Alamos Scientific Laboratory, 1950).
7. Paul Boyer, *By the Bomb's Early Light: American Thought and Culture at the Dawn of the Atomic Age* (Chapel Hill: University of North Carolina Press, 1994); Robert J. Lifton and Greg Mitchell, *Hiroshima in America: A Half Century of Denial* (New York: HarperCollins, 1995).
8. Lifton and Mitchell, *Hiroshima in America*, 59.
9. Wilfred Burchett, *Shadows of Hiroshima* (London: Verso, 1983), 35.
10. H. N. Friesen, *A Perspective on Atmospheric Nuclear Tests in Nevada: Fact Book* (Las Vegas, NV: DOE, 1992).

11. Barton C. Hacker, *Elements of Controversy: The Atomic Energy Commission and Radiation Safety in Nuclear Weapons Testing, 1947–1974* (Berkeley: University of California Press, 1994).

12. Richard G. Hewlett and Francis Duncan, *Atomic Shield, 1947–1952*, vol. 2, *A History of the Atomic Energy Commission* (University Park: Pennsylvania State University Press, 1969).

13. Richard L. Garwin and Georges Charpak, *Megawatts and Megatons: A Turning Point in the Nuclear Age?* (New York: Knopf, 2001).

14. Friesen, *Perspective*, 13.

15. *Allen et al. v. United States*, 588 F. Supp. 247 (D. Utah 1984).

16. John G. Fuller, *The Day We Bombed Utah: America's Most Lethal Secret* (New York: New American Library, 1984).

17. Stafford Warren to Major General Groves, "Report on Test II at Trinity, July 1945," July 21, 1945, NTA Accession No. NV0026157. By 1952, Stafford Warren had taken a position as the dean of the school of medicine at the University of California, Los Angeles. Groves had retired from the US Army, in part, due to political pressure stemming from his outspoken pursuit of near-absolute military control of atomic technology.

18. The 1990 RECA bill encompasses select areas of Clark County, Nevada, that are not listed in table 4.2, and, therefore, the population total for Nevada is slightly undercounted. Note too that only Mohave County, Arizona, north of the Grand Canyon is included in the RECA legislation. In turn, the total population displayed for Mohave County is overinflated.

19. Clark County, Nevada, grew from 48,289 people to 127,016 between 1950 and 1960, and Salt Lake County grew from 274,895 to 383,035 over this period. Neither were included in the proposed RECA bill of 1979 or in the bill signed into law in 1991; US Census Bureau, *U.S. Census of Population: 1960*, vol. 1, *Characteristics of the Population* (Washington, DC: US Government Printing Office, 1963).

20. Valerie L. Kuletz, *The Tainted Desert: Environmental and Social Ruin in the American Desert* (New York: Routledge, 1998).

21. Charles Pinderhughes, "Toward a New Theory of Internal Colonialism," *Socialism and Democracy* 25, no. 1 (2011): 236.

22. Eric Frohmberg, Robert Goble, Virginia Sanchez, and Dianne Quigley, "The Assessment of Radiation Exposures in Native American Communities from Nuclear Weapons Testing in Nevada," *Risk Analysis* 20, no. 1 (2000): 101–111.

23. Kuletz, *Tainted Desert*, 8.

24. Evelyn Nakano Glenn, "Settler Colonialism as Structure: A Framework for Comparative Studies of U.S. Race and Gender Formation," *Sociology of Race and Ethnicity* 1, no. 1 (2015): 54–74.

25. DNA, "Operation Tumbler-Snapper, 1952."

26. Lifton and Mitchell, *Hiroshima in America*; Beverly Ann Deepe Keever, *News Zero: The "New York Times" and the Bomb* (Monroe, ME: Common Courage, 2004).

27. Boyer, *By the Bomb's Early Light*.
28. William L. Laurence, "Millions on TV See Explosion That Rocks Desert like Quake," *New York Times*, March 18, 1953, 1, 12.
29. Max Weber, *Economy and Society*, vol. 1, ed. Guenther Roth and Claus Wittich (Berkeley: University of California Press, 1978).
30. Scott Kirsch, "Watching the Bombs Go Off: Photography, Nuclear Landscapes, and Spectator Democracy," *Antipode* 29, no. 3 (1997): 227–255.
31. Philip L. Fradkin, *Fallout: An American Nuclear Tragedy* (Tucson: University of Arizona Press, 1989), 101. Note that the last airdrop of the Tumbler-Snapper series, Dog, was conducted on May 1.
32. Fradkin; Miller, *Under the Cloud*.
33. US Census Bureau, *U.S. Census of Population*.
34. Howard Ball, *Justice Downwind: America's Atomic Testing Program in the 1950s* (New York: Oxford University Press, 1986).
35. "S.L. A-Dust Declared Harmless," *Deseret News*, May 8, 1952.
36. "Atoms in the Dust," *Deseret News*, May 9, 1952.
37. Ball, *Justice Downwind*.
38. Utah Governor J. Bracken Lee to Gordon Dean, May 9, 1952, NTA Accession No. NV0704806. Note that Lee frames radiation exposure in an additive, cumulative manner.
39. Miller, *Under the Cloud*.
40. Friesen, *Perspective*; Lynn R. Anspaugh and Bruce W. Church, "Historical Estimates of External Gamma Exposure and Collective External Gamma Exposure from Testing at the Nevada Test Site. I. Test Series through HARDTACK II, 1958," Lawrence Livermore National Laboratory and DOE Nevada Operations Office, December 1985, NTA Accession No. NV0061768.
41. Miller, *Under the Cloud*.
42. Harold L. Beck, Irene K. Helfer, Andre Bouville, and Mona Dreicer, "Estimates of Fallout in the Continental U.S. from Nevada Weapons Testing Based on Gummed-Film Monitoring Data," *Health Physics* 59, no. 5 (1990): 565–576.
43. Thomas L. Shipman to Alvin Graves, "Nevada Cattle," August 8, 1952, NTA Accession No. NV0078840.
44. Thomas L. Shipman to Alvin Graves, "Nevada Cattle," August 23, 1952, NTA Accession No. NV0032992.
45. Thomas L. Shipman to Alvin Graves, "Nevada Cattle," August 6, 1952, NTA Accession No. ALLA0008408.
46. AEC, *Report of Committee on Operational Future of Nevada Proving Grounds* (Santa Fe Operations Office, May 11, 1953), 3, 13, NTA Accession No. NV0128723.
47. AEC.
48. AEC, 14.
49. AEC, 5.
50. Ball, *Justice Downwind*.
51. AEC, "Report of Committee," 7, 27.

52. AEC, 24.

53. Charles Perrow, "A Society of Organizations," *Theory and Society* 20, no. 6 (1991): 725–762.

54. Diane Vaughan, "The Dark Side of Organizations: Mistake, Misconduct, and Disaster," *Annual Review of Sociology* 25 (1999): 271–305.

55. Barry A. Turner, "The Organizational and Interorganizational Development of Disasters," *Administrative Science Quarterly* 21, no. 3 (1976): 378–397; Barry A. Turner and Nick F. Pidgeon, *Man-Made Disasters*, 2nd ed. (Oxford, UK: Butterworth-Heinemann, 1997).

56. Turner, "Development of Disasters," 381.

57. Edgar H. Schein, *Organizational Culture and Leadership* (New York: Wiley, 2004).

58. Diane Vaughan, "Rational Choice, Situated Action, and the Social Control of Organizations," *Law & Society Review* 32, no. 1 (1998): 37.

59. Schein, *Organizational Culture.*

60. Lee Clarke, "I'm Warning You," *The Fukushima Dai-Ichi Accident*, ed. Peter Bernard Ladkin, Christoph Goeker, and Bernd Sieker (Zürich: Lit Verlag, 2013), 76.

61. Turner, "Development of Disasters."

62. Karl E. Weick, "Foresights of Failure: An Appreciation of Barry Turner," *Journal of Contingencies and Crisis Management* 6, no. 2 (1998): 74.

63. Turner, "Development of Disasters"; Turner and Pidgeon, *Man-Made Disasters*; Karl. E. Weick, *Sensemaking in Organizations* (Thousand Oaks, CA: Sage, 1995).

64. Lee Clarke, "The Disqualification Heuristic: When Do Organizations Misperceive Risk?," *Research in Social Problems and Public Policy* 5 (1993): 289–312.

65. Brian Toft and Simon Reynolds, *Learning from Disasters: A Management Approach*, 3rd ed. (Leicester, UK: Perpetuity, 2005).

66. Turner, "Development of Disasters."

67. Weick, "Foresights of Failure," 74.

68. Maurizio Catino, *Organizational Myopia: Problems of Rationality and Foresight in Organizations* (New York: Cambridge University Press, 2014).

69. Robert P. Gephart Jr., "Making Sense of Organizationally Based Environmental Disasters," *Journal of Management* 10, no. 2 (1984): 205–225.

70. Karl E. Weick, Kathleen M. Sutcliffe, and David Obstfeld, "Organizing and the Process of Sensemaking," *Organization Science* 16, no. 4 (2005): 411.

71. Gephart, "Making Sense," 213.

72. Robert P. Gephart Jr., "Normal Risk: Technology, Sense Making, and Environmental Disasters," *Organization & Environment* 17, no. 1 (2004): 22.

73. Ulrich Beck, *Risk Society: Towards a New Modernity* (Thousand Oaks, CA: Sage, 1992).

74. John Downer, "'737-Cabriolet': The Limits of Knowledge and the Sociology of Inevitable Failure," *American Journal of Sociology* 117, no. 3 (2011): 725–762; Charles Perrow, *Normal Accidents: Living with High-Risk Technologies* (Princeton, NJ: Princeton University Press, 1999).

75. Rob Nixon, *Slow Violence and the Environmentalism of the Poor* (Cambridge, MA: Harvard University Press, 2011).

76. Ulrich Beck, *Ecological Politics in an Age of Risk* (Cambridge, UK: Polity, 1995), 86.

77. Peter Harries-Jones, *A Recursive Vision: Ecological Understanding and Gregory Bateson* (Toronto: University of Toronto Press, 1995), 8.

CHAPTER 5. DIRTY HARRY AND THE MATERIAL-DISCURSIVE BIND, 1953

1. William R. Freudenburg and Margarita Alario, "Weapons of Mass Distraction: Magicianship, Misdirection, and the Dark Side of Legitimation," *Sociological Forum* 22, no. 2 (2007): 146–173.

2. AEC, "Continental Weapons Tests . . . Public Safety," March 31, 1953, NTA Accession No. NV0317129.

3. AEC, 4, 18, 51.

4. *Allen et al. v. United States*, 588 F. Supp. 247 (D. Utah 1984).

5. Sherry Cable, Thomas E. Shriver, and Donald W. Hastings, "The Silenced Majority: Quiescence and Government Social Control on the Oak Ridge Nuclear Reservation," *Research in Social Problems and Public Policy* 7 (1999): 78; see also Tamara L. Mix, Sherry Cable, and Thomas E. Shriver, "Social Control and Contested Environmental Illness: The Repression of Ill Nuclear Weapons Workers," *Human Ecology Review* 16, no. 2 (2009): 172–183.

6. AEC, "Atomic Test Effects in the Nevada Test Site Region," March 7, 1955, NTA Accession No. NV0018294; AEC, "Atomic Tests in Nevada," March 31, 1957, NTA Accession No. NV0006372. Pamphlets were also disseminated in a more limited manner in parts of California and Arizona.

7. *Allen et al.*, 588 F. Supp. 247.

8. Lee Clarke, *Mission Improbable: Using Fantasy Documents to Tame Disaster* (Chicago, IL: The University of Chicago Press, 1999), 4.

9. Clarke, 2 (emphasis in original).

10. Clarke, 66.

11. Michel Foucault, *Power/Knowledge: Selected Interviews and Other Writings, 1972–1977*, ed. Colin Gordon (New York: Vintage, 1980).

12. AEC, "Public Relations Plan Operations Buster and Jangle," July 24, 1951, 13, NTA Accession No. NV0128200.

13. Raymond Murphy, "Disaster or Sustainability: The Dance of Human Agents with Nature's Actants," *Canadian Review of Sociology and Anthropology* 41, no. 3 (2004): 253, 261.

14. Andrew Kirk, "Rereading the Nature of Atomic Doom Towns," *Environmental History* 17, no. 3 (2012): 638.

15. Joseph Masco, *The Nuclear Borderlands: The Manhattan Project in Post–Cold War New Mexico* (Princeton, NJ: Princeton University Press, 2006).

16. Val Peterson, "Panic: The Ultimate Weapon?," *Collier's Weekly*, August 21, 1953, 100.

17. Kirk, "Nature of Atomic Doom Towns," 636.
18. Harold L. Beck, "External Radiation Exposure to the Population of the Continental U.S. from Nevada Weapons Tests and Estimates of Deposition Density of Radionuclides that Could Significantly Contribute to Internal Radiation Exposure via Ingestion," Appendix E, *Report on the Feasibility of a Study of the Health Consequences to the American Population from Nuclear Weapons Tests Conducted by the United States and Other Nations*, Volume 2, Centers for Disease Control and Prevention and National Cancer Institute, May 2005.
19. Lieutenant Colonel Tom D. Collison, "Report to the Test Director: Radiological Safety Operation," June 1953, NTA Accession No. NV0000079.
20. Richard L. Miller, *Under the Cloud: The Decades of Nuclear Testing* (The Woodlands, TX: Two Sixty, 1991); AEC, "Radioactive Debris from Operations Upshot and Knothole," Health and Safety Laboratory, New York Operations Office, June 25, 1954, vi, NTA Accession No. NV0011545.
21. Testimony of Ken Bulloch, in US Congress, Subcommittee on Oversight and Investigations of the House Committee on Interstate and Foreign Commerce, Health and Scientific Research Subcommittee of the Labor and Human Resources Committee, and the Senate Committee on the Judiciary, *Health Effects of Low-Level Radiation*, joint hearing, 96th Congress, 1st session, April 19, 1979, vol. 1, 227.
22. PHS, "Report of Public Health Service Activities in the Offsite Monitoring Program, Nevada Proving Ground—Spring 1953," December 31, 1953, NTA Accession No. NV0064563.
23. Collison, "Report to the Test Director."
24. Monroe A. Holmes, "Compiled Report on Co-operative Field Survey of Sheep Deaths in S.W. Utah (Cedar City)," December 31, 1953, NTA Accession No. NV0020585.
25. Lieutenant Colonel N. M. Lulejian, "Radioactive Fall-Out from Atomic Bombs," Air Research and Development Command (Baltimore, MD), November 1953, 36, NTA Accession No. NV0400160.
26. Lulejian, 18.
27. AEC, "Report of the Committee to Study Nevada Proving Grounds" (Tyler Committee), February 1, 1954, NTA Accession No. NV0061646.
28. Harold L. Beck, Irene K. Helfer, Andre Bouville, and Mona Dreicer, "Estimates of Fallout in the Continental U.S. from Nevada Weapons Testing Based on Gummed-Film Monitoring Data," *Health Physics* 59, no. 5 (1990): 565–576.
29. Miller, *Under the Cloud*, 166.
30. Defense Nuclear Agency, "Shot Simon: A Test of the Upshot-Knothole Series, 25 April 1953," January 13, 1982.
31. Philip L. Fradkin, *Fallout: An American Nuclear Tragedy* (Tucson: University of Arizona Press, 1989), 106.
32. Miller, *Under the Cloud*.
33. Robert J. List, "The Transport of Atomic Debris from Operation Upshot-Knothole," US Weather Bureau, June 25, 1954, NTA Accession No. NV0005743.

34. John C. Bugher to Brig. Gen. Kenneth E. Fields, "Rainout in the Troy, New York Area," May 14, 1953, NTA Accession No. NV0030140.

35. Herbert M. Clark, "The Occurrence of an Unusually High-Level Radioactive Rainout in the Area of Troy, N.Y.," *Science* 119, no. 3097 (1954): 619–622; Bill Heller, *A Good Day Has No Rain* (Albany, NY: Whitston, 2003).

36. Miller, *Under the Cloud*, 170.

37. Heller, *Good Day*; Ernest J. Sternglass, *Secret Fallout: Low-Level Radiation from Hiroshima to Three-Mile Island* (New York: McGraw-Hill, 1981).

38. Clark, "Occurrence."

39. Sternglass, *Secret Fallout*, 14.

40. Miller, *Under the Cloud*. Fallout data from the gummed-film network were composed of disintegration rate units, a measure of activity, which then were converted to roentgen equivalent units.

41. Heller, *Good Day*; Sternglass, *Secret Fallout*.

42. Heller, *Good Day*.

43. Clark, "Occurrence"; E. J. Kilcawley, H. M. Clark, H. L. Ehrlich, W. J. Kelleher, H. E. Schultz, and N. L. Krascella, "Measurement of Radioactive Fallout in Reservoirs," *Journal of the American Water Works Association* 46, no. 11 (1954): 1101–1111.

44. AEC, "AEC Meeting Minutes No. 862," May 13, 1953, NTA Accession No. NV0072744.

45. Clark, "Occurrence."

46. Kilcawley et al., "Measurement of Radioactive Fallout."

47. Heller, *Good Day*; Matthew W. Wald, "Radiation from 1953 Nuclear Test Fell on Albany," *New York Times*, May 2, 1982, sec. 1, 48. Miller observes that the counties hardest hit include Albany (NY), Columbia (NY), Fulton (NY), Rensselaer (NY), Saratoga (NY), Schenectady (NY), Warren (NY), Washington (NY), Addison (VT), and Bennington (VT). Richard L. Miller, *The U.S. Atlas of Nuclear Fallout*, vol. 1, *Total Fallout, 1951–1962* (The Woodlands, TX: Legis Books, 2000). Note that roentgen equivalent man (rem) is a measure of dosage employed to measure the potential health effects of ionizing radiation.

48. Beck et al., "Estimates of Fallout."

49. Merril Eisenbud and John H. Harley, "Radioactive Dust from Nuclear Detonations," *Science* 117, no. 3033 (1953): 141–147; Leland B. Taylor, "History of Air Force Atomic Cloud Sampling, Volume I," Air Force Systems Command, January 1963, NTA Accession No. NV0410473.

50. Lulejian, "Radioactive Fall-Out."

51. Taylor, "Atomic Cloud Sampling."

52. Emory Jerry Jessee, "Radiation Ecologies: Bombs, Bodies, and Environment during the Atmospheric Nuclear Weapons Testing Period, 1942–1965" (PhD diss., Montana State University, 2013).

53. Eisenbud and Harley, "Radioactive Dust."

54. Merril Eisenbud, "The First Years of the Atomic Energy Commission New York Operations Office Health and Safety Laboratory," *Environment International* 20,

no. 5 (1994): 561–571; Jessee, "Radiation Ecologies." Note that the nationwide fallout monitoring network began in 1951 with forty-five stations and by 1953 had expanded to ninety-five continental stations and twenty-six stations outside the United States.

55. Eisenbud, "First Years," 567.

56. Jessee, "Radiation Ecologies."

57. List, "Transport of Atomic Debris," ix.

58. List, 53.

59. List, 71.

60. List, 63.

61. Jessee, "Radiation Ecologies," 324.

62. Beck et al., "Estimates of Fallout"; H. N. Friesen, "A Perspective on Atmospheric Nuclear Tests in Nevada: Fact Book," DOE, August 1985, NTA Accession No. NV0041479.

63. Lieutenant Colonel R. P. Campbell to Brigadier General Kenneth Fields, "Radioactive Fallout from Upshot-Knothole," May 8, 1953, 4, NTA Accession No. NV0030143.

64. Defense Nuclear Agency, "Operation Upshot-Knothole 1953" (Washington, DC, 1981).

65. Fradkin, *Fallout*; Mary D. Wammack, "Chain Reaction: The Tragedy of Atomic Governance" (PhD diss., University of Nevada, Las Vegas, 1998).

66. Fradkin, *Fallout*, 3.

67. J. C. Clark to C. L. Tyler, "Weather Briefings from 16 March–15 June 1953," Los Alamos Scientific Laboratory, NTA Accession No. NV0039348. The quotes are from the weather briefings for May 17 and May 18.

68. Fradkin, *Fallout*, 3.

69. Clark, "Weather Briefings," verbatim from the May 18 weather briefing (emphasis added). Participants mentioned include Howard L. Andrews of the National Institutes of Health; John C. Clark, Gordon Dunning, Paul H. Fackler of the US Air Force; Lester Machta of the US Weather Bureau; Karl Z. Morgan of the Oak Ridge National Laboratory; Oliver R. Placak of the PHS; and Carroll L. Tyler.

70. Clark, 177.

71. PHS, "Report of Public Health Service."

72. John G. Fuller, *The Day We Bombed Utah: America's Most Lethal Secret* (New York: New American Library, 1984).

73. William S. Johnson to Lt. Colonel Tom D. Collison, "Upshot-Knothole Report of Offsite Rad-Safe Group and Monitoring Data Logbook Shot Harry," May 24, 1953, NTA Accession No. NV0000012; DOE, "Discussions with Frank Butrico Monitor at St. George, Utah, May 1953," deposition at Nevada Operations Office, Las Vegas, NV, August 14, 1980, NTA Accession No. NV0011593.

74. Fuller, *Day We Bombed Utah*.

75. Fradkin, *Fallout*, 13.

76. Johnson, "Upshot-Knothole Report of Offsite Rad-Safe Group"; DOE, "Discussions with Frank Butrico."

77. *Allen et al.*, 588 F. Supp. 247.

78. Frank A. Butrico to William S. Johnson, "Report on the Sequence of Events Occurring in St. George, Utah, as a Result of the Detonation of Shot IX," May 30, 1953, in US Congress, Subcommittee on Oversight and Investigations of the House Committee on Interstate and Foreign Commerce, *Low-Level Radiation Effects on Health*, 96th Congress, 1st session, April 23, May 24, and August 1, 1979, 781–784. Hereafter cited as *LLREH*.

79. DOE, "Discussions with Frank Butrico." See also Janice Perry, "Health Official Claims His Signature Forged," United Press International, September 22, 1982.

80. Butrico, "Report on the Sequence."

81. Johnson, "Upshot-Knothole Report of Offsite Rad-Safe Group." See log entry for May 19.

82. Carol B. Thompson, Richard D. McArthur, and Stan W. Hutchinson, "Development of the Town Data Base: Estimates of Exposure Rates and Times of Fallout Arrival near the Nevada Test Site," DOE, September 1994, NTA Accession No. NV0058984. Distance is measured as the crow flies and assumes straight-line trajectories.

83. DOE, "Discussions with Frank Butrico."

84. Butrico, "Report on the Sequence," 781.

85. Virgil E. Quinn, V. Doyle Urban, and Norman C. Kennedy, "Analysis of Upshot-Knothole 9 (Harry) Radiological and Meteorological Data," National Oceanic and Atmospheric Administration, April 30, 1981, NTA Accession No. NV0015206.

86. DOE, "Discussions with Frank Butrico"; DOE, "Proceedings of the Offsite Monitors Workshop," vol. 1, June 25, 1980, NTA Accession No. NV0012755.

87. DOE, "Discussions with Frank Butrico," 2.

88. DOE, "Proceedings of the Offsite Monitors," vol. 1, 144.

89. *Allen et al.*, 588 F. Supp. 247.

90. Stewart L. Udall, *The Myths of August: A Personal Exploration of Our Tragic Affair with the Atom* (New York: Pantheon Books, 1994), 229.

91. Gordon R. Dunning, "Protective and Remedial Measures Taken Following Three Incidents of Fallout" (paper presented at the annual meeting of the Professional Association for Radiation Protection, Interlaken, Switzerland, May 1, 1968), 1, NTA Accession No. NV0410529. Note that Butrico's report was classified at the time.

92. Udall, *Myths of August*, 228 (emphasis in original).

93. *Allen et al.*, 588 F. Supp. at 391.

94. *Allen et al.*, 588 F. Supp. at 391–392.

95. William S. Johnson to Alvin C. Graves, "Information to be Provided by Offsite Monitoring Group for Upshot-Knothole," September 26, 1952, NTA Accession No. NV0122662.

96. DOE, "Proceedings of the Offsite Monitors," vol. 1, 50.

97. *Allen et al.*, 588 F. Supp. 247.

98. Collison, "Report to the Test Director," 123.

99. Johnson, "Upshot-Knothole Report of Offsite Rad-Safe Group." See the section titled "Summary of Roadblock Operations at Mesquite, Nevada," 4.

100. DOE, "Discussions with Frank Butrico," 3.

101. Williams S. Johnson, Jean McClelland, and Clarence P. Skillern, "Monitoring of Cow's Milk for Fresh Fission Products Following an Atomic Detonation," Los Alamos Scientific Laboratory, October 1953, 3, NTA Accession No. NV0520836.

102. Johnson et al., 3.

103. DOE, "Proceedings of the Offsite Monitors," vol. 1, 47.

104. DOE, "Proceedings of the Offsite Monitors Workshop," vol. 2, June 26, 1980, 1, NTA Accession No. NV0011585.

105. Gordon M. Dunning, "Effects of Nuclear Weapons Testing," *Scientific Monthly* 81, no. 6 (1955): 266, 267.

106. Gordon M. Dunning, "Health Aspects of Nuclear Weapons Testing," AEC, January 1964, 11, NTA Accession No. NV0402696.

107. Heller, *Good Day*.

108. US Department of Agriculture, "Defense Against Radioactive Fallout on the Farm," June 1957, 9, 5 (the latter quote refers to the human body), NTA Accession No. NV0100801.

109. Steven L. Simon, Andre Bouville, and Harold L. Beck, "The Geographic Distribution of Radionuclide Deposition across the Continental US from Atmospheric Nuclear Testing," *Journal of Environmental Radioactivity* 74, nos. 1–3 (2004): 91–105.

110. AEC, "AEC Meeting Minutes No. 865, Subject: Fallout Resulting from May 19 Shot," May 21, 1953, NTA Accession No. NV0072212; AEC, "Fourteenth Semiannual Report of the Atomic Energy Commission," July 31, 1953, NTA Accession No. NV0727877.

111. Udall, *Myths of August*.

112. *Allen et al.*, 588 F. Supp. 247. Stewart Udall refers to data adjustment based on shielding and biological repair as the "Dunning factors" and insists that Gordon Dunning had an undue influence in their creation (*Myths of August*, 232). For discussion of shielding and biological repair, see Gordon M. Dunning, "Criteria for Evaluating Gamma Radiation Exposures from Fallout Following Nuclear Detonations," *Radiology* 66, no. 4 (1956): 585–594.

113. Elijah D. Dickson, "Experimental Shielding Evaluation of the Radiation Protection Provided by Residential Structures" (PhD diss., Oregon State University, 2013).

114. Linsey McGoey, "The Logic of Strategic Ignorance," *British Journal of Sociology* 63, no. 3 (2012): 553–576.

115. *Allen et al.*, 588 F. Supp. at 381 (emphasis in original).

116. *Allen et al.*, 588 F. Supp. at 379.

117. O. R. Placak, M. W. Carter, R. A. Gilmore, Roscoe H. Goeke, and Charles L. Weaver, "Operation Plumbbob Offsite Radiological Safety Report," US Public Health Service, December 1957, NTA Accession No. NV0020578. See also J. Sanders, O. R. Placak, and M. W. Carter, "Offsite Radiological Safety Plan (Teapot Operation) and Report of Off-Site Radiological Safety Activities," March 1956, NTA Accession No. NV0120739.

118. *Allen et al.*, 588 F. Supp. at 374.

119. "A.E.C. Denies Rays Killed Utah Sheep," *New York Times*, January 17, 1954, 46.

120. "Nevada Atom Test Affects Utah Area," *New York Times*, May 20, 1953, 8.

121. Gladwin Hill, "Atom Test Studies Show Area Is Safe," *New York Times*, May 25, 1953, 21. Note that Grable, the atomic cannon shot, was not the final test, as the AEC elected to add an eleventh shot, code-named Climax, a sixty-one-kiloton airdrop.

122. DOE, *United States Nuclear Tests: July 1945 through September 1992* (Las Vegas: Nevada Operations Office, 2000).

123. Gladwin Hill, "Cannon Fires Atomic Shell: Target 7 Miles Away Blasted," *New York Times*, May 26, 1953, 1, 18.

124. Gladwin Hill, "Mightiest Atom Blast of Tests Unleashed on Nevada Desert," *New York Times*, June 5, 1953, 1.

125. "A-Blasts and Weather," *Deseret News*, May 21, 1953, 16A; "We're Getting Annoyed," *Salt Lake Tribune*, May 21, 1953; "Atomic Blasts," *Las Vegas Review-Journal*, May 24, 1953, 4.

126. Campbell, "Radioactive Fallout."

127. A. Costandina Titus, *Bombs in the Backyard: Atomic Testing and American Politics*, 2nd ed. (Reno: University of Nevada Press, 2001), 93.

128. Richard G. Hewlett and Jack M. Holl, *Atoms for Peace and War, 1953–1961: Eisenhower and the Atomic Energy Commission* (Berkeley: University of California Press, 1989).

129. AEC, "AEC Meeting Minutes No. 862," 3. "The Chairman" refers to Gordon Dean.

130. AEC, "Memo for the President, Subject: AEC Has Made a Review of Possible Locations for the Required Additional Atomic Weapons Site," National Security Council, December 18, 1953, NTA Accession No. NV0304388.

131. AEC, "AEC Meeting Minutes No. 865," 2–3.

132. Fradkin, *Fallout*.

133. Note that Borst's letter and John C. Bugher's letter in response are enclosed in "Health and Safety Aspects of Continental Tests," NTA Accession No. NV0072033.

134. *Allen et al.*, 588 F. Supp. at 378.

135. "A-Cloud Dangers in S.L. Studied: University Nuclear Expert Finds High Rays Concentration," *Deseret News*, March 26, 1953.

136. "The Safest Way," *Deseret News*, March 27, 1953.

137. "Effects of Atom Blasts on Southern Utah Discussed by U of U Student," *Iron County Record*, May 7, 1953.

138. Howard Ball, *Justice Downwind: America's Atomic Testing Program in the 1950s* (New York: Oxford University Press, 1986), 70.

139. Senator Arthur V. Watkins to AEC Chairman Gordon Dean, May 23, 1953, NTA Accession No. NV0018853.

140. Frank A. Butrico to William S. Johnson, "Discussion with Representative Stringfellow in St. George Utah," May 28, 1953, NTA Accession No. NV0040984.

141. Fradkin, *Fallout*, 20.

142. Fradkin, 20.

143. AEC, "AEC Meeting Minutes No. 875," June 10, 1953, NTA Accession No. NV0072203.

144. AEC, "AEC Meeting Minutes No. 888," July 15, 1953, 2, NTA Accession No. NV0072196.

145. Ball, *Justice Downwind*.

146. Fradkin, *Fallout*, 82.

147. Fradkin, 111.

148. Hewlett and Holl, *Atoms for Peace*.

149. *Allen et al.*, 588 F. Supp. at 389.

150. Andrew Pickering, *The Cybernetic Brain: Sketches of Another Future* (Chicago: University of Chicago Press, 2010).

151. Andrew Pickering, "Science as Theatre: Gordon Pask, Cybernetics and the Arts," *Cybernetics & Human Knowing* 14, no. 4 (2007): 44.

152. Andrew Pickering, "In Our Place: Performance, Dualism, and Islands of Stability," *Common Knowledge* 23, no. 3 (2017): 381–395.

153. Pickering, 393.

154. Pickering, *Cybernetic Brain*.

155. Gilbert G. Germain, *A Discourse on Disenchantment: Reflections on Politics and Technology* (Albany: State University of New York Press, 1993).

156. Jacob Darwin Hamblin, *Arming Mother Nature: The Birth of Catastrophic Environmentalism* (New York: Oxford University Press, 2013).

157. James Gleick, *Chaos: Making a New Science* (New York: Penguin Books, 1991), 16, 23.

158. Gleick, 24.

159. Sarah Alisabeth Fox, *Downwind: A People's History of the Nuclear West* (Lincoln: University of Nebraska Press, 2014); Carole Gallagher, *American Ground Zero* (Cambridge, MA: MIT Press, 1993).

160. Lynn R. Anspaugh and Bruce W. Church, "Historical Estimates of External Gamma Exposure and Collective External Gamma Exposure from Testing at the Nevada Test Site. I. Test Series through HARDTACK II, 1958," Lawrence Livermore National Laboratory and DOE Nevada Operations Office, December 1985, NTA Accession No. NV0061768.

161. Collison, "Report to the Test Director."

162. Anspaugh and Church, "Historical Estimates."

163. Simon, Bouville, and Beck, "Geographic Distribution." Note that this includes all detonations in the South Pacific and in Nevada.

164. AEC, "Radioactive Debris"; Beck et al., "Estimates of Fallout."

165. Simon, Bouville, and Beck, "Geographic Distribution."

166. The total explosive yield of the Upshot-Knothole series was approximately 252 kilotons and Plumbbob comprised around 344 kilotons.

CHAPTER 6. DEAD SHEEP AND THE FLUIDITY OF FALLOUT

1. AEC, "Report on Sheep Losses Adjacent to the Nevada Proving Grounds," January 6, 1954, 1, NTA Accession No. NV0020422.

2. AEC, "Report on Sheep Deaths in Cedar City in Spring of 1953," press release, January 8, 1954, NTA Accession No. NV0141987; AEC, "Fourteenth Semiannual Report of the Atomic Energy Commission," July 31, 1953, 52, NTA Accession No. NV0727877.

3. AEC, "Meeting of Livestockmen and AEC Officials, January 13, 9:30 a.m., Fire House, Conference Room, Cedar City, Utah (sheep)," January 13, 1954, NTA Accession No. NV0025557; Barton C. Hacker, *Elements of Controversy: The Atomic Energy Commission and Radiation Safety in Nuclear Weapons Testing, 1947–1974* (Berkeley: University of California Press, 1994).

4. Robert N. Proctor, "Agnotology: A Missing Term to Describe the Cultural Production of Ignorance (and Its Study)," in *Agnotology: The Making and Unmaking of Ignorance*, ed. Robert N. Proctor and Londa Schiebinger (Stanford, CA: Stanford University Press, 2008), 1–36.

5. Jan Wille Wieland, "Willful Ignorance," *Ethical Theory and Moral Practice* 20 (2017): 106.

6. Alexander Sarch, "Willful Ignorance in Law and Morality," *Philosophy Compass* 13, no. 5 (2018): 2.

7. Linsey McGoey, "The Logic of Strategic Ignorance," *British Journal of Sociology* 63, no. 3 (2012): 553–576; Proctor, "Agnotology"; Steve Rayner, "Uncomfortable Knowledge: The Social Construction of Ignorance in Science and Environmental Policy Discourses," *Economy and Society* 41, no. 1 (2012): 107–125.

8. McGoey, "Logic of Strategic Ignorance," 571, 556.

9. Eugene A. Rosa, "Metatheoretical Foundations for Post-normal Risk," *Journal of Risk Research* 1, no. 1 (1998): 15–44.

10. Naomi Oreskes and Erik M. Conway, *Merchants of Doubt* (New York: Bloomsbury, 2010), 34.

11. Ulrich Beck, *World at Risk* (Malden, MA: Polity, 2009), 116.

12. Frederick R. Karl, *Franz Kafka: Representative Man* (New York: Ticknor and Fields, 1991).

13. Torben Beck Jorgensen, "Weber and Kafka: The Rational and the Enigmatic Bureaucracy," *Public Administration* 90, no. 1 (2012): 194–210.

14. Beck, *World at Risk*; Olga Kuchinskaya, *The Politics of Invisibility: Public Knowledge about Radiation Health Effects after Chernobyl* (Cambridge, MA: MIT Press, 2014).

15. Kuchinskaya, *Politics of Invisibility*.
16. Proctor, "Agnotology."
17. Beck, *World at Risk*, 115 (emphasis in original).
18. Bruno Latour, *Reassembling the Social: An Introduction to Actor-Network Theory* (New York: Oxford University Press, 2005), 245.
19. Proctor, "Agnotology."
20. Barton C. Hacker, "'Hotter than a $2 Pistol': Fallout, Sheep, and the Atomic Energy Commission, 1953–1986," in *The Atomic West*, ed. Bruce Hevly and John M. Findlay (Seattle: University of Washington Press, 1998), 157–175.
21. Robert Thompsett to R. E. Cole, "Possible Radiation of Animals," n.d., NTA Accession No. NV0025957; Monroe A. Holmes, "Preliminary Report of Radioassays of Selected Tissue Specimens of Sheep from the Cedar City, Utah Area," n.d., NTA Accession No. NV0032618; Arthur Wolff to William Hadlow, June 10, 1953, NTA Accession No. NV0000385; Robert J. Veenstra to William Allare, June 17, 1953, NTA Accession No. NV0020400; William Hadlow to Monroe A. Holmes, July 10, 1953, NTA Accession No. NV0001282. See also George Spendlove's comments in "Dead Utah Sheep Show Radiation," *Deseret News and Telegram*, July 10, 1953, 1A.
22. James G. Terrill Jr., "Interim Report on the Investigation of the Deaths of Sheep in Areas Affected by Atomic Fall-out," August 21, 1953, NTA Accession No. NV0025963.
23. Monroe A. Holmes, "Data on Sheep Loss—Cedar City, Utah, Area," June 17, 1953, NTA Accession No. NV0020399.
24. F. H. Melvin, Bureau of Animal Industry, to Dr. Simms, June 8, 1953, NTA Accession No. NV0020403.
25. Holmes, "Data on Sheep Losses."
26. Holmes. See also AEC, "Statement on Sheep, NPG," press release, June 9, 1953, NTA Accession No. NV0033564; Hacker, *Elements of Controversy*.
27. Thompsett, "Possible Radiation of Animals"; Holmes, "Preliminary Report of Radioassays"; Veenstra to Allare, June 17, 1953; Paul B. Pearson, "Livestock Losses around Test Site," June 21, 1953, NTA Accession No. NV0025948; Monroe A. Holmes, "Compiled Report," December 31, 1953, NTA Accession No. NV0020585.
28. Hacker, *Elements of Controversy*.
29. Testimony of Jack Pace, in US Congress, Subcommittee on Oversight and Investigations of the House Committee on Interstate and Foreign Commerce, Health and Scientific Research Subcommittee of the Labor and Human Resources Committee, and the Senate Committee on the Judiciary, *Health Effects of Low-Level Radiation*, joint hearing, 96th Congress, 1st session, April 19, 1979, vol. 1, 230. Hereafter cited as *HELLR*.
30. Testimony of Stephen L. Brower, in *HELLR*, vol. 1, 232–243.
31. Stephen L. Brower to Monroe A. Holmes, August 28, 1953, NTA Accession No. NV0001268; George Spendlove to Stephen L. Brower, September 2, 1953, NTA Accession No. NV0001267; Monroe A. Holmes to Stephen L. Brower,

September 21, 1953, NTA Accession No. NV0001269; James H. Steele to George A. Spendlove, October 12, 1953, in *HELLR*, vol. 2, 1617.

32. Stephen L. Brower to Utah Governor Scott M. Matheson, *HELLR*, vol. 1, 567.

33. AEC, "AEC Meeting Minutes No. 875," June 10, 1953, item 10, NTA Accession No. NV0072203.

34. Hacker, *Elements of Controversy*; Carroll L. Tyler, "Livestock and Mining Matters, Nevada Proving Grounds Area," 26 June 1953, NTA Accession No. NV0000337.

35. Pearson, "Livestock Losses around Test Site."

36. Hacker, *Elements of Controversy*, 110.

37. Pearson, "Livestock Losses around Test Site," 10.

38. AEC, "AEC Meeting Minutes No. 877," June 17, 1953, NTA Accession No. NV0072198.

39. Veenstra to Allare, June 17, 1953.

40. Hadlow to Holmes, July 10, 1953; Wolff to Hadlow, June 10, 1953.

41. Thompsett, "Possible Radiation of Animals," 3.

42. Veenstra to Allare, June 17, 1953, 2.

43. Warren B. Earl to John I. Curtis, June 11, 1953, NTA Accession No. NV0001080; Vernon Metcalf to Monroe A. Holmes, July 10, 1953, NTA Accession No. NV0001288.

44. Morse Salisbury, "Statement that Might Be Issued If Authoritative Sources Criticize NPG Operations," July 29, 1953, NTA Accession No. NV0404706. See also Shelby Thompson, "NPG Operations Statement," August 3, 1953, NTA Accession No. NV0072194.

45. Hacker, *Elements of Controversy*.

46. AEC, "AEC Meeting Minutes No. 875"; AEC, "AEC Meeting Minutes No. 877"; AEC, "AEC Meeting Minutes No. 884," July 7, 1953, NTA Accession No. NV0072034.

47. Philip L. Fradkin, *Fallout: An American Nuclear Tragedy* (Tucson: The University of Arizona Press, 1989), 151.

48. Fradkin; Brower to Holmes, August 28, 1953; Spendlove to Brower, September 2, 1953; Holmes to Brower, September 21, 1953; Steele to Spendlove, October 12, 1953. See also Brower's testimony before congressional investigators, in *HELLR*, vol. 1, 238–240.

49. J. H. Rust, B. F. Trum, and C. L. Comar, "Report of Farm Animal Survey at Nevada Test Site," June 16, 1953, 5, NTA Accession No. NV0014132.

50. J. H. Rust to Paul B. Pearson, July 21, 1953, 3, NTA Accession No. NV0014178.

51. Bernard Trum, "Special for AEC," University of Tennessee–AEC Agricultural Research Program, September 7, 1953, NTA Accession No. NV0001271. Note that Arthur Wolff subsequently offered a pointed critique of a number of conclusions in Trum's report. See Arthur Wolff to James G. Terrill, October 13, 1953, NTA Accession No. NV0001265.

52. Terrill, "Interim Report," 3–4.

53. Terrill, 5.

54. AEC, "Notes Taken at Meeting of Atomic Energy Commission, State Health Department, Public Health Department, Livestockmen and Others, Held at the City and County Building, Cedar City, Utah, August 9, 1953, Beginning at Approximately 8:30 a.m.," August 9, 1953, 1, NTA Accession No. NV0001276.

55. Monroe A. Holmes to James H. Steele, "Report on Conference Held at Los Alamos, New Mexico, on Utah Sheep Deaths and Experimentally Induced Beta-Radiation Burns on Sheep," November 9, 1953, NTA Accession No. NV0107322. For Bernard Trum's perspective, see Trum to C. Lushbaugh, November 4, 1953, NTA Accession No. NV0014114. Poisonous plants were considered but were quickly ruled out; see W. T. Huffman, US Department of Agriculture, to H. W. Schoening, June 23, 1953, NTA Accession No. NV0025523.

56. AEC, "Notes Taken at Meeting." In regard to the range conditions, see also L. A. Stoddart, "Report on Livestock Conditions Adjacent to the Las Vegas Bombing Range," June 22, 1953, NTA Accession No. NV0020406.

57. AEC, "Notes Taken at Meeting," 2.

58. AEC, 2.

59. AEC, 3.

60. AEC, 4–5 (emphasis in original).

61. AEC.

62. Warren B. Earl to John I. Curtis, June 11, 1953, NTA Accession No. NV0001080; Vernon Metcalf to Monroe Holmes, July 10, 1953, NTA Accession No. NV0001288.

63. Terrill, "Interim Report"; Joe B. Sanders, "Investigation of Livestock Conditions, Ely-Eureka Areas," August 24, 1953, NTA Accession No. NV0000334.

64. AEC, "Notes Taken at Meeting," 7.

65. Terrill, "Interim Report"; Paul B. Pearson to Carroll L. Tyler, "Report on Sheep Losses," November 5, 1953, NTA Accession No. NV0014059; Paul B. Pearson to John C. Bugher, "Report on Meetings in Utah on Sheep Losses," January 19, 1954, NTA Accession No. NV0025979.

66. Terrill, "Interim Report," 5.

67. AEC, "Notes Taken at Meeting," 2.

68. Jim Meigs, "Why *Titanic* Still Matters," *Popular Mechanics*, April 2012, 59.

69. Fradkin, *Fallout*, 149.

70. Testimony of Stephen L. Brower, in *HELLR*, vol. 1, 240–241.

71. Arthur Wolff, "Report of Trip to Los Alamos, October 26–29," n.d. [1953], NTA Accession No. NV0020415; F. H. Melvin to Monroe A. Holmes, October 29, 1953, NTA Accession No. NV0001261; Holmes, "Report on Conference"; Hacker, *Elements of Controversy*.

72. Stewart L. Udall, *The Myths of August: A Personal Exploration of Our Tragic Affair with the Atom* (New York: Pantheon Books), 1994. See also Testimony of Harold A. Knapp, in *HELLR*, vol. 1, 342–352.

73. Gordon Dunning, "Los Alamos Conference on Livestock Losses," October 27, 1953. NTA Accession No. NV0404967. The following are indicated as having signed: Bernard Trum, Joe Sanders, Monroe Holmes, Arthur Wolff, Clarence

Lushbaugh, Robert Thompsett, and F. H. Melvin. The Dunning statements were at odds with the empirical evidence but congruent with, at times word for word, what AEC officials were stating in regard to the sheep losses within internal memoranda prior to the October 27 meeting in Los Alamos. See N. H. Woodruff to Carroll L. Tyler, "Sheep Losses around Cedar City," October 16, 1953, NTA Accession No. NV0404968.

74. Earl to Curtis, June 11, 1953; Metcalf to Holmes, July 10, 1953; Terrill, "Interim Report"; Sanders, "Investigation of Livestock."

75. Gordon M. Dunning, "Developing Critical Thinking through Elementary Science," *School Science and Mathematics* 51, no. 1 (1951): 61–63; Dunning, "Evaluation of Critical Thinking," *Science Education* 38, no. 3 (1954): 191–211; Dunning, "Critical Thinking and Research," *Science Education* 40, no. 2 (1956): 83–86.

76. Holmes, "Report on Conference."

77. Wolff, "Report of Trip."

78. Melvin to Holmes, October 29, 1953.

79. Gordon M. Dunning, "Response to Congressional Committee Hearings Held in 1979," September 10, 1984, 1, 9, NTA Accession No. NV0041905.

80. Gordon Dunning to Morse Salisbury, "Alleged Radiation Damage to Sheep," November 3, 1953, 1, NTA Accession No. NV0014061.

81. Joe B. Sanders to James E. Reeves, "Livestock Losses Vicinity of NPG," October 30, 1953, NTA Accession No. NV0000363; Paul B. Pearson to John C. Bugher, "Conference at Los Alamos on Beta Burns on Sheep," November 4, 1953, NTA Accession No. NV0014063; Pearson, "Report on Sheep Losses"; Paul B. Pearson to John C. Bugher, "Conference with Dr. George A. Spendlove," November 5, 1953, NTA Accession No. NV0014060.

82. Fradkin, *Fallout*.

83. L. K. Bustad, S. Marks, N. L. Dockum, D. R. Kallwarf, and H. A. Kornberg, "A Comparative Study of Hanford and Utah Range Sheep," Hanford Atomic Products Operation, November 30, 1953, NTA Accession No. NV0025066.

84. R. Jeffrey Smith, "Scientists Implicated in Atom Test Deception," *Science* 218, no. 4572 (1982): 545–547; *Bulloch et al. v. United States*, 95 F.R.D. 123 (D. Utah 1982).

85. Bustad et al., "Comparative Study," 4; Smith, "Scientists Implicated."

86. Monroe A. Holmes to George A. Spendlove, "Observations and Comments on Draft of Paul Pearson's Final Report on Sheep Losses, Dated December 16th, 1953," December 30, 1953, 2, NTA Accession No. NV0000330. See also Holmes, "Compiled Report."

87. Pearson, "Report on Meetings."

88. Holmes, "Report on Conference"; Huffman to Schoening, June 23, 1953.

89. AEC, "Meeting of Livestockmen"; AEC, "Report on Sheep Losses"; AEC, "Report on Sheep Deaths."

90. Monroe A. Holmes to S. C. Ingrahan, "Narrative Report on Continued Investigations of Sheep Deaths in S.W. Utah from Unknown Sources," April 14, 1954, NTA Accession No. NV0000327; Monroe A. Holmes to George A. Spendlove,

"Continued Investigation of Utah Sheep Deaths from Unknown Etiology," June 9, 1954, NTA Accession No. NV0020429.

91. Karl, *Franz Kafka*.

92. Franz Kafka, *The Zürau Aphorisms*, trans. Michael Hofman and Geoffrey Brock (New York: Schocken, 2006), 53.

93. Udall, *Myths of August*, 211.

CHAPTER 7. RESPITE, RECONFIGURATION, AND OPERATION
TEAPOT, 1954–1955

1. Terry Tempest Williams, *Refuge: An Unnatural History of Family and Place* (New York: Vintage Books, 1992), 5.

2. Wallace Stegner, *Mormon Country* (Lincoln: University of Nebraska Press, 1970), 34.

3. Williams, *Refuge*, 286.

4. Traci Brynne Voyles, *Wastelanding: Legacies of Uranium Mining in Navajo Country* (Minneapolis: University of Minnesota Press, 2015).

5. Valerie L. Kuletz, *The Tainted Desert: Environmental and Social Ruin n the American West* (New York: Routledge, 1998).

6. AEC, "Report of the Committee to Study Nevada Proving Grounds" (Tyler Committee), February 1, 1954, NTA Accession No. NV0061646.

7. AEC, 40.

8. Colonel Ben Holzman and Dr. Lester Machta, "Weather Forecasting Service at the Nevada Proving Grounds," January 7, 1954, 1–2, NTA Accession No. NV0339402.

9. AEC, "Report of the Committee."

10. AEC, 17, 36.

11. AEC, 7.

12. AEC, 32.

13. Thomas P. Hughes, "The Evolution of Large Technological Systems," in *The Social Construction of Technological Systems*, ed. Wiebe E. Bijker, Thomas Parke Hughes, and Trevor Pinch (Cambridge, MA: MIT Press, 1987), 54.

14. AEC, "Report of the Committee," 12.

15. AEC, 50, 19.

16. Captain Howard L. Andrews, "Residual Radioactivity Associated with the Testing of Nuclear Devices within the Continental Limits of the United States," National Institutes of Health, September 13, 1953, 7, 8–9 (emphasis added), NTA Accession No. NV0125353.

17. Hughes, "Evolution," 56.

18. Erving Goffman, *Frame Analysis: An Essay on the Organization of Experience* (New York: Harper Colophon, 1974).

19. DOE, *A Perspective on Atmospheric Nuclear Tests in Nevada*, DOE/NV-296 (Rev. 2) (Las Vegas: Nevada Operations Office, 1995).

20. Robert J. List, "World-Wide Fallout from Operation Castle," US Weather Bureau, 17 May 1955, NTA Accession No. NV0039820.

21. Gerald J. DeGroot, *The Bomb* (Cambridge, MA: Harvard University Press, 2006); Robert Jacobs, "The Bravo Test and the Death and Life of the Global Ecosystem in the Early Anthropocene," *Asia-Pacific Journal* 13, no. 29 (2015): 1–17.

22. Jacob Darwin Hamblin and Linda M. Richards, "Beyond the Lucky Dragon: Japanese Scientists and Fallout Discourse in the 1950s," *Historia Scientiarum* 25, no. 1 (2015): 36–56; Jacobs, "Bravo Test."

23. Samuel Glasstone, *The Effects of Atomic Weapons* (Los Alamos, NM: Los Alamos Scientific Laboratory, 1964).

24. Jacobs, "Bravo Test," 1.

25. DeGroot, *Bomb*; Jacobs, "Bravo Test"; Hamblin and Richards, "Beyond the Lucky Dragon."

26. Jacob Darwin Hamblin, *Arming Mother Nature: The Birth of Catastrophic Environmentalism* (New York: Oxford University Press, 2013).

27. Ralph E. Lapp, "Strontium Limits in Peace and War," *Bulletin of the Atomic Scientists* 12, no. 8 (1956): 287–289, 287.

28. Laura A. Bruno, "The Bequest of the Nuclear Battlefield: Science, Nature, and the Atom during the First Decade of the Cold War," *Historical Studies in the Physical and Biological Sciences* 33, no. 2 (2003): 237–260; Ronald E. Doel, "Constituting the Postwar Earth Sciences: The Military's Influence on the Environmental Sciences in the USA after 1945," *Social Studies of Science* 33, no. 5 (2003): 635–666; Paul N. Edwards, "Entangled Histories: Climate Science and Nuclear Weapons Research," *Bulletin of the Atomic Scientists* 68, no. 4 (2012): 28–40; Rachel Rothschild, "Environmental Awareness in the Atomic Age: Radioecologists and Nuclear Technology," *Historical Studies in the Natural Sciences* 43, no. 4 (2013): 492–530; Spencer R. Weart, "Global Warming, Cold War, and the Evolution of Research Plans," *Historical Studies in the Physical and Biological Sciences* 27, no. 2 (1997): 319–356.

29. Jacobs, "Bravo Test," 12.

30. Hamblin, *Arming Mother Nature*.

31. Edwards, "Entangled Histories," 29.

32. AEC, "Atomic Test Effects in the Nevada Test Site Region," March 7, 1955, iii, NTA Accession No. NV0018294.

33. Charles Thorpe, "Against Time: Scheduling, Momentum, and Moral Order at Wartime Los Alamos," *Journal of Historical Sociology* 17, no. 1 (2004): 31–55.

34. Sheldon Ungar, "Moral Panics, the Military-Industrial Complex, and the Arms Race," *Sociological Quarterly* 31, no. 2 (1990): 165–185.

35. AEC, "Report of the Committee," 55; AEC, "Atomic Test Effects," 5.

36. AEC, "Atomic Test Effects," 23.

37. AEC, "Report of the Committee," 46–50.

38. AEC, "Atomic Test Effects," 22.

39. Philip L. Fradkin, *Fallout: An American Nuclear Tragedy* (Tucson: The University of Arizona Press, 1989), 119.

40. Fradkin, 26.
41. AEC, *Atomic Tests in Nevada: The Story of AEC's Continental Proving Ground,* color film, 28 minutes, 1955. All quotes from the film refer to this source.
42. Barton C. Hacker, *Elements of Controversy: The Atomic Energy Commission and Radiation Safety in Nuclear Weapons Testing, 1947–1974* (Berkeley: University of California Press, 1994).
43. Hacker; Richard L. Miller, *Under the Cloud: The Decades of Nuclear Testing* (The Woodlands, TX: Two Sixty, 1991).
44. Sarah Alisabeth Fox, *Downwind: A People's History of the Nuclear West* (Lincoln: University of Nebraska Press, 2014).
45. J. B. Sanders, O. R. Placak, and M. W. Carter, "Report of Offsite Radiological Safety Activities, Operation Teapot, Nevada Test Site, Spring 1955," December 31, 1955, 14–15, NTA Accession No. NV0016360.
46. AEC, "Invitations to Open Teapot Event—Suggested Members of the Offsite Communities as Possible Observers," December 31, 1955, NTA Accession No. NV0150103.
47. AEC, "Physicians of Southwestern Utah and Southern Nevada Invited to Conference in St. George, Utah on Feb. 12," press release, February 7, 1955, NTA Accession No. NV0336270.
48. Hacker, *Elements of Controversy.*
49. Clinton P. Anderson with Milton Viorst, *Outsider in the Senate* (New York: World, 1970).
50. AEC, "AEC Meeting Minutes No. 1062," February 23, 1955, 3, 4, NTA Accession No. NV0072872.
51. AEC, 6.
52. AEC, 7–8.
53. Miller, *Under the Cloud.*
54. Sanders, Placak, and Carter, "Offsite Radiological Safety Activities."
55. Miller, *Under the Cloud,* 224, 225.
56. Sanders, Placak, and Carter, "Offsite Radiological Safety Activities."
57. Miller, *Under the Cloud,* 224; Lynn R. Anspaugh and Bruce W. Church, "Historical Estimates of External Gamma Exposure and Collective External Gamma Exposure from Testing at the Nevada Test Site. I. Test Series through HARDTACK II, 1958," Lawrence Livermore National Laboratory and DOE Nevada Operations Office, December 1985, NTA Accession No. NV0061768.
58. Miller, *Under the Cloud,* 231.
59. Miller.
60. Sanders, Placak, and Carter, "Offsite Radiological Safety Activities"; Miller, *Under the Cloud.*
61. Tanfer Emin Tunc, "Eating in Survival Town: Food in 1950s Atomic America," *Cold War History* 15, no. 2 (2015): 181.
62. Joseph Masco, "'Survival Is Your Business': Engineering Ruins and Affect in Nuclear America," *Cultural Anthropology* 23, no. 2 (2008): 361–398, 374.

63. W. J. Lloyd, J. L. Gear, A. H. Stevenson, and R. L. Corsbie, "Operation Teapot, Project 35.5, Effects of a Nuclear Explosion on Records and Records Storage Equipment (February–May 1955)," April 30, 1956, NTA Accession No. NV0051120; E. Roland McConnell, George O. Sampson, and John R. Shari, "Operation Teapot, Nevada Test Site, February–May 1955, Project 32.2A, The Effect of Nuclear Explosions on Commercially Packaged Beverages," January 24, 1957, NTA Accession No. NV0011597; Ebe R. Shaw and Frank P. McNea, "Exposure of Mobile Homes and Emergency Vehicles to Nuclear Explosions," FCDA, July 31, 1957, NTA Accession No. NV0051295.

64. McConnell, Sampson, and Shari, "Commercially Packaged Beverages," 14.

65. Masco, "Survival Is Your Business"; Tunc, "Eating in Survival Town"; Laura McEnaney, *Civil Defense Begins at Home: Militarization Meets Everyday Life in the Fifties* (Princeton, NJ: Princeton University Press, 2000).

66. Anspaugh and Church, "Historical Estimates."

67. Sanders, Placak, and Carter, "Offsite Radiological Safety Activities," 21.

68. Fradkin, *Fallout*.

69. "Don't Hurry Back," *Deseret News*, May 13, 1955.

CHAPTER 8. *BULLOCH ET AL. V. UNITED STATES*

1. Utah State Agricultural College, "The Effect of the Level of Nutrition on the Pathology and Productivity of Sheep," n.d., Agricultural Experiment Station, NTA Accession No. NV0001243.

2. Sarah Alisabeth Fox, *Downwind: A People's History of the Nuclear West* (Lincoln: University of Nebraska Press, 2014).

3. Monroe Holmes articulated the mismatch between the study and what he encountered. See Monroe Holmes to Dr. Wayne Binns, April 16, 1954, NTA Accession No. NV0001229; Monroe Holmes to Arthur M. Wolff, May 24, 1954, in US Congress, Subcommittee on Oversight and Investigations of the House Committee on Interstate and Foreign Commerce, Health and Scientific Research Subcommittee of the Labor and Human Resources Committee, and the Senate Committee on the Judiciary, *Health Effects of Low-Level Radiation*, joint hearing, 96th Congress, 1st session, April 19, 1979, vol. 2, 1839. Hereafter cited as *HELLR*.

4. Fox, *Downwind*.

5. AEC, "Meeting of Livestockmen and AEC Officials, January 13, 9:30 a.m., Fire House, Conference Room, Cedar City, Utah (sheep)," January 13, 1954, NTA Accession No. NV0025557.

6. *Bulloch et al. v. United States*, 145 F. Supp. 824 (D. Utah 1956).

7. Philip L. Fradkin, *Fallout: An American Nuclear Tragedy*, (Tucson, AZ: The University of Arizona Press, 1989), 156.

8. Major Grant Kuhn, "Trip Report of Major Grant Kuhn, March 9–22, 1955," March 25, 1955, NTA Accession No. NV0070612.

9. Monroe A. Holmes to Samuel C. Ingraham, "Narrative Report on Continued Investigations of Sheep Deaths in S.W. Utah from Unknown Source," April 14,

1954, NTA Accession No. NV0000327; see also Stephen L. Brower to Monroe A. Holmes, April 13, 1954, NTA Accession No. NV0001248.

10. Bernard F. Trum, "Report of Lt. Col. Bernard F. Trum, March 31 to April 20, 1955," April 30, 1955, NTA Accession No. NV0014242.

11. Fradkin, *Fallout*.

12. Bernard F. Trum to Robert J. Veenstra, March 25, 1955, NTA Accession No. NV0027280.

13. Trum, "Report," 3 (emphasis added).

14. Robert J. Veenstra to Bernard F. Trum, April 7, 1955, 1, NTA Accession No. NV0014055.

15. Robert J. Veenstra to Bernard F. Trum (handwritten note), April 7, 1955, NTA Accession No. NV0014287.

16. Trum, "Report," 12.

17. Fradkin, *Fallout*, 158.

18. Bernard F. Trum to R. E. Thompsett, May 9, 1955, NTA Accession No. NV0070613.

19. Trum, "Report."

20. Bernard F. Trum to Llewellyn O. Thomas, May 12, 1955, 3, 4 (emphasis in original), NTA Accession No. NV0027279.

21. Trum to Thomas, 1.

22. R. E. Thompsett to R. E. Cole, "Possible Radiation of Animals," n.d., NTA Accession No. NV0025957; Robert Veenstra to William Allare, June 17, 1953, NTA Accession No. NV0020400; Trum to Thomas.

23. Trum to Thomas, 4, 5.

24. Trum to Thomas, 5.

25. Trum to Thomas, 6.

26. Trum to Thomas; For a synopsis of initial field and laboratory reports illustrating that the sheep were exposed to substantial levels of radioactivity, see Dan S. Bushnell, "Trial Memorandum," in *HELLR*, vol. 1, 601–623.

27. Thompsett, "Possible Radiation of Animals," 3.

28. Veenstra to Allare, June 17, 1953, 2.

29. Monroe A. Holmes to James H. Steele, "Compiled Report on Co-operative Field Survey of Sheep Deaths in S.W. Utah (Cedar City)," December 31, 1953, 14, NTA Accession No. NV0020585.

30. Thompsett, "Possible Radiation of Animals."

31. Monroe A. Holmes to James H. Steele, "Report on Conference Held at Los Alamos, New Mexico, on Utah Sheep Deaths and Experimentally Induced Beta-Radiation Burns on Sheep," November 9, 1953, 3, NTA Accession No. NV0107322.

32. Stewart L. Udall, *The Myths of August: A Personal Exploration of Our Tragic Cold War Affair with the Atom* (New York: Pantheon, 1994), 212.

33. Deposition of R. E. Thompsett, March 14, 1956, in *HELLR*, vol. 1, 850–863.

34. Deposition of Monroe A. Holmes, September 1, 1956, in *HELLR*, vol. 1, 910.

35. Closing argument in *Bulloch et al. v. United States*, 145 F. Supp. 824, October 1–2, 1956, in *HELLR*, vol. 1, 1013–1014. Note that John J. Finn was an attorney representing the government alongside Llewellyn O. Thomas and Donald R. Fowler.

36. *HELLR*, vol. 1, 1051–1052.

37. *HELLR*, vol. 1, 564–565.

38. Judge A. Sherman Christensen, findings of the court in *Bulloch et al. v. United States*, 145 F. Supp. 824, October 2, 1956, in *HELLR*, vol. 1, 965.

39. Fox, *Downwind*, 92 (emphasis in original).

40. Fox; Carole Gallagher, *American Ground Zero* (Cambridge, MA: MIT Press, 1993).

41. Daniel Ford, *The Cult of the Atom* (New York: Simon and Schuster, 1982).

42. Stephen L. Brower to Governor Scott M. Matheson, February 14, 1979, in *HELLR*, vol. 1, 568.

43. *Bulloch et al. v. United States*, 95 F.R.D. 123 (D. Utah 1982).

44. Howard Ball, *Justice Downwind: America's Atomic Testing Program in the 1950s* (New York: Oxford University Press, 1986); A. Costandina Titus, *Bombs in the Backyard: Atomic Testing and American Politics*, 2nd ed. (Reno: University of Nevada Press, 2001); Deborah Wollen, "Reformation of the Burden of Proof," *Natural Resources Journal* 26, no. 2 (1986): 377–389.

45. L. K. Bustad, S. Marks, N. L. Dockum, D. R. Kallwarf, and H. A. Kornberg, "A Comparative Study of Hanford and Utah Range Sheep," Hanford Atomic Products Operation, November 30, 1953, 1, NTA Accession No. NV0025066.

46. Ball, *Justice Downwind*.

47. Sidney Marks, Norman L. Dockum, and Leo K. Bustad, "Histopathology of the Thyroid Gland of Sheep in Prolonged Administration of I[131]," *American Journal of Pathology* 33, no. 2 (1957): 228.

48. *Bulloch et al. v. United States*, 763 F.2d 1115, 1124 (10th Cir., May 22, 1985).

49. Leo K. Bustad, *Compassion: Our Last Great Hope*, 2nd ed. (Renton, WA: Delta Society, 1996), 27.

50. *Bulloch et al.*, 763 F.2d 1115.

51. Philip W. Allen, Lester Machta, and Kenneth N. Nagler, "Transport of Radioactive Debris from Operations Buster and Jangle," Armed Forces Special Weapons Project, March 15, 1952, NTA Accession No. NV0316159; Gaelen Felt to A. C. Graves, "Jangle Fallout Problems," June 28, 1951, NTA Accession No. NV0062743; H. F. Schulte to Dr. T. L. Shipman, "Preliminary Report on Buster-Jangle Fall-Out Program," Los Alamos Scientific Laboratory, December 15, 1951, NTA Accession No. NV0122093.

52. Robert J. List, "The Transport of Atomic Debris from Operation Upshot-Knothole," US Weather Bureau, June 25, 1954, NTA Accession No. NV0005743; PHS, "Report of Public Health Service Activities in the Offsite Monitoring Program, Nevada Proving Ground—Spring 1953," December 31, 1953, NTA Accession No. NV0064563.

53. AEC, "Report of the Committee to Study Nevada Proving Grounds" (Tyler Committee), February 1, 1954, 40 (emphasis added), NTA Accession No. NV0061646.

54. Lieutenant Colonel N. M. Lulejian, "Radioactive Fall-Out from Atomic Bombs," Air Research and Development Command (Baltimore, MD), November 1953, 36, NTA Accession No. NV0400160.

55. Closing argument in *Bulloch et al. v. United States*, 145 F. Supp. 824, in *HELLR*, vol. 1, 1013–1014.

56. AEC, "Notes Taken at Meeting of Atomic Energy Commission, State Health Department, Public Health Department, Livestockmen and Others, Held at the City and County Building, Cedar City, Utah, August 9, 1953, Beginning at Approximately 8:30 a.m.," August 9, 1953, NTA Accession No. NV0001276; AEC, "Meeting of Livestockmen and AEC Officials, January 13, 9:30 a.m., Fire House, Conference Room, Cedar City, Utah (sheep)," January 13, 1954; "Statements of a Panel of Utah Citizens to Discuss the Impact of Radiation on Sheep," in *HELLR*, vol. 1, 227–243.

57. Judge A. Sherman Christensen, findings of the court in *Bulloch et al. v. United States*, 145 F. Supp. 824 (October 2, 1956), in *HELLR*, vol. 1, 965.

58. Testimony in *Bulloch et al. v. United States*, 145 F. Supp. 824 (1956), in *HELLR*, vol. 1, 1045, 1032.

59. Udall, *Myths of August*, 216.

60. AEC, "Report on Sheep Losses Adjacent to the Nevada Proving Grounds," January 6, 1954, NTA Accession No. NV0020422.

61. Mary Dickson, "Living and Dying with Fallout," *Dialogue: A Journal of Mormon Thought* 37, no. 2 (2004): 1–35; Fox, *Downwind*; Fradkin, *Fallout*; Gallagher, *American Ground Zero*.

62. R. Jeffrey Smith, "Atom Bomb Tests Leave Infamous Legacy," *Science* 218, no. 4569 (1982): 266.

63. Harold A. Knapp, "Sheep Deaths in Utah and Nevada Following the 1953 Nuclear Tests," in *HELLR*, vol. 1, 287.

64. Testimony of Harold A. Knapp, in US Congress, Subcommittee on Oversight and Investigations of the House Committee on Interstate and Foreign Commerce, *Low-Level Radiation Effects on Health*, 96th Congress, 1st session, April 23, May 24, and August 1, 1979, 292–307. Hereafter cited as *LLREH*.

65. Testimony of Harold A. Knapp, in *LLREH*, 296; Knapp, "Sheep Deaths."

66. AEC, "Notes Taken at Meeting," 3.

67. Testimony of Harold A. Knapp, in *LLREH*, 293.

68. Testimony of Harold A. Knapp, in *HELLR*, vol. 1, 344.

69. Smith, "Atom Bomb Tests," 268.

70. Knapp, "Sheep Deaths"; testimony of Harold A. Knapp, in *LLREH*.

71. R. Jeffrey Smith, "Scientists Implicated in Atom Test Deception," *Science* 218, no. 4572 (1982): 545–547.

72. Knapp, "Sheep Deaths," 291.

73. Testimony of Harold A. Knapp, in *LLREH*, 292–307.

74. Knapp, 297.

75. Fradkin, *Fallout*, 147.

76. Lee Clarke, *Worst Cases: Terror and Catastrophe in the Popular Imagination* (Chicago: University of Chicago Press, 2006).

77. Testimony of Charles Dunham, in US Congress, Subcommittee on Radiation, JCAE, *The Nature of Radioactive Fallout and Its Effects on Man*, 85th Congress, 1st session, part I, May 27–29 and June 3, 1957, 15.

CHAPTER 9. OPERATION PLUMBBOB

1. Sarah Alisabeth Fox, *Downwind: A People's History of the Nuclear West* (Lincoln: University of Nebraska Press, 2014); Carole Gallagher, *American Ground Zero* (Cambridge, MA: MIT Press, 1993).

2. Philip L. Fradkin, *Fallout: An American Nuclear Tragedy* (Tucson: University of Arizona Press, 1989).

3. US Congress, Subcommittee on Oversight and Investigations of the House Committee on Interstate and Foreign Commerce, *Low-Level Radiation Effects on Health*, 96th Congress, 1st Session, April 23, May 24, and August 1, 1979, 39. Hereafter cited as *LLREH*.

4. *LLREH*, 35 (emphasis added).

5. *LLREH*, 18.

6. Captain Howard L. Andrews, "Residual Radioactivity Associated with the Testing of Nuclear Devices within the Continental Limits of the United States," National Institutes of Health, September 13, 1953, 9, NTA Accession No. NV0125353.

7. Margaret Jones Patterson and Robert H. Russell, *Behind the Lines: Case Studies in Investigative Reporting* (New York: Columbia University Press, 1986), 160.

8. Gordon Eliot White, "Deaths High in Utah Fallout Area," *Deseret News*, August 12, 1977.

9. Fradkin, *Fallout*; Patterson and Russell, *Behind the Lines*; Statement of Joseph L. Lyon, MD, in US Congress, Subcommittee of the Committee on Appropriations, US Senate, *Radioactive Fallout from Nuclear Testing at the Nevada Test Site, 1950–60*, 105th Congress, 1st session, October 1, 1997, 17–19. Hereafter cited as *RFNT*.

10. Joseph L. Lyon, Melville R. Klauber, John W. Gardner, and King S. Udall, "Childhood Leukemias Associated with Fallout from Nuclear Testing," *New England Journal of Medicine* 300, no. 8 (1979): 397–402.

11. Joseph L. Lyon, "Nuclear Weapons Testing and Research Efforts to Evaluate Health Effects on Exposed Populations in the United States," *Epidemiology* 10, no. 5 (1999): 558.

12. Lyon et al., "Childhood Leukemias."

13. Fradkin, *Fallout*, 221.

14. Statement of Joseph L. Lyon, in *RFNT*, 21–22. Note that Charles E. Land of the National Cancer Institute critiqued the Lyon et al. study in the *New England Journal of Medicine*. The statistician Herman Chernoff then examined the study by Lyon et al. and Land's critique for the US General Accounting Office. Chernoff concluded, "In short, the Lyon et al. paper is basically sound. Although they do not claim it, I feel that fallout probably caused an increase in leukemia." Charles

E. Land, "The Hazards of Fallout or of Epidemiological Research?," *New England Journal of Medicine* 300, no. 8 (1979): 431; Herman Chernoff, "Report for the United States Accounting Office," Department of Mathematics, Massachusetts Institute of Technology, 1979, 30.

15. Edward S. Weiss, "Leukemia Mortality in Southwestern Utah: 1950–1964," PHS, September 14, 1965, reprinted in US Congress, Subcommittee on Oversight and Investigations of the House Committee on Interstate and Foreign Commerce, Health and Scientific Research Subcommittee of the Labor and Human Resources Committee, and the Senate Committee on the Judiciary, *Health Effects of Low-Level Radiation*, joint hearing, 96th Congress, 1st session, April 19, 1979, vol. 2, 2191–2216; Lyon, "Nuclear Weapons Testing."

16. Weiss, "Leukemia Mortality."

17. Lyon, in *RFNT*, 17–18.

18. Judson Hardy, "Meeting of PHS-AEC Representatives: re. Utah Study," September 2, 1965, NTA Accession No. NV0018768; Gordon Dunning to Dwight Ink, "PHS Announcement on Study of Possible Fallout Effects in Utah," August 31, 1965, NTA Accession No. NV0004479; Dwight A. Ink to Luther L. Terry, Surgeon General, PHS, September 10, 1965, NTA Accession No. NV0161830; Dwight Ink to AEC Commissioners, "USPS Epidemiology Studies in Southwestern Utah," September 9, 1965, NTA Accession No. NV0018102.

19. Lyon, "Nuclear Weapons Testing," 557.

20. Patterson and Russell, *Behind the Lines*.

21. Lyon, in *RFNT*, 19–24.

22. Walter Stevens, Duncan C. Thomas, Joseph L. Lyon, John E. Till, Richard A. Kerber, Steven L. Simon, Ray D. Lloyd, Naima Abd Elghany, and Susan Preston-Martin, "Leukemia in Utah and Radioactive Fallout from the Nevada Test Site: A Case-Control Study," *Journal of the American Medical Association* 264, no. 5 (1990): 585–591.

23. Richard A. Kerber, John E. Till, Steven L. Simon, Joseph L. Lyon, Duncan C. Thomas, Susan Preston-Martin, Marvin L. Rallison, Ray D. Lloyd, and Walter Stevens, "A Cohort Study of Thyroid Disease in Relation to Fallout from Nuclear Weapons Testing," *Journal of the American Medical Association* 270, no. 17 (1993): 2076–2082.

24. Charles E. Land, Frank W. Mckay, and Stella G. Machado, "Childhood Leukemia and Fallout from the Nevada Nuclear Tests," *Science* 223, no. 4632 (1984): 139–144; Stella G. Machado, Charles E. Land, and Frank W. Mckay, "Cancer Mortality and Radioactive Fallout in Southwestern Utah," *American Journal of Epidemiology* 125, no. 1 (1987): 44–61; R. D. Lloyd, "NTS Fallout-Induced Cancer in Southwestern Utah," *Health Physics* 72, no. 6 (1997): 938–940.

25. Sandra Steingraber, *Living Downstream: An Ecologist Looks at Cancer and the Environment* (Reading, MA: Addison-Wesley, 1997), 59.

26. Fox, *Downwind*; Gallagher, *American Ground Zero*.

27. Mary Dickson, "Living and Dying with Fallout," *Dialogue: A Journal of Mormon Thought* 37, no. 2 (2004): 1–34; Fox, *Downwind*; Gallagher, *American Ground Zero*.

28. Ian Hodder, *Entangled: An Archaeology of the Relationships between Humans and Things* (Malden, MA: Wiley-Blackwell, 2012), 8.

29. Andrew Pickering, *The Cybernetic Brain: Sketches of Another Future* (Chicago: University of Chicago Press, 2010).

30. Stacy Alaimo, *Bodily Natures: Science, Environment, and the Material Self* (Bloomington: Indiana University Press, 2010), 11, 19–20.

31. Alaimo, 11; Steingraber, *Living Downstream*.

32. Steve Mattewman, "Risk Society Revisited, Again," *Thesis Eleven* 128, no. 1 (2015): 142.

33. Alaimo, *Bodily Natures*.

34. Ulrich Beck, *World at Risk* (Malden, MA: Polity, 2009).

35. Alaimo, *Bodily Natures*, 19.

36. Rachel Carson, *Silent Spring* (Boston: Houghton Mifflin, 1962).

37. Steingraber, *Living Downstream*.

38. Kristen Iversen, *Full Body Burden: Growing Up in the Nuclear Shadow of Rocky Flats* (New York: Crown, 2013), 198.

39. Terry Tempest Williams, *Refuge: An Unnatural History of Family and Place* (New York: Vintage Books, 1991), 281.

40. Beck, *World at Risk*.

41. Ulrich Beck, *World Risk Society* (Malden, MA: Polity, 1999), 70.

42. Ulrich Beck, *Risk Society: Towards a New Modernity* (Thousand Oaks, CA: Sage, 1992).

43. Richard G. Hewlett and Jack M. Holl, *Atoms for Peace and War, 1953–1961: Eisenhower and the Atomic Energy Commission* (Berkeley: University of California Press, 1989), 279.

44. H. L. Nieburg, "The Eisenhower AEC and Congress: A Study in Executive-Legislative Relations," *Midwest Journal of Political Science* 6, no. 2 (1962): 115–148.

45. US Congress, Subcommittee on Radiation, Joint Committee on Atomic Energy, *The Nature of Radioactive Fallout and Its Effects on Man*, 85th Congress, 1st session, part I, May 27–29 and June 3, 1957. Hereafter cited as *NRFEM*.

46. Chet Holifield, "Congressional Hearings on Radioactive Fall-out," *Bulletin of the Atomic Scientists* 14, no. 1 (1958): 53.

47. Statement of Dr. Willard F. Libby, in *NRFEM*, part II, June 4–7, 1957, 1210.

48. Laura A. Bruno, "Bequest of the Nuclear Battlefield: Science, Nature, and the Atom during the First Decade of the Cold War," *Historical Studies in the Physical and Biological Sciences* 33, no. 2 (2003): 237–260.

49. Libby, in *NRFEM*, part II, 1211.

50. Libby, 1211.

51. Willard F. Libby, "Radioactive Fallout," *Proceedings of the National Academy of Sciences* 43, no. 8 (1957): 758–775.

52. W. F. Libby, "Distribution and Effects of Fall-out," *Bulletin of the Atomic Scientists* 14, no. 1 (1958): 27–30.

53. Libby, "Radioactive Fallout," 759, 769 (emphasis in original).

54. Statement of Dr. Lester Machta, in *NRFEM*, part I, 141–161; L. Machta, R. J. List, and L. F. Hubert, "World-Wide Travel of Atomic Debris," *Science* 124, no. 3220 (1956): 474–477.

55. Machta, in *NRFEM*, part I, 141–161.

56. Libby, in *NRFEM*, part II, 1216.

57. Libby, 1216.

58. W. F. Libby, "Radioactive Strontium Fallout," *Proceedings of the National Academy of Sciences* 42, no. 6 (1956): 365–390; Willard F. Libby, "Current Research Findings on Radioactive Fallout," *Proceedings of the National Academy of Sciences* 42, no. 12 (1956): 945–962; Libby, "Radioactive Fallout"; Libby, "Distribution and Effects."

59. W. F. Libby and C. E. Palmer, "Stratospheric Mixing from Radioactive Fallout," *Journal of Geophysical Research* 65, no. 10 (1960): 3307–3317.

60. P. Fabian, W. F. Libby, and C. E. Palmer, "Stratospheric Residence Time and Interhemispheric Mixing of Strontium 90 from Fallout in Rain," *Journal of Geophysical Research* 73, no. 12 (1968): 3611–3616.

61. Bruno Latour, *Science in Action* (Cambridge, MA: Harvard University Press, 1987), 93.

62. Chet Holifield, "Who Should Judge the Atom?," *Saturday Review*, August 3, 1957, 36.

63. Clinton P. Anderson with Milton Viorst, *Outsider in the Senate* (New York: World, 1970).

64. Andrew Pickering, "New Ontologies," in *The Mangle in Practice: Science, Society, and Becoming*, ed. Andrew Pickering and Keith Guzik (Durham, NC: Duke University Press, 2008), 1–14.

65. The quote in the subhead is from AEC, "Atomic Test Effects in the Nevada Test Site Region," March 7, 1955, 32, NTA Accession No. NV0018294; AEC, "Continental Weapons Tests . . . Public Safety," March 31, 1953, NTA Accession No. NV0317129; AEC, "Atomic Tests in Nevada," March 31, 1957, NTA Accession No. NV0006372. The pamphlets were converted to plain text files (.txt) and then uploaded into WordSmith Tools, version 6. The analysis draws from the discourse theory of Ernesto Laclau and Chantal Mouffe, *Hegemony and Socialist Strategy: Towards a Radical Democratic Politics*, 2nd ed. (New York: Verso, 2001).

66. Michael Stubbs, *Words and Phrases: Corpus Studies of Lexical Semantics* (Oxford, UK: Blackwell, 2001).

67. Note that this span is widely used in corpus linguistics.

68. AEC, "Continental Weapons Tests," 18.

69. AEC, "Atomic Tests in Nevada," 48.

70. AEC, "Continental Weapons Tests," 18.

71. AEC, 32.

72. AEC, 16.

73. Jeannie Massie and Inara Gravitis, "Safety Experiments November 1955–March 1958," DNA, August 2, 1982, NTA Accession No. NV0767858; DOE, "Plutonium Dispersal Tests at the Nevada Test Site," August 2013.

74. Annie Jacobsen, *Area 51: An Uncensored History of America's Top Secret Military Base* (New York: Little, Brown, 2011).

75. The quote is by Dr. Robert Pendleton, a University of Utah scientist, in UPI, "Atomic 'Safety Tests' in the 1950s Showered Utah with Plutonium," *Washington Post*, March 3, 1979, A6.

76. Iversen, *Full Body Burden*, 173.

77. Massie and Gravitis, "Safety Experiments."

78. Jacobsen, *Area 51*. Note that in 1981, remediation efforts were undertaken to clean up Area 13.

79. Paul Jacobs, "Clouds from Nevada," *Reporter*, May 16, 1957, 10–29. For a compendium of internal memoranda in reference to Paul Jacobs, see AEC, "Correspondence Concerning Paul Jacobs from The Reporter Magazine (Letters, Memos, Dosage Reports, 1955–1958)," n.d., NTA Accession No. NV0002029.

80. Paul Jacobs, "The Little Cloud That Got Away," *Reporter*, April 3, 1958, 18–19. See also Jacobs, "Precautions Are Being Taken by Those Who Know," *Atlantic Monthly*, February 1971.

81. Massie and Gravitis, "Safety Experiments."

82. E. P. Hardy, P. W. Krey, and H. L. Volchok, "Plutonium Fallout in Utah," AEC Health and Safety Laboratory, July 1972, NTA Accession No. NV0012878.

83. Edward P. Hardy Jr., "Plutonium in Soil Northeast of the Nevada Test Site," AEC Health and Safety Laboratory, July 1976, NTA Accession No. NV0018140.

84. James V. Cizdziel, Vernon F. Hodge, and Scott H. Faller, "Plutonium Anomalies in Attic Dust and Soils at Locations Surrounding the Nevada Test Site," *Chemosphere* 37, no. 6 (1998): 1164, 1165.

85. Mary Turner, Mark Rudin, James Cizdziel, and Vernon Hodge, "Excess Plutonium in Soil near the Nevada Test Site, USA," *Environmental Pollution* 125, no. 2 (2003): 193.

86. Harold L. Beck, Irene K. Helfer, Andre Bouville, and Mona Dreicer, "Estimates of Fallout in the Continental U.S. from Nevada Weapons Testing Based on Gummed-Film Monitoring Data," *Health Physics* 59, no. 5 (1990): 565–576. Note that Upshot-Knothole had the highest collective population exposure to external gamma radiation. See Bernard Shleien, "External Radiation Exposure to the Off-site Population from Nuclear Tests at the Nevada Test Site between 1951 and 1970," *Health Physics* 41, no. 2 (1981): 243–254.

87. Richard L. Miller, *The U.S. Atlas of Nuclear Fallout 1951–1962*, vol. 1, *Total Fallout* (Woodlands, TX: Two Sixty, 2001).

88. Fradkin, *Fallout*, 125.

89. Terrence R. Fehner and F. G. Gosling, *Battlefield of the Cold War: The Nevada Test Site*, vol. 1 (Washington, DC: DOE, 2006).

90. K. P. Ferlic, "Fallout: Its Characteristics and Management," Defense Nuclear Agency, December 1983; Samuel Glasstone and Philip J. Dolan, *The Effects of Nuclear Weapons*, 3rd ed. (Washington, DC: DOE, 1977). Note that Virgil E. Quinn suggests that the Boltzmann hot spot may be due to an error in measurement and,

thus, did not occur. Quinn, "Analysis of Meteorological and Radiological Data for Selected Fallout Episodes," *Health Physics*, 59, no. 5 (1990): 577–592.

91. Summary-Analysis of Hearings, in US Congress, Subcommittee on Radiation, JCAE, *Fallout from Nuclear Weapons Tests*, 86th Congress, 1st session, May 5–8, 1959, 6.

92. Fehner and Gosling, *Battlefield of the Cold War*, 168.

93. AEC, "Operation Plumbbob: Technical Summary of Military Effects, Programs 1–9," August 15, 1962, NTA Accession No. NV0014657.

94. US Air Force, *Operation Plumbbob Military Effects Studies*, 1957, National Archives and Records Administration.

95. Richard L. Miller, *Under the Cloud: The Decades of Nuclear Testing* (The Woodlands, TX: Two Sixty, 1991).

96. Beck et al., "Estimates of Fallout"; Harold L. Beck and Burton G. Bennett, "Historical Overview of Atmospheric Nuclear Weapons Testing and Estimates of Fallout in the Continental United States," *Health Physics* 82, no. 5 (2002): 591–608.

97. Jan DeBlieu, *Wind: How the Flow of Air Has Shaped Life, Myth, and the Land* (Boston: Houghton Mifflin, 1998), 65.

98. Loudon S. Wainwright, "The Heroic Disarming of Diablo: Atomic Engineers Make a Suspense-Laden Climb," *Life*, September 16, 1957, 134.

99. Miller, *Under the Cloud*, 267.

100. Wainwright, "Heroic Disarming."

101. H. N. Friesen, "A Perspective on Atmospheric Nuclear Tests in Nevada: Fact Book," DOE, August 1985; Beck et al., "Estimates of Fallout"; Beck and Bennett, "Historical Overview."

102. Miller, *Under the Cloud*.

103. US Air Force, *Project Genie*, 1957, National Archives and Records Administration. The Air Force officers included Lieutenant Colonel Frank P. Ball, Major Norman Bodinger, Colonel Sidney Bruce, Major John Hughes, and Don Lutrel. The videographer was George Yoshitake.

104. Fehner and Gosling, *Battlefield of the Cold War*.

105. Iain Thomson, "Did Speeding American Manhole Cover Beat Sputnik into Space?," *The Register*, July 16, 2015, www.theregister.com; Eric Schulman, *A Briefer History of Time: From the Big Bang to the Big Mac* (New York: W. H. Freeman, 1999).

106. Schulman, *Briefer History of Time*.

107. Robert R. Brownlee of Los Alamos Scientific Laboratory was tasked with developing underground detonations and was a key figure behind the Pascal-B shot. See Brownlee, "Learning to Contain Underground Nuclear Explosions," June 2002, http://nuclearweaponarchive.org.

108. Miller, *Under the Cloud*, 282.

109. P. S. Harris, C. Lowery, A. G. Nelson, S. Obermiller, W. J. Ozeroff, and E. Weary, "Shot Smoky: A Test of the Plumbbob Series," Defense Nuclear Agency, May 31, 1981.

110. Miller, *Under the Cloud*.
111. Gordon M. Dunning, "Fallout from Nuclear Tests at the Nevada Test Site," AEC, May 1959, NTA Accession No. NV0705727.
112. F. Ward Whicker, Thomas B. Kirchner, Lynn R. Anspaugh, and Yook C. Ng, "Ingestion of Nevada Test Site Fallout: Internal Dose Estimates," *Health Physics*, 71, no. 4 (1996): 481.
113. K. H. Larson, J. W. Neel, H. A. Hawthorne, H. M. Mork, R. H. Rowland, L. Baurmash, R. G. Lindberg, J. H. Olafson, and B. W. Kowalewsky, "Distribution, Characteristics, and Biotic Availability of Fallout, Operation Plumbbob," July 26, 1966, 40, NTA Accession No. NV0519242.
114. Larson et al.
115. DOE, "Proceedings of the Offsite Monitors Workshop," vol. 3, June 27, 1980, 28, NTA Accession No. NV0011586.
116. Glyn G. Caldwell, Delle B. Kelley, and Clark W. Heath Jr., "Leukemia among Participants in Military Maneuvers at a Nuclear Bomb Test," *Journal of the American Medical Association* 244, no. 14 (1980): 1575–1578; Glyn G. Caldwell, Matthew M. Zack, Michael T. Mumma, Henry Falk, Clark W. Heath Jr., John E. Till, Heidi Chen, and John D. Boice Jr., "Mortality among Military Participants at the 1957 Plumbbob Nuclear Weapons Test Series and on Leukemia among Participants at the Smoky Test," *Journal of Radiological Protection* 36, no. 3 (2016): 474–489.
117. ACHRE, "Final Report of the President's Advisory Committee on Human Radiation Experiments" (Washington, DC: US Government Printing Office, 1995), 482.
118. Miller, *Under the Cloud*, 284.
119. Miller, 291.
120. Clifford T. Honicker, "Premeditated Deceit: The Atomic Energy Commission against Joseph August Sauter" (master's thesis, University of Tennessee at Knoxville, 1987).

CHAPTER 10. BABY TEETH, PROJECT SUNSHINE, AND A MORATORIUM ON TESTING, 1958–1962

1. Herman M. Kalckar, "An International Milk Teeth Radiation Census," *Nature* 182, no. 4631 (1958): 283.
2. W. K. Wyant Jr., "50,000 Baby Teeth," *Nation*, June 13, 1959, 535.
3. Petra Goedde, *The Politics of Peace: A Global Cold War History* (New York: Oxford University Press, 2019).
4. Caroline Jack and Stephanie Steinhardt, "Atomic Anxiety and the Tooth Fairy: Citizen Science in the Midcentury Midwest," *Appendix* 2, no. 4 (2014).
5. William Krasner, "Baby Tooth Survey: First Results," *Environment* 55, no. 2 (1961): 18–24; Louise Zibold Reiss, "Strontium-90 Absorption in Deciduous Teeth," *Science* 134, no. 3491 (1961): 1669–1673.
6. H. T. Blumenthal, "Strontium 90 in Children," *Scientist and Citizen* 6, nos. 9–10 (1964): 3–7.

7. Quote from a 1993 interview in the *Chicago Tribune* and referenced in Daniel Lewis, "Saw an Earth at Risk and Let the World Know," *New York Times*, October 2, 2012, A1.

8. Karen Barad, *Meeting the Universe Halfway: Quantum Physics and the Entanglement of Matter and Meaning* (Durham, NC: Duke University Press, 2007), 151.

9. AEC, "Report on Project Gabriel," Division of Biology and Medicine, July 1954, NTA Accession No. NV0708119.

10. Eileen Welsome, *The Plutonium Files: America's Secret Medical Experiments in the Cold War* (New York: Dial, 1999), 299.

11. AEC, "Rand Sunshine Project," Division of Biology and Medicine, December 30, 1953, 1, NTA Accession No. NV0407554.

12. Emory Jerry Jessee, "Radiation Ecologies: Bombs, Bodies, and Environment during the Atmospheric Nuclear Weapons Testing Period, 1942–1965" (PhD diss., Montana State University, 2013), 158.

13. RAND Corporation, "Project Sunshine: Worldwide Effects of Atomic Weapons," August 6, 1953, 5, NTA Accession No. NV0407582.

14. AEC, "Rand Sunshine Project," 9.

15. AEC, "Biophysics Conference," January 18, 1955, NTA Accession No. NV0727441. See also discussion of Project Sunshine in ACHRE, "Final Report of the President's Advisory Committee on Human Radiation Experiments," (Washington, DC: US Government Printing Office, 1995). Note that the 1955 Biophysics Conference report was declassified from "restricted data" status in 1995 due to the efforts of ACHRE.

16. AEC, "Biophysics Conference," 6.

17. AEC, 3, 81; see also ACHRE, "Final Report."

18. AEC, "Thirty-Sixth Meeting of the General Advisory Committee to the U.S. Atomic Energy Commission," meeting minutes, August 17–19, 1953, 17, NTA Accession No. NV0058974.

19. AEC, "Biophysics Conference," 12, 190.

20. AEC, "Biophysics Conference"; AEC, "Rand Sunshine Project," 9.

21. AEC, "Biophysics Conference," 14.

22. AEC, 50, 23, 51.

23. AEC, 64–65 (emphasis added).

24. Welsome, *Plutonium Files*.

25. ACHRE, "Final Report"; US General Accounting Office, "Information on DOE's Human Tissue Analysis Work," May 1995.

26. AEC, "Biophysics Conference," 14.

27. AEC, "Report on Project Gabriel."

28. Ulrich Beck, *Risk Society: Towards a New Modernity* (Thousand Oaks CA: Sage, 1992), 68.

29. Merril Eisenbud, "Global Distribution of Strontium-90 from Nuclear Detonations," *Scientific Monthly* 84, no. 5 (1957): 237–244, 237.

30. ACHRE, "Final Report."

31. GAO, "Human Tissue Analysis Work."

32. ACHRE, "Final Report."

33. Welsome, *Plutonium Files.*

34. Gerald Markowitz, "'A Little of the Buchenwald Touch': America's Secret Radiation Experiments," *Reviews in American History* 28 (2000): 601–606; Welsome, *Plutonium Files.*

35. Joseph Masco, *The Nuclear Borderlands: The Manhattan Project in Post–Cold War New Mexico* (Princeton, NJ: Princeton University Press, 2006), 29 (emphasis in original).

36. Michael Marder, *The Chernobyl Herbarium: Fragments of an Exploded Consciousness* (London: Open Humanities Press, 2016), 24.

37. DOE, *A Perspective on Atmospheric Nuclear Tests in Nevada* (Las Vegas: Nevada Operations Office, 1995).

38. Richard G. Hewlett and Jack M. Holl, *Atoms for Peace and War, 1953–1961: Eisenhower and the Atomic Energy Commission* (Berkeley: University of California Press, 1989).

39. Toshihiro Higuchi, *Political Fallout: Nuclear Weapons Testing and the Making of a Global Environmental Crisis* (Stanford, CA: Stanford University Press, 2020).

40. Gerard J. DeGroot, *The Bomb: A Life* (Cambridge, MA: Harvard University Press, 2005)., 254.

41. DOE, *Perspective.*

42. Steven L. Simon, Andre Bouville, and Harold L. Beck, "The Geographic Distribution of Radionuclide Deposition across the Continental US from Atmospheric Nuclear Testing," *Journal of Environmental Radioactivity* 74, no. 1–3 (2004): 91–105.

43. Sarah Alisabeth Fox, *Downwind: A People's History of the Nuclear West* (Lincoln: University of Nebraska Press, 2014).

44. Gordon M. Dunning, "Effects of Nuclear Weapons Testing," *Scientific Monthly* 81, no. 6 (1955): 265–270.

45. Testimony of Harold A. Knapp, in US Congress, Subcommittee on Oversight and Investigations of the House Committee on Interstate and Foreign Commerce, *Low-Level Radiation Effects on Health*, 96th Congress, 1st session, April 23, May 24, and August 1, 1979, 292–307. Hereafter cited as *LLREH.*

46. Report by Gordon M. Dunning, in US Congress, Subcommittee on Radiation, JCAE, *The Nature of Radioactive Fallout and Its Effects on Man*, 85th Congress, 1st session, May 27–29 and June 3, 1957, part I, 235.

47. Report by Gordon M. Dunning, in US Congress, Subcommittee on Radiation, JCAE, *Biological and Environmental Effects of Nuclear War*, 86th Congress, 1st session, June 22–26, 1959, part I, 445. Hereafter cited as *BEENW.*

48. Testimony of E. B. Lewis, in *BEENW*, part II, 1552.

49. Harold Knapp to Charles L. Dunham, June 27, 1963, 11, NTA Accession No. NV0004884. Note that this document contains as an attachment, Gordon M. Dunning to N. H. Woodruff, Director of Operational Safety, "Comments on

'Iodine-131 in Fresh Milk and Human Thyroids Following a Single Deposition of Nuclear Test Fallout,' by Dr. Harold Knapp," June 14, 1963, in which Dunning comments on Knapp's report and Knapp has inserted his counterresponses.

50. H. A. Knapp, "Iodine-131 in Fresh Milk and Human Thyroids Following a Single Deposition of Nuclear Test Fallout," AEC Division of Biology and Medicine, June 1, 1953, NTA Accession No. NV0001758.

51. Testimony of Harold A. Knapp, in *LLREH*, 305.

52. Harold Knapp to Charles L. Dunham, "Transmittal of Report on Radioiodine," September 13, 1962, 1, NTA Accession No. NV0138706.

53. Ralph E. Lapp, "Nevada Test Fallout and Radioiodine in Milk," *Science* 137, no. 3532 (1962): 756–758.

54. Knapp, "Transmittal."

55. Charles L. Dunham to H. A. Knapp, "Draft Document: 'Average and Above Average Doses to the Thyroids of Children in the United States from Radioiodine from Nuclear Weapons Tests,'" October 24, 1962, 1, NTA Accession No. NV0004911.

56. Testimony of Harold A. Knapp, in US Congress, Subcommittee on Oversight and Investigations of the House Committee on Interstate and Foreign Commerce, Health and Scientific Research Subcommittee of the Labor and Human Resources Committee, and the Senate Committee on the Judiciary, *Health Effects of Low-Level Radiation*, joint hearing, 96th Congress, 1st session, April 19, 1979, vol. 1, 343. Hereafter cited as *HELLR*.

57. Barton C. Hacker, *Elements of Controversy: The Atomic Energy Commission and Radiation Safety in Nuclear Weapons Testing, 1947–1974* (Berkeley: University of California Press, 1994).

58. Testimony of Harold A. Knapp, in *HELLR*, vol. 1, 351.

59. Knapp to Dunham, June 27, 1963, 4, 15, 10, 16, 17, 9 (emphasis in original), NTA Accession No. NV0004884.

60. Knapp, "Iodine-131 in Fresh Milk," vii, viii, ix.

61. Testimony of Harold A. Knapp, in *LLREH*, 292–307.

62. E. J. Bloch, Assistant General Manager for Operations, to Gordon M. Dunning, "Recommendations of the Ad Hoc Working Group on Radioiodine in the Environment" (Working Group), April 25, 1963, 1, NTA Accession No. NV0153315.

63. Harold A. Knapp, "Iodine-131 in Fresh Milk and Human Thyroids Following a Single Deposition of Nuclear Test Fallout," *Nature* 202 (4932): 534–537.

64. Philip L. Fradkin, *Fallout: An American Nuclear Tragedy* (Tucson: University of Arizona Press, 1989).

65. Harold A. Knapp to L. Joe Deal, Assistant Director for Field Operations, DOE, January 5, 1979, reprinted in in *HELLR*, vol. 1, 271, 274.

66. Testimony of Harold A. Knapp, *LLREH*.

67. Charles Perrow, *Normal Accidents: Living with High-Risk Technologies* (Princeton, NJ: Princeton University Press, 1999), 14.

68. In the early years of atmospheric testing, measurement of radioiodine in milk was laborious, but widespread availability of gamma spectrometry by the late

1950s made assessment easier. See Merril Eisenbud and M. E. Wrenn, "Biological Disposition of Radioiodine—A Review," *Health Physics* 9, no. 12 (1963): 1133–1139. As Harold Knapp illustrated, however, it was possible to employ proportional scaling, and this could have been done years earlier had it been a priority within the AEC.

69. James J. Nickson, "Dosimetric and Protective Considerations for Radioactive Iodine," *Journal of Clinical Endocrinology & Metabolism* 8, no. 9 (1948): 721–731; J. B. Trunnell, "The Treatment of Human Thyroid Disease with Radioactive Iodine," *Transactions of the New York Academy of Sciences* 11, no. 6 (1949): 195–201; Sidney C. Werner, Edith H. Quimby, and Charlotte Schmidt, "Radioactive Iodine, I-131, in the Treatment of Hyperthyroidism," *American Journal of Medicine* 7, no. 6 (1949): 731–740.

70. J. E. Rall, "The Role of Radioactive Iodine in the Diagnosis of Thyroid Disease," *American Journal of Medicine* 20, no. 5 (1956): 729 (emphasis added).

71. Arthur B. Schneider and Elaine Ron, "Radiation and Thyroid Cancer: Lessons from a Half Century of Study," in *Diseases of the Thyroid*, 2nd ed., ed. Lewis E. Braverman (Totowa, NJ: Humana, 2003), 239–262.

72. H. Miller and R. S. Weetch, "The Excretion of Radioactive Iodine in Human Milk," *Lancet* 266, no. 6998 (1955): 1013; C. L. Comar, Bernard F. Trum, U. S. G. Kuhn III, R. H. Wasserman, M. M. Nold, and J. C. Schooley, "Thyroid Radioactivity after Nuclear Weapons Tests," *Science* 126, no. 3262 (1957): 16–18.

73. Schneider and Ron, "Radiation and Thyroid Cancer," 240.

74. Pat Ortmeyer and Arjun Makhijani, "Worse than We Knew," *Bulletin of the Atomic Scientists* 53, no. 6 (1997): 46–50.

75. Testimony of Harold A. Knapp, in *LLREH*, 303.

76. Ulrich Beck, *Risk Society: Towards a New Modernity* (Thousand Oaks, CA: Sage, 1992), 65 (emphasis in original).

77. Thomas D. Beamish, *Silent Spill: The Organization of an Industrial Crisis* (Cambridge, MA: MIT Press, 2002).

78. Barry A. Turner, "The Organizational and Interorganizational Development of Disasters," *Administrative Science Quarterly* 21, no. 3 (1976): 381.

79. Lee Clarke, *Worst Cases: Terror and Catastrophe in the Popular Imagination* (Chicago: University of Chicago Press, 2006.)

80. NCI, *Estimated Exposures and Thyroid Doses Received by the American People from Iodine-131 Fallout Following Nevada Atmospheric Nuclear Bomb Tests* (Bethesda, MD: NCI, 1997), www.cancer.gov.

81. IEER, *Fact Sheet on Fallout Report and Related Maps* (Takoma Park, MD: IEER, 2002), http://ieer.org.

82. CDC and NCI, *A Feasibility Study of the Health Consequences to the American Population of Nuclear Weapons Tests Conducted by the United States and Other Nations* (Atlanta: CDC, 2002), www.cdc.gov. Note that this study was limited to analysis of open-air testing between 1951 and 1962.

83. IEER, *Fact Sheet*.

84. For more information, see Steven L. Simon, André Bouville, and Charles E. Land, "Fallout from Nuclear Weapons Tests and Cancer Risks." *American Scientist* 94 (January–February 2006): 48–57.

85. IEER, *Fact Sheet*.

CHAPTER 11. SEDAN, *SILENT SPRING*, AND THE REENCHANTMENT OF NATURE

1. Scott Kirsch, *Proving Grounds: Project Plowshare and the Unrealized Dream of Nuclear Earthmoving* (New Brunswick, NJ: Rutgers University Press, 2005).

2. AEC, "Statement on the Results of Project Sedan," July 7, 1962, NTA Accession No. NV0137578; John S. Kelly, "Moving Earth and Rock with a Nuclear Device," *Science* 138, no. 3536 (1962): 50–51.

3. Edward Teller, "We're Going to Work Miracles," *Popular Mechanics* 113, no. 3 (1960): 97–101, 278, 280, 282.

4. Carolyn Merchant, "'The Violence of Impediments:' Francis Bacon and the Origins of Experimentation," *Isis* 99 (2008): 731–760.

5. Morris Berman, *The Reenchantment of the World* (Ithaca, NY: Cornell University Press, 1981); Merchant, "Violence of Impediments."

6. Richard Tarnas, *The Passion of the Western Mind* (New York: Ballantine Books, 1991), 273.

7. DOE, *United States Nuclear Tests: July 1945 through September 1992*, DOE/NV-209 (Rev. 15) (Las Vegas: Nevada Operations Office, 2000).

8. Kirsch, *Proving Grounds*; M. D. Nordyke and M. M. Williamson, *The Sedan Event* (Livermore, CA: Lawrence Radiation Laboratory, University of California, 1965).

9. Kirsch, *Proving Grounds*.

10. Dan O'Neill, *The Firecracker Boys* (New York: St. Martin's, 1994), 253.

11. Philip L. Fradkin, *Fallout: An American Nuclear Tragedy* (Tucson: University of Arizona Press, 1989); Kirsch, *Proving Grounds*; Richard L. Miller, *Under the Cloud: The Decades of Nuclear Testing* (The Woodlands, TX: Two Sixty, 1991).

12. CDC and NCI, *A Feasibility Study of the Health Consequences to the American Population of Nuclear Weapons Tests Conducted by the United States and Other Nations* (Atlanta: CDC, 2002), www.cdc.gov, appendix E. Note that Sedan is not categorized as an atmospheric detonation but a cratering experiment.

13. Ralph H. Lutts, "Chemical Fallout: Rachel Carson's *Silent Spring*, Radioactive Fallout, and the Environmental Movement," *Environmental Review* 9, no. 3 (1985): 210–225; Linda Lear, *Rachel Carson: Witness for Nature* (Boston: Houghton Mifflin, 2009); William Souder, *On a Farther Shore: The Life and Legacy of Rachel Carson* (New York: Broadway Books, 2012).

14. Rachel Carson to Dorothy Freeman, February 1, 1958, in *Letters of the Century: America 1900–1999*, ed. Lisa Grunwald and Stephen J. Adler (New York: Dial, 1999), 407.

15. Edward Teller with Allen Brown, *The Legacy of Hiroshima* (New York: Doubleday, 1962), 84.

16. William Leiss, *The Domination of Nature* (Montreal: McGill-Queen's University Press, 1994), xviii.

17. Rachel Carson, *Silent Spring* (Boston: Houghton Mifflin, 1962).

18. The quotation in the subhead is from Teller, "We're Going to Work Miracles."

19. David F. Noble, *The Religion of Technology* (New York: Penguin Books, 1999), 67.

20. Carolyn Merchant, *The Death of Nature: Women, Ecology, and the Scientific Revolution* (New York: Harper, 1980).

21. Paolo Rossi, *Francis Bacon: From Magic to Science* (New York: Routledge, 2009). As Merchant notes, Bacon employed distinctively masculine rhetorical images to articulate the conduct of empirical inquiry and feminine metaphors in describing the character of nature so acted on (*Death of Nature*).

22. O'Neill, *Firecracker Boys*.

23. Teller, *Legacy of Hiroshima*, 56.

24. William H. Berman and Lee M. Hydeman, "Nuclear Explosions for Peaceful Purposes," *Natural Resources Journal* 1, no. 1 (1961): 1–22; Teller, "We're Going to Work Miracles."

25. Experiments outside Nevada include Project Gnome in 1961 in southeastern New Mexico, which comprised detonation in an underground salt deposit to gauge the viability of generating heat for electricity production as well as rare radioisotopes; Project Gasbuggy in 1967 in northwestern New Mexico, which was an underground detonation to enhance the recoverability of natural gas deposits; Project Rulison two years later in western Colorado, which was also designed to enhance the recoverability of natural gas; and Project Rio Blanco in 1973 in Colorado, which also entailed a detonation deep underground to test the viability of enhanced natural gas recovery.

26. Kirsch, *Proving Grounds*; O'Neill, *Firecracker Boys*.

27. Ralph E. Lapp, "The 'Humanitarian' H-Bomb," *Bulletin of the Atomic Scientists* 12, no. 7 (1956): 261–264.

28. Edward Teller and Albert L. Latter, *Our Nuclear Future* (New York: Criterion Books, 1958), 172.

29. Robert A. Divine, *Blowing on the Wind: The Nuclear Test Ban Debate, 1954–1960* (New York: Oxford University Press, 1978); Toshihiro Higuchi, "'Clean' Bombs: Nuclear Technology and Nuclear Strategy in the 1950s," *Journal of Strategic Studies* 29, no. 1 (2006): 85; Katherine Magraw, "Teller and the 'Clean Bomb' Episode," *Bulletin of the Atomic Scientists* 44, no. 4 (1988): 32–37.

30. Teller, *Legacy of Hiroshima*.

31. Edward Teller and Albert L. Latter, "The Compelling Need for Nuclear Tests," *Life*, February 10, 1958, 72.

32. O'Neill, *Firecracker Boys*.

33. Dan O'Neill, "Project Chariot: How Alaska Escaped Nuclear Excavation," *Bulletin of the Atomic Scientists* 45, no. 10 (1989): 28.

34. Kirsch, *Proving Grounds*; O'Neill, *Firecracker Boys*.

35. Kirsch, *Proving Grounds*.

36. CDC and NCI, *Feasibility*.

37. Pat Ortmeyer and Arjun Makhijani, "Worse than We Knew," *Bulletin of the Atomic Scientists* 53, no. 6 (1997): 46–50.

38. NCI, *Estimated Exposures and Thyroid Doses Received by the American People from Iodine-131 in Fallout Following Nevada Atmospheric Nuclear Bomb Tests* (Washington, DC: NCI, 1997), chap. 2, table 2.2.

39. DOE, *United States Nuclear Tests*.

40. Phil Patton, "Journey to the End of the World," *Esquire*, July 1995, 128.

41. Max Weber, "Science as a Vocation," in *From Max Weber: Essays in Sociology*, ed. and trans. H. H. Gerth and C. Wright Mills (New York: Oxford University Press, 1946), 143.

42. Berman, *Reenchantment of the World*, 28.

43. Berman; Merchant, *Death of Nature*.

44. Paul B. Sears, "Ecology—a Subversive Subject," *BioScience* 14, no. 7 (1964): 11–13; Paul Shepard, "Introduction: Ecology and Man—a Viewpoint," in *The Subversive Science: Essays Toward an Ecology of Man*, ed. Paul Shepard and Daniel McKinley (Boston: Houghton Mifflin, 1969), 1–10.

45. Sears, "Ecology," 12.

46. Carson, *Silent Spring*, 6.

47. Rob Nixon, *Slow Violence and the Environmentalism of the Poor* (Cambridge, MA: Harvard University Press, 2011), 2.

48. Nixon.

49. Raymond Murphy, "Disaster or Sustainability: The Dance of Human Agents with Nature's Actants," *Canadian Review of Sociology and Anthropology* 41, no. 3 (2004): 249–266.

50. Carson, *Silent Spring*, 13.

51. Carson, 13.

52. Merchant, *Death of Nature*.

53. Merchant, "Violence of Impediments," 155.

54. Merchant, *Death of Nature*.

55. Moody E. Prior, "Bacon's Man of Science," *Journal of the History of Ideas* 15, no. 3 (1954): 348–370.

56. Karen Armstrong, *The Battle for God: A History of Fundamentalism* (New York: Random House, 2000), 71.

57. Carson, *Silent Spring*, 297.

58. Weber, "Science as a Vocation," 139.

59. Rachel Carson, *Lost Woods: The Discovered Writing of Rachel Carson*, ed. Linda Lear (Boston: Beacon, 1998), 94. The quotation is from a speech she gave upon receiving the John Burroughs Medal in New York City, 1952.

60. Carson.

61. Max Weber, *Economy and Society*, vol. 1, ed. Guenther Roth and Claus Wittich (Berkeley: University of California Press, 1978).

62. Carson, *Silent Spring*, 12.

63. Oliver Stone and Peter Kuznick, *The Untold History of the United States* (New York: Simon and Schuster, 2012); Derek Mead, "The U.S.'s Insane Attempt to Build a Harbor with a Two Megaton Nuclear Bomb," *Motherboard*, August 9, 2012, https://motherboard.vice.com.

64. Ulrich Beck, *Risk Society: Towards a New Modernity* (Thousand Oaks, CA: Sage, 1992).

65. Merchant, "Violence of Impediments."

66. Carson, *Silent Spring*.

67. David A. Kirsch, "Project Plowshare: The Cold War Search for a Peaceful Nuclear Explosive," in *Science, Values, and the American West*, ed. Stephen Tchudi (Reno: Nevada Humanities Committee, 1997), 191–222.

68. Carson, *Silent Spring*, 12–13.

69. The quotation is from the CBS Reports television program *The Silent Spring of Rachel Carson*, airing in April 1963.

CONCLUSION

1. Eric Rutkow, *American Canopy: Trees, Forests, and the Making of a Nation* (New York: Scribner, 2012).

2. Rutkow.

3. Rutkow, 4.

4. Mary Shelley, *Frankenstein; or, The Modern Prometheus* (New York: Tom Doherty, 1988), 38.

5. Langdon Winner, *Autonomous Technology: Technics-out-of-Control as a Theme in Political Thought* (Cambridge, MA: MIT Press, 1977), 313. Note that Winner is referencing a quote from Milton's *Paradise Lost* that was included on the title page of Shelley's novel.

6. Andrew Pickering, *The Cybernetic Brain: Sketches of Another Future* (Chicago: University of Chicago Press, 2010).

7. Todd R. La Porte and Daniel S. Metlay, "Hazards and Institutional Trustworthiness: Facing a Deficit of Trust," *Public Administration Review* 56, no. 4 (1996): 342.

8. Klaus Becker, "Radiophobia: A Serious but Curable Mental Disorder," Japan Atomic Energy Institute Conference, March 2005, 154–158, https://inis.iaea.org; Jack De Ment, "Radiophobia: A New Psychological Syndrome," *Western Journal of Surgery, Obstetrics, and Gynecology* 59, no. 11 (1951): viii–x, 602; Michael Myslobodsky, "The Origin of Radiophobias," *Perspectives in Biology and Medicine* 44 no. 4 (2001): 543–555; Spencer R. Weart, *The Rise of Nuclear Fear* (Cambridge, MA: Harvard University Press, 2012).

9. DOE, *United States Nuclear Tests: July 1945 through September 1992* (Las Vegas: Nevada Operations Office, 2000).

10. Matt Wray, "A Blast from the Past: Preserving and Interpreting the Atomic Age," *American Quarterly* 58, no. 2 (2006): 469.

11. Patrick W. McCray, "Viewing America's Bomb Culture: The Atomic Testing Museum," *Public Historian* 28, no. 1 (2006): 153.

12. Michael A. Elliott, "Our Memorials, Ourselves," *American Quarterly* 63, no. 1 (2011): 229–240; Martin Harwit, *Exhibit Denied: Lobbying the History of the Enola Gay* (New York: Springer-Verlag, 1996).

13. Benedict Giamo, "The Myth of the Vanquished: The Hiroshima Peace Memorial Museum," *American Quarterly* 55, no. 4 (2003): 724.

14. Nevada Test Site Historical Foundation, "Information Regarding NTSHF," accessed October 1, 2015, www.atomictestingmuseum.org.

15. Chip Ward, *Canaries on the Rim: Living in the Downwind West* (New York: Verso, 1999), 41.

16. Wallace Stegner, *The Sound of Mountain Water* (Lincoln: University of Nebraska Press, 1980), 201.

17. John C. Van Dyke, *The Desert* (Baltimore: Johns Hopkins University Press, 1999), 43.

INDEX

ABOUT THE AUTHOR

JAMES C. RICE is Professor of Sociology at New Mexico State University, specializing in environmental and organizational sociology. His research and teaching focuses on power, inequality, and the production of environmental risk in society.